Stochastic Processes and Their Applications

Stochastic Processes and Their Applications

Frank E. Beichelt and L. Paul Fatti

London and New York

First published 2002 by Taylor & Francis
11 New Fetter Lane, London EC4P 4EE

Simultaneously published in the USA and Canada
by Taylor & Francis Inc,
29 West 35th Street, New York, NY 10001

Taylor & Francis is an imprint of the Taylor & Francis Group

This book has been produced from camera-ready copy supplied by the authors
Printed and bound in Great Britain by
TJ International Ltd, Padstow, Cornwall

Titel der Originalausgabe: Beichelt, Frank: Stochastische Prozesse Für Ingenieure

Translation arranged with the approval of the publisher B.G. Teubner
from the original German edition into English.

British Library Cataloguing in Publication Data
A catalogue record for this book is available from the British Library

Library of Congress Cataloging in Publication Data
A catalog record for this book has been requested

ISBN 0-415-27232-7

Contents

Preface ix

Symbols and Abbreviations xi

1 Probability Theory 1
1.1 Random Events and their Probabilities 1
1.2 Random Variables 4
1.2.1 Discrete Random Variables 4
1.2.2 Continuous Random Variables 6
1.2.3 Nonnegative Random Variables 8
1.3 Random Vectors 13
1.3.1 Two-Dimensional Random Vectors 13
1.3.2 *n*-Dimensional Random Vectors 20
1.4 Sums and Sequences of Random Variables 22
1.5 Transformations of Probability Distributions 31
1.5.1 *z*-Transformation 32
1.5.2 Laplace-Transformation 34
Exercises 37

2 Stochastic Processes 43
2.1 Introduction 43
2.2 Characteristics of Stochastic Processes 47
2.3 Properties of Stochastic Processes 49
2.4 Special StochasticProcesses 53
2.4.1 Continuous-Time Stochastic Processes 53
2.4.2 Stationary Discrete-Time Stochastic Processes 60
Exercises 67

3 Poisson Processes 69
3.1 Homogeneous Poisson Process 69
3.1.1 Definition and Properties 69
3.1.2 Poisson Process and Uniform Distribution 77
3.2 Inhomogeneous Poisson Process 85
3.2.1 Definition and Properties 85
3.2.2 Minimal Repair 89
Exercises 97

4 Renewal Processes **99**

4.1 Foundations 99

4.2 Renewal Function 102

4.2.1 Renewal Equations 102

4.2.2 Bounds for the Renewal Function 107

4.3 Recurrence Times 111

4.4 Asymptotic Behaviour 114

4.5 Stationary Renewal Processes 118

4.6 Alternating Renewal Processes 120

4.7 Cumulative Stochastic Processes 125

4.8 Regenerative Stochastic Processes 131

Exercises 134

5 Discrete-Time Markov Chains **137**

5.1 Foundations and Examples 137

5.2 Classification of States 145

5.2.1 Closed Sets of States 145

5.2.2 Equivalence Classes 147

5.2.3 Periodicity 150

5.2.4 Recurrence and Transience 152

5.3 Limit Theorems and Stationary Distribution 158

5.4 Birth- and Death Processes 163

Exercises 165

6 Continuous-Time Markov Chains **169**

6.1 Foundations 169

6.2 Kolmogorov's Differential Equations 173

6.3 Stationary State Probabilities 183

6.4 Construction of Markovian Systems 187

6.5 Erlang's Phase Method 189

6.6 Birth- and Death Processes 191

6.6.1 Time-Dependent State Probabilities 191

6.6.2 Stationary State Probabilities 202

6.7 Sojourn Times 207

6.8 Applications in Queueing Theory 209

6.8.1 Introduction 209

6.8.2 Loss Systems 212

6.8.3 Waiting Systems 216

6.8.4 Waiting-Loss-Systems 219

6.8.5 Special Single-Server Queueing Systems 223

6.8.6 Networks of Queueing Systems 229

6.9 Semi-Markov Processes 241

Exercises 250

7 Wiener Processes **257**

7.1 Definition and Properties 257

7.2 First Passage Times 263
7.3 Transformations of the Wiener Process 269
7.3.1 Elementary Transformations 269
7.3.2 Ornstein-Uhlenbeck Process 271
7.3.3 Wiener-Process with Drift 272
7.3.4 Integral Transformations 285
Exercises 289

8 Spectral Analysis of Stationary Processes 293
8.1 Foundations 293
8.2 Processes with Discrete Spectrum 295
8.3 Processes with Continuous Spectrum 299
8.3.1 Spectral Representation of the Covariance Function 299
8.3.2 Spectral Representation of Stationary Processes 308
Exercises 310

Appendix 1 Landau Order Symbol 312

Appendix 2 Dirac Delta Function 313

Answers to Selected Exercises 314

References 319

Index 323

PREFACE

This book is an introduction to stochastic processes and their applications for students in engineering, industrial statistics, science, operations research, business and public policy analysis and finance. It provides theoretical foundations for modeling time-dependent random phenomena as they occur in physics, electronics, computer science and biology. Likewise, it takes into account applications in the field of operations research, in particular in queueing, maintenance and reliability theory as well as in financial markets. Through numerous, mostly science and engineering-based examples, the subject is represented in a comprehensible, practically oriented way. Hence the book is also suitable for self-study. As a non-measure theoretic introduction to stochastic processes, its study only requires a knowledge of calculus and elementary probability theory which should be familiar to students after having finished their undergraduate studies in mathematics and statistics. However, to make the book attractive to mathematically interested readers as well, some important proofs and theoretically challenging examples and exercises have been included. Solutions to most of the exercises can be found in an appendix or the exercises are given together with their solutions. Those sections, examples or exercises marked with a * symbol are either theoretically more difficult or of less practical importance. The chapters are organized in such a way that reading a chapter usually requires a knowledge of the previous ones.

A mathematically rigorous treatment of Wiener processes (Brownian motion processes) and the spectral analysis of stationary processes, dealt with in chapters 7 and 8, respectively, is not possible without a basic knowledge of measure theory, Fourier analysis, and the theory of generalized functions. Therefore, these chapters sometimes present heuristically motivated explanations and formulas instead of mathematically exact concepts and derivations. This approach does not detract from the main purpose of this book, which is to enable readers to apply stochastic modeling in their own field. This book generally does not deal with the data analysis aspects of stochastic processes. It can be anticipated that readers will use statistical software for tackling numerical problems. However, after having studied our book they will be able to work more creatively with such software and write their own analysis programs. The text may also serve as a basis for preparing senior undergraduate and graduate level courses.

The authors wish to thank their Honours students of the year 2000, who checked and tested the exercises and gave many valuable hints. In particular, we acknowledge the contributions of H. Christoforou, T. Levin, V. Nkwambi and D. Tugendhaft. Finally, the authors would like to thank the staff of Gordon and Breach Science Publishers for their constructive cooperation.

The book is a thoroughly checked and revised translation of the German original published by B. G. Teubner, Stuttgart in 1997 under the title *Stochastische Prozesse für Ingenieure.*

The numerous suggestions of German-speaking readers helped very much to prepare this English edition. Many thanks for their support as well.

Further helpful comments on this book are very welcome and should be sent to the authors: University of the Witwatersrand, Department of Statistics and Actuarial Science, WITS 2050, Johannesburg, RSA.
E-mail: 010frank@cosmos.wits.ac.za; fatti@stats.wits.ac.za.

Frank E. Beichelt and L. Paul Fatti

Symbols and Abbreviations

□ ■ ●	symbols after an example, a theorem, a definition
$f(t) \equiv c$	$f(t) = c$ for all $t \in \mathbf{T}$
$f * g$	convolution of two functions f and g
$f^{*(n)}$	n-th convolution power of f
$\hat{f}(s)$, $L\{f\}$	Laplace-transform of a function f
$o(x)$	Landau order symbol
$\delta(x)$	Dirac delta function

Probability Theory

$X,\ Y,\ Z$	random variables
$E(X)$, $Var(X)$	expected (mean) value, variance of X
$f_X(x)$, $F_X(x)$	probability density function, (cumulative probability) distribution function of X
$\lambda(x)$, $\Lambda(x)$	failure rate, integrated failure rate (hazard function)
$N(\mu, \sigma^2)$	normal random variable with expected value μ and variance σ^2
$\phi(x)$, $\Phi(x)$	probability density function, distribution function of a standard normal random variable
$f_{\mathbf{X}}(x_1, x_2, \ldots, x_n)$	joint probability density function of $\mathbf{X} = (X_1, X_2, \ldots, X_n)$
$F_{\mathbf{X}}(x_1, x_2, \ldots, x_n)$	joint distribution function of $\mathbf{X} = (X_1, X_2, \ldots, X_n)$
$Cov(X, Y)$	covariance between X and Y
$\rho(X, Y)$	correlation coefficient of X and Y
$M(z)$	z-transform (moment generating function) of a discrete random variable or its probability distribution

Stochastic Processes

$\{X(t), t \in \mathbf{T}\}$, $\{X_t, t \in \mathbf{T}\}$	continuous-time, discrete-time stochastic process with parameter space $\mathbf{T}$
$\mathbf{Z}$	state space of a stochastic process
$f_t(x)$, $F_t(x)$	probability density, distribution function of $X(t)$
$f_{t_1,t_2,\ldots,t_n}(x_1, x_2, \ldots, x_n)$, $F_{t_1,t_2,\ldots,t_n}(x_1, x_2, \ldots, x_n)$	joint probability density function, distribution function of $(X(t_1), X(t_2), \ldots, X(t_n))$
$m(t)$	trend function of a stochastic process
$C(s,t)$	covariance function of a stochastic process
$C(\tau)$	covariance function of a stationary process
$\rho(s,t)$	correlation function of a stochastic process

$\{N(t), t \geq 0\}$	point process, counting process
$\{T_1, T_2, ...\}$	pulse process
$H(t),\ H_1(t)$	renewal function of an ordinary, delayed renewal processs
$R(t),\ V(t)$	backward, forward recurrence time
$A(t),\ A$	point, stationary (long-run) availability
$L(x)$	first-passage time of level x
$p_{ij},\ p_{ij}^{(m)}$	one-step, m-step transition probabilities of a homogeneous, discrete-time Markov chain
$\mathbf{P}$	transition matrix
$f_{ij}^{(n)}$	first-passage probability; probability that, starting from state i, the first transition into state j occurs at time $t = n$
f_{ij}	probability that, starting in state i, the process ever makes a transition into state j
$p_{ij}(t)$	transition probabilities of homogeneous, continuous-time Markov chain
$q_{ij},\ q_i$	conditional, unconditional transition rates (transition intensities) of a homogeneous, continuous-time Markov chain
$\{\pi_i; i \in \mathbf{Z}\}$	stationary state distribution of a homogeneous Markov chain
$\lambda_j,\ \mu_j$	birth, death rates
$\lambda,\ \mu$	arrival rate, service rate (section 6.8)
μ_i	expected sojourn time of a semi-Markov process in state i
μ	drift parameter of a Wiener process with drift
ρ	correlation coefficient; traffic intensity (section 6.8)
η	degree of server utilisation
W	waiting time in a queueing system
L	random lifetime, queue length
$L(a,b)$	first-passage time of level $\min(a, b)$
$\{W(t), t \geq 0\}$	Wiener-process with drift
$\{Z(t), t \geq 0\}$	white noise
ω	circular frequency
$\phi,\ \Phi$	phase, random phase
$s(\omega),\ S(\omega)$	spectral density, spectral function
$\mathbf{S}$	spectrum $\mathbf{S} = \{\omega;\ s(\omega) > 0\}$
w	band width, $\mathrm{w} = \sup_{\omega \in \mathbf{S}} \omega - \inf_{\omega \in \mathbf{S}} \omega$

1 Probability Theory

1.1 Random Events and Their Probabilities

Random events occur in connection with *random experiments*. A random experiment is characterized by two properties:

1) repetitions of one and the same experiment, even if carried out under identical conditions, generally have different outcomes, and

2) the set of all possible outcomes of the experiment is known.

Hence, the outcome of a single random experiment may not be predicted with certainty. Thus, it only makes sense to carry out random experiments if they are repeated sufficiently frequently under identical conditions, in order to identify and quantify *stochastic* (*statistical*) *regularities*. Examples of random experiments are:

1) Tossing a coin. The possible outcomes are "*Heads*" and "*Tails*".

2) Tossing a die. The possible outcomes are the numbers 1, 2, ..., 6.

3) Counting the number of vehicles arriving at a filling station per day.

4) Counting the number of e-mails arriving in a firm per day.

5) Counting the number of faults occuring per unit time in a computer network. The possible outcomes are, as in the previous two random experiments, $0, 1, \ldots$.

6) Recording the life lengths of electronic parts.

7) Recording the maximum fluctuation of stock prices per unit time. The possible outcomes are, as in the random experiment 6, any real numbers in $[0, \infty)$.

The following concepts refer to the same random experiment. A possible outcome a of the experiment is called an *elementary* (*simple*) *event*. The set of all elementary events is called *space of elementary events* or *sample space*. Here and in what follows the sample space is denoted by **M**. A sample space is *discrete* if it is a finite or countably infinite set. A *random event* (briefly: *event*) A is a subset of **M**. An event A is said to have occured if the outcome a of the random experiment is an element of A, i.e., if $a \in A$. Let A and B be two events. Then the set-theoretic operations *intersection* "$\cap$"and *union* "$\cup$" can be interpreted in the following way: $A \cap B$ is the event that both A and B occur and $A \cup B$ is the event that A or B (or both) occur. If $A \subseteq B$, i.e. if A is a subset of B, then the occurence of A implies the occurence of B. $A \setminus B$ is the set of all those elementary events which are elements of A, but not of B. Hence, $A \setminus B$ is the event that A occurs but not B. The event

$\overline{A} = \mathbf{M} \setminus A$ is the *complement of* A. Thus, if A occurs, then not $\overline{A}$ and vice versa. Let $A_1, A_2, ..., A_n$ be a sequence of events. Then the *rules of de Morgan* hold:

$$\overline{\bigcup_{i=1}^{n} A_i} = \bigcap_{i=1}^{n} \overline{A}_i, \quad \overline{\bigcap_{i=1}^{n} A_i} = \bigcup_{i=1}^{n} \overline{A}_i. \tag{1.1}$$

In particular, if $n = 2$, $A_1 = A$ and $A_2 = B$,

$$\overline{A \cup B} = \overline{A} \cap \overline{B}, \quad \overline{A \cap B} = \overline{A} \cup \overline{B}. \tag{1.2}$$

The empty set $\emptyset$ is the *impossible event*, since, not containing any elementary events, it can never occur. By definition, **M** contains all elementary events so that it must always occur. Hence **M** is called the *certain event*. Two events A and B are called *mutually exclusive* if their joint occurence is impossible, i.e. if $A \cap B = \emptyset$. In this case the occurence of A implies that B does not occur and vice versa. In particular, A and $\overline{A}$ are mutually exclusive.

Let $\mathcal{M}$ be the set of all events which can occur when carrying out the random experiment. Further, let $P = P(\cdot)$ be a function on $\mathcal{M}$ with the following properties:

I) $P(\emptyset) = 0, \quad P(\mathbf{M}) = 1.$

II) For any event A, $0 \le P(A) \le 1$.

III) For any sequence of pairwise mutually exclusive events $A_1, A_2, ...$, i.e. $A_i \cap A_j = \emptyset$ for $i \ne j$,

$$P\left(\bigcup_{i=1}^{\infty} A_i\right) = \sum_{i=1}^{\infty} P(A_i). \tag{1.3}$$

The number $P(A)$ is the *probability* of the event A. $P(A)$ characterizes the degree of certainty of the occurence of A. This interpretation of the probability is justified by the following implications of the the properties I) to III).

1) $P(\overline{A}) = 1 - P(A)$.

2) If $A \subseteq B$, then $P(A) \le P(B)$.

3) If A and B are mutually exclusive events, i.e. if $A \cap B = \emptyset$, then

$$P(A \cup B) = P(A) + P(B).$$

4) For any events A and B,

$$P(A \cup B) = P(A) + P(B) - P(A \cap B).$$

Hint: It is assumed that all events which arise from applying the operations $\cap$, $\cup$, $\subseteq$ and $\setminus$ to any subsets of $\mathcal{M}$ are also elements of $\mathcal{M}$.

The probabilites of random events are usually unknown. However, they can be estimated by their *relative frequencies*. If, in a series of n repetitions of the same random experiment the event A has been observed $m = m(A)$ times, then the relative frequency of A is given by

$$\hat{p}_n(A) = \frac{m(A)}{n}.$$

Generally, the relative frequency of A tends to $P(A)$ as n increases:

$$\lim_{n\to\infty} \hat{p}_n(A) = P(A).$$

Therefore, the probability of A can be estimated with any required level of accuracy from its relative frequency by sufficient repetitions of the random experiment.

Two events A and B are called *independent* if

$$P(A \cap B) = P(A)P(B). \tag{1.4}$$

The events $A_1, A_2, ..., A_n$ are *independent* if for any subset $\{A_{i_1}, A_{i_2}, \cdots, A_{i_k}\}$ of $\{A_1, A_2, ..., A_n\}$, $k \le n$,

$$P(A_{i_1} \cap A_{i_2} \cap ... \cap A_{i_k}) = P(A_{i_1})P(A_{i_2}) \cdots P(A_{i_k}).$$

In particular, if $k = n$, then the independence of the A_i implies

$$P(A_1 \cap A_2 \cap ... \cap A_n) = P(A_1)P(A_2) \cdots P(A_n). \tag{1.5}$$

Let A and B be two events with $P(B) > 0$. Then the *conditional probability* of A given B is

$$P(A|B) = \frac{P(A \cap B)}{P(B)}. \tag{1.6}$$

$\{A_1, A_2, ..., A_n\}$ is called an *exhaustive set* of events if

$$\bigcup_{i=1}^{n} A_i = \mathbf{M}.$$

Let $\{A_1, A_2, ..., A_n\}$ be an exhaustive and mutually exclusive set of events. Then the formula

$$P(B) = \sum_{i=1}^{n} P(B|A_i)P(A_i) \tag{1.7}$$

is called *total probability rule* and the formula

$$P(A_i|B) = \frac{P(B|A_i)}{\sum_{i=1}^{n} P(B|A_i)P(A_i)}, \quad i = 1, 2, \cdots, n.$$

is called *Bayes' theorem*.

1.2 Random Variables

Given a random experiment with sample space **M**, a *random variable* X is a real function on **M**: $X = X(a)$, $a \in \mathbf{M}$. Thus, a random variable associates a number with each elementary event, i.e. with each outcome of the random experiment. Any value which X can assume is called a *realization* of X. The set of all realizations of X is called the *range* of X. Because it is random which elementary event will be outcome of the random experiment, the resulting realization of X is also random. By introducing a random variable X one passes from the sample space **M** of a random experiment to the range $\{X(a),\ a \in \mathbf{M}\}$ of X, which is a set of real numbers. This transition makes sense if the elementary events are not real numbers so that a direct quantitative evaluation of the random experiment is not possible. However, in many cases the elementary events are themselves real numbers, in which case they can be considered to be realizations of a random variable. Hence a random variable may be interpreted as the (random) outcome of a random experiment the elementary events of which are real numbers.

A random variable X is completely characterized by its (*cumulative*) *distribution function* $F(x)$:

$$F(x) = P(X \le x).$$

Thus, $F(x)$ is the probability of the random event that X assumes a realization which is less than or equal to x. For $a < b$,

$$P(a < X \le b) = F(b) - F(a). \tag{1.8}$$

Any distribution function $F(x)$ has the following properties:

1) $F(-\infty) = 0, \quad F(+\infty) = 1.$

2) $F(x)$ is nondecreasing in x.

Conversely, every function $F(x)$ satisfying these two properties can be considered to be the distribution function of a random variable X. The distribution function of a random variable characterizes its *probability distribution.*

1.2.1 Discrete Random Variables

The range of a *discrete random variable* is a finite or countably infinite set. Examples of discrete random variables were given in section 1.1 (examples 2 to 5).

Let $\{x_0, x_1, x_2, \cdots\}$ be the range of X, $x_i \le x_k$ for $i < k$. Further, let p_i be the probability of the random event that X assumes the realization x_i:

$$p_i = P(X = x_i), \quad i = 0, 1, 2, \cdots$$

The set $\{p_0, p_1, ...\}$ characterizes the probability distribution of X. Since X must assume one of its realizations,

$$\sum_{i=1}^{\infty} p_i = 1.$$

Conversely, any sequence of nonnegative numbers $\{p_0, p_1, ...\}$ satisfying this condition can be considered to be the probability distribution of a discrete random variable.

The distribution function of X is

$$F(x) = \begin{cases} 0 & \text{for } x < x_0 \\ \Sigma_{i=0}^{k} p_i & \text{for } x_k \le x < x_{k+1}, \quad k = 0, 1, 2, \cdots \end{cases}.$$

If the range of X is finite and if x_n is the greatest realization of X, then this definition has to be supplemented by

$$F(x) = 1 \text{ for } x_n \le x.$$

$F(x)$ is a piecewise constant function with jumps of size p_i at $x = x_i - 0$. Therefore,

$$p_i = F(x_i) - F(x_i - 0); \quad i = 0, 1, 2, \cdots$$

Thus, given $\{p_0, p_1, ...\}$ the distribution function of X can be constructed and, vice versa, given the distribution function of X, the probabilities $p_i = P(X = x_i)$ can be obtained.

The *expected value* (*mean value*) $E(X)$ and the *variance* $Var(X)$ of X are given by

$$E(X) = \Sigma_{i=0}^{\infty} x_i p_i$$

and

$$Var(X) = \Sigma_{i=0}^{\infty} (x_i - E(X))^2 p_i,$$

respectively, provided the sums exists. The arithmetic mean of n realizations of X obtained from n independent repetitions of the random experiment tends to $E(X)$ as n tends to infinity. The variance $Var(X)$ is the expected squared deviation of X from its expected value $E(X)$.

In particular, let X be a *binary random variable* with range $\{0, 1\}$ and probability distribution $\{P(X = 1) = p, P(X = 0) = 1 - p\}$. Then,

$$E(X) = p \quad \text{and} \quad Var(X) = p(1 - p). \tag{1.9}$$

The *n th moment* μ_n of X is the expected value of X^n:

$$\mu_n = E(X^n) = \Sigma_{i=0}^{\infty} x_i^n p_i; \quad n = 0, 1, 2, ...$$

Table 1.1 Probability distributions of discrete random variables X

Distribution	*Range of X*	$p_i = P(X = x_i)$	$E(X)$	$Var(X)$
uniform distribution	$x_1, x_2, \cdots, x_n$	$p_i = \frac{1}{n},\ n < \infty$	$\frac{1}{n}\sum_{i=1}^{n} x_i$	$\frac{1}{n}\sum_{i=1}^{n}(x_i - E(X))^2$
geometric distribution	$x_i = i$ $(i = 1, 2, \cdots)$	$p(1-p)^{i-1}$ $(0 < p < 1)$	$\frac{1}{p}$	$\frac{1-p}{p^2}$
binomial distribution	$x_i = i$ $(i = 0, 1, \cdots, n)$	$\binom{n}{i} p^i (1-p)^{n-i}$ $(0 < p < 1)$	np	$np(1-p)$
negative binomial distribution	$x_i = i$ $(i = r, r+1, \cdots)$	$\binom{i-1}{r-1} p^r (1-p)^{i-r}$ $(r < \infty,\ 0 < p < 1)$	$\frac{r}{p}$	$\frac{r(1-p)}{p^2}$
Poisson distribution	$x_i = i$ $(i = 0, 1, \cdots)$	$\frac{\lambda^i}{i!} e^{-\lambda}$ $(0 < \lambda < \infty)$	λ	λ

1.2.2 Continuous Random Variables

A random variable X with distribution function $F(x)$ is called *continuous* if $F(x)$ has a first derivative

$$f(x) = \frac{dF(x)}{dx}.$$

$f(x)$ is called *probability density function* of X (briefly: *probability density* or only *density*). According to its definition,

$$F(x) = \int_{-\infty}^{x} f(u)\,du.$$

In particular,

$$\int_{-\infty}^{+\infty} f(x)\,dx = F(\infty) = 1.$$

Conversely, every nonnegative function $f(x)$ satisfying this condition can be considered to be the probability density of a random variable X. As with its distribution function, a continuous random variable is also completely characterized by its probability density.

Expected value (*mean value*) $E(X)$ and *variance* $Var(X)$ of X are defined by

$$E(X) = \int_{-\infty}^{+\infty} x\, f(x)\,dx \quad \text{and} \quad Var(X) = \int_{-\infty}^{+\infty} (x - E(X))^2 f(x)\,dx,$$

provided the integrals exist.

The *n th moment* of X is

$$\mu_n = E(X^n) = \int_{-\infty}^{+\infty} x^n f(x)\,dx; \quad n = 0, 1, 2, \ldots,$$

provided the integrals exist. A useful representation of $Var(X)$ is

$$Var(X) = E(X^2) - (E(X))^2 = \mu_2 - \mu_1^2. \tag{1.10}$$

This relationship also holds for discrete random variables. For a continuous random variable X the probability (1.8) can be written in the following form:

$$P(a < X \le b) = \int_a^b f(x)\,dx$$

The range of X coincides with the set of all those x for which $f(x) > 0$.

Table 1.2 Probability distributions of continuous random variables X

Distribution	*Range of X*	$f(x)$	$E(X)$	$Var(X)$
Uniform distribution over [c,d]	$c \le x \le d$ $(c < d)$	$\frac{1}{d-c}$	$\frac{c+d}{2}$	$\frac{1}{12}(d-c)^2$
Exponential distribution	$x \ge 0$	$\lambda e^{-\lambda x}$ $(\lambda > 0)$	$\frac{1}{\lambda}$	$\frac{1}{\lambda^2}$
Gamma distribution	$x \ge 0$	$\frac{\lambda^\alpha}{\Gamma(\alpha)} x^{\alpha-1} e^{-\lambda x}$ $(\alpha > 0,\ \lambda > 0)$	$\frac{\alpha}{\lambda}$	$\frac{\alpha}{\lambda^2}$
Beta distribution in [0,1]	$0 \le x \le 1$	$\frac{1}{B(\alpha,\beta)} x^{\alpha-1}(1-x)^{\beta-1}$ $(\alpha > 0,\ \beta > 0)$	$\frac{\alpha}{\alpha+\beta}$	$\frac{\alpha\beta}{(\alpha+\beta+1)(\alpha+\beta)^2}$
Erlang distribution	$x > 0$	$\lambda \frac{(\lambda x)^{n-1}}{(n-1)!} e^{-\lambda x}$ $(\lambda > 0,\ n = 1, 2, \cdots)$	$\frac{n}{\lambda}$	$\frac{n}{\lambda^2}$
Normal (Gauss-) distribution	$-\infty < x < +\infty$	$\frac{1}{\sqrt{2\pi}\,\sigma} e^{-\frac{(x-\mu)^2}{2\sigma^2}}$ $(-\infty < \mu < +\infty,\ \sigma > 0)$	μ	σ^2

Comment The densities have the given functional forms over the ranges specified in the second column. Elsewhere they are identically zero. The Gamma function $\Gamma(x)$ and Beta function $B(x, y)$ are defined by

$$\Gamma(x) = \int_0^\infty u^{x-1} e^{-u}\,du \quad \text{and} \quad B(x, y) = \int_0^1 u^{x-1}(1-u)^{y-1}\,du.$$

1.2.3 Nonnegative Random Variables

If X is a nonnegative discrete random variable in the range $\{0, 1, ...\}$ with probability distribution $\{p_i = P(X = i);\ i = 0, 1, ...\}$, then its expected value

$$E(X) = \sum_{i=1}^{\infty} i\, p_i < \infty$$

can be written in the form

$$E(X) = \sum_{i=1}^{\infty} \sum_{k=i}^{\infty} p_k = \sum_{i=1}^{\infty} P(X \ge i)\,. \tag{1.11}$$

In the remainder of this section, X is assumed to be a continuous nonnegative random variable with distribution function $F(x)$ and density $f(x)$. Since both functions $F(x)$ and $f(x)$ are equal to 0 for $x < 0$, it is sufficient to define them for $x \ge 0$. An analogous formula to (1.11) for $E(X)$ given that $E(X) < \infty$ can be obtained by partial integration:

$$E(X) = \lim_{t\to\infty} \int_0^t x\, f(x)\, dx = \lim_{t\to\infty} \left[t\,F(t) - \int_0^t F(x)\, dx \right]$$

$$= \lim_{t\to\infty} \int_0^t (F(t) - F(x))\, dx$$

Hence,

$$E(X) = \int_0^{\infty} (1 - F(x))\, dx \tag{1.12}$$

An important example of a nonnegative random variable is the lifetime of technical systems, i.e. the time span to its first failure. (In this context, a failure is assumed to occur instantanously.) The terminology used in what follows refers to this example. Consequently, $F(x)$ is called *failure probability* and

$$\overline{F}(x) = 1 - F(x)$$

survival probability because $F(x)$ and $\overline{F}(x)$ are the respective probabilities that the system does or does not fail in $[0, x]$.

Let us now consider the distribution function of the residual lifetime of a system which has already worked for t time units without failing (Figure 1). This conditional failure probability will be denoted by $F_t(x)$:

$$F_t(x) = P(X - t \le x \,|\, X > t)$$

Hence, according to (1.6),

$$F_t(x) = \frac{P(X - t \le x \cap X > t)}{P(X > t)}$$

$$= \frac{P(t < X \le t + x)}{P(X > t)}$$

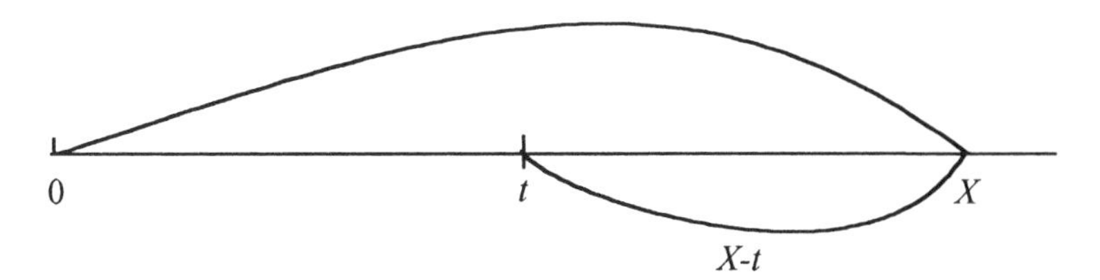

Figure 1.1 Illustration of the residual lifetime

Formula (1.8) then yields the desired result:

$$F_t(x) = \frac{F(t+x) - F(t)}{\overline{F}(t)}. \tag{1.13}$$

The corresponding *conditional survival probability* $\overline{F}_t(x) = 1 - F_t(x)$ is given by

$$\overline{F}_t(x) = \frac{\overline{F}(t+x)}{\overline{F}(t)} . \tag{1.14}$$

Example 1.1 (*uniform distribution*) Let the random variable X be uniformly distributed over [0, T]. Then its density and distribution function are (see Table 1.2)

$$f(x) = \begin{cases} 1/T & \text{for } \; 0 \le x \le T, \\ 0, & \text{elsewhere}, \end{cases} \qquad F(x) = \begin{cases} 0 & \text{for } \; x < 0, \\ x/T & \text{for } \; 0 \le x \le T, \\ 1 & \text{for } \; T < x. \end{cases}$$

Hence, the conditional failure probability is

$$F_t(x) = \frac{x}{T-t}; \quad 0 \le t < T, \; 0 \le x \le T - t.$$

Thus, the residual lifetime after the time point t is uniformly distributed over the interval $[0, T-t]$.

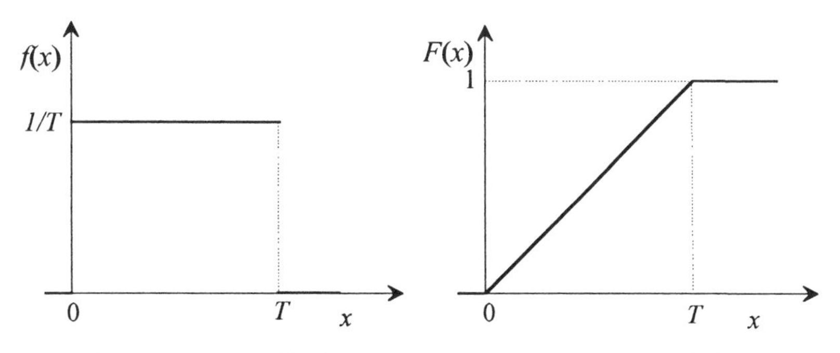

Figure 1.2 Density and distribution function of a random variable being uniformly distributed over [0, T]

Obviously, the conditional failure probability is increasing with increasing t. Specifically, for $T = 100$ the system fails in the interval (50, 60] with probability 1/10. However, if it is known that the system has not failed in the interval [0, 40], then it fails in the interval [50, 60] with probability

$$F_{40}(20) - F_{40}(10) = \frac{20}{100-40} - \frac{10}{100-40} \approx 0.17$$ □

Example 1.2 ***(exponential distribution)*** Let the random lifetime X of a system be exponentially distributed with parameter λ, i.e. its density and distribution function are (see Table 1.2)

$$f(x) = \lambda e^{-\lambda x}, \quad x \geq 0, \quad F(x) = 1 - e^{-\lambda x}, \quad x \geq 0.$$

Given t, the corresponding conditional failure probability is

$$F_t(x) = \frac{(1 - e^{-\lambda(t+x)}) - (1 - e^{-\lambda t})}{e^{-\lambda t}} = 1 - e^{-\lambda x}, \quad x \geq 0. \tag{1.15}$$

Thus, the residual lifetime of the system has the same distribution function as the lifetime of a new system: it is exponentially distributed with parameter λ. The exponential distribution is the only continuous probability distribution which has this so-called *memoryless property* or *lack of memory property*. Consequently, the age of a system with exponential lifetime has no influence on its future failure behaviour. Or, equivalently, if the system has not failed in the interval $[0, t]$, then, with respect to its failure behaviour in $[t, \infty)$, it is "as good as new". Electronic hardware often has this property after the "early failure time period".

The relationship (1.15) can also be written in the form

$$\overline{F}(t+x) = \overline{F}(t)\,\overline{F}(x). \tag{1.16}$$

It can be shown that the exponential distribution is the only one which satisfies (1.16). □

The practical background of the conditional failure probabilit motivates the following definition:

Definition 1.1 A system is *aging* in the interval $[t_1, t_2]$, $t_1 < t_2$, if and only if for an arbitrary, but fixed x and for increasing t, $t_1 \leq t \leq t_2$, the conditional failure probability $F_t(x)$ (conditional survival probability $\overline{F}_t(x)$) is increasing (decreasing). ●

The following considerations provide another approach to modeling the aging behaviour of systems: When the conditional failure probability $F_t(\Delta t)$ of a system in the interval $[t, t+\Delta t]$ is considered relative to the length Δt of this interval, one obtains a conditional failure probability per unit time $F_t(\Delta t)/\Delta t$, that is, a "failure probability rate". As $\Delta t \to 0$, this rate tends to a function $\lambda(t)$, which gives information on the instantaneous tendency of the system to fail:

$$\lambda(t) = \lim_{\Delta t \to 0} \frac{1}{\Delta t} F_t(\Delta t) = \lim_{\Delta t \to 0} \frac{F(t+\Delta t) - F(t)}{\Delta t} \cdot \frac{1}{\overline{F}(t)} \tag{1.17}$$

Hence,

$$\lambda(t) = f(t) / \overline{F}(t)$$

Integration on both sides of this relationship yields

$$F(x) = 1 - e^{-\int_0^x \lambda(t)\, dt}$$

The function $\lambda(t)$ is called *failure rate*. The *integrated failure rate* or the *hazard function* is given by

$$\Lambda(x) = \int_0^x \lambda(t)\, dt$$

By making use of $\Lambda(x)$ we obtain

$$F(x) = 1 - e^{-\Lambda(x)}, \qquad \overline{F}(x) = e^{-\Lambda(x)}, \tag{1.18}$$

$$F_t(x) = 1 - e^{-[\Lambda(t+x) - \Lambda(t)]}, \qquad \overline{F}_t(x) = e^{-[\Lambda(t+x) - \Lambda(t)]} \tag{1.19}$$

(1.18) implies an important property of the failure rate:

A system ages in the interval $[t_1, t_2]$, $t_1 < t_2$, if and only if its failure rate is increasing in this interval.

Example 1.3 (*Weibull distribution*) A random variable X has a *Weibull distribution* with parameters β and θ if it has density

$$f(x) = \frac{\beta}{\theta}\left(\frac{x}{\theta}\right)^{\beta-1} e^{-(x/\theta)^\beta}; \quad x \geq 0; \ \beta > 0, \ \theta > 0$$

(Figure 1.3). Hence its distribution function is

$$F(x) = 1 - e^{-(x/\theta)^\beta}, \quad x \geq 0$$

The parameter θ is a *scale parameter*. It is equal to 1 if θ is the measurement unit for x. Thus, θ is unessential for characterizing the Weibull distribution. The structure of the distribution function immediately yields the corresponding hazard function:

$$\Lambda(x) = \left(\frac{x}{\theta}\right)^\beta, \quad x \geq 0$$

By differentiation,

$$\lambda(x) = \frac{\beta}{\theta}\left(\frac{x}{\theta}\right)^{\beta-1}$$

Consequently, a system with a Weibull-distributed lifetime ages if and only if $\beta > 1$.

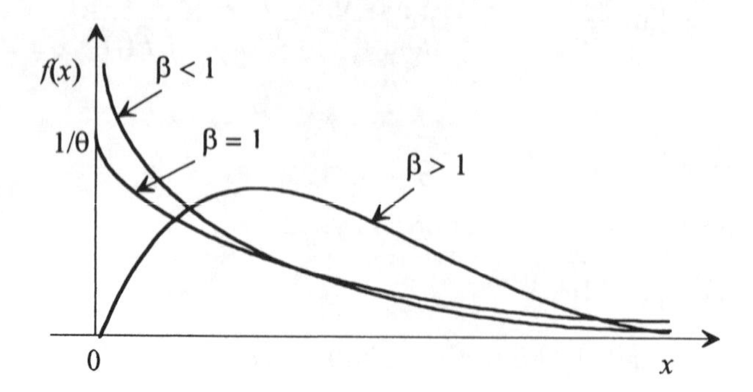

Figure 1.3 Densities of the Weibull distribution

From (1.12), the expected value of X is

$$E(X) = \int_0^\infty e^{-\left(\frac{x}{\theta}\right)^\beta} dx = \theta\Gamma\left(\frac{1}{\beta}+1\right).$$

This result shows once more that θ is a scale parameter. Special cases of the Weibull distribution are the exponential distribution ($\beta = 1$) and the *Rayleigh distribution* ($\beta = 2$). Analogously to the exponential distribution, the distribution function of a Weibull distributed random variable is sometimes written in the form

$$F(x) = 1 - e^{-\lambda x^\beta}, \quad x \geq 0.$$

Thus, $\lambda = (1/\theta)^\beta$. A disadvantage of this representation is that λ is not a scale parameter. □

For many applications, the following property of the failure rate $\lambda(x)$ is important:

$$P(X - x \leq \Delta x | X > x) = \lambda(x)\Delta x + o(\Delta x). \tag{1.20}$$

$o(h)$ is *Landau's order symbol* with respect to $h \to 0$, i.e. any function of h satisfying

$$\lim_{h\to 0} \frac{o(h)}{h} = 0$$

(see Appendix 1). Therefore, for Δx sufficiently small, $\lambda(x)\Delta x$ is approximately the probability of a system failure in $(x, x+\Delta x]$ provided that it has operated in $[0, x]$ without failing. This property of the failure rate can be used for statistical estimation: At time $t = 0$ a specified number of systems with independent, identically distributed lifetimes start working. Then the failure rate of these systems in the interval $(x, x+\Delta x]$ is approximately equal to the number of systems having failed in $(x, x+\Delta x]$ divided by the number of systems which are still operating at time x.

1.3 Random Vectors

Outcomes of random experiments frequently have to be described by more than one random variable to obtain the required information. In these cases, the outcomes are vectors $(X_1, X_2, ..., X_n)$, the components X_i of which are random variables. $(X_1, X_2, ..., X_n)$ is called a *random vector* or a *multidimensional random variable.* If, after having carried out the random experiment, $X_i = x_i;\ i = 1, 2, ..., n;$ then $(x_1, x_2, ..., x_n)$ is called a *realization* of $(X_1, X_2, ..., X_n)$. Generally, the probabilistic treatment of random vectors cannot be done by considering their components separately from each other, they have to be considered together. This requires the introduction of the *joint* or *common probability distribution* of random vectors. The joint probability distribution of $(X_1, X_2, ..., X_n)$ contains both complete information on the probability distributions of all the components X_i and complete information on their statistical dependence.

1.3.1 Two-Dimensional Random Vectors

This section considers random vectors with two components $X_1 = X$ and $X_2 = Y$. Random vectors with discrete and continuous random variables will be dealt with separately from each other, although -as in case of one-dimensional random variables- the important concepts and properties are usually identical.

Discrete components Let X and Y be two discrete random variables with respective ranges $\{x_0, x_1, ...\}$ and $\{y_0, y_1, ...\}$ and respective probability distributions $\{p_i = P(X = x_i);\ i = 0, 1, ...\}$ and $\{q_j = P(Y = y_j);\ j = 0, 1, ...\}$. Furthermore, let

$$r_{ij} = P(X = x_i \cap Y = y_j).$$

The set of probabilities $\left\{r_{ij};\ i, j = 0, 1, ...\right\}$ is the *joint* or *two-dimensional probability distribution* of the random vector (X, Y). The individual probability distributions of X and Y are referred to as the *marginal distributions* of the joint distribution of (X, Y). The following relationship holds between the joint probability distribution and its marginal distributions:

$$p_i = \sum_{j=0}^{\infty} r_{ij}, \quad q_j = \sum_{i=0}^{\infty} r_{ij}; \quad i, j = 0, 1, ... \tag{1.21}$$

The conditional probability of $X = x_i$ given $Y = y_j$ is, according to (1.6),

$$P(X = x_i | Y = y_j) = \frac{r_{ij}}{q_j}$$

The sets

$$\left\{\frac{r_{ij}}{q_j};\ i=0,1,...\right\} \text{ and } \left\{\frac{r_{ij}}{p_i};\ j=0,1,...\right\}$$

are the *conditional distributions of* X *given* $Y=y_j$ and *of* Y *given* $X=x_i$, respectively. Hence, the corresponding conditional expected values are

$$E(X|Y=y_j)=\sum_{i=0}^{\infty} x_i \frac{r_{ij}}{q_j} \quad \text{and} \quad E(Y|X=x_i)=\sum_{j=0}^{\infty} y_j \frac{r_{ij}}{p_i}.$$

Thus, the conditional expected value of X given Y is

$$E(E(X|Y))=\sum_{j=0}^{\infty} E(X|Y=y_j)P(Y=y_j)=\sum_{j=0}^{\infty}\sum_{i=0}^{\infty} x_i \frac{r_{ij}}{q_j} q_j.$$

Because the roles of X and Y can be exchanged, (1.21) yields

$$E(E(X|Y))=E(X) \quad \text{and} \quad E(E(Y|X))=E(Y). \tag{1.22}$$

The expected value of the sum $X+Y$ is

$$E(X+Y)=\sum_{i=0}^{\infty}\sum_{j=0}^{\infty}(x_i+y_j)r_{ij}=\sum_{i=0}^{\infty} x_i\sum_{j=0}^{\infty} r_{ij}+\sum_{j=0}^{\infty} y_j\sum_{i=0}^{\infty} r_{ij}.$$

Thus, in view of (1.21)

$$E(X+Y)=E(X)+E(Y). \tag{1.23}$$

The random variables X and Y are said to be *independent* if

$$r_{ij}=p_i\cdot q_j;\quad i,j=0,1,...;$$

i.e. X and Y are independent if the random events "$X=x_i$" and "$Y=y_j$" are independent for all $i,j=0,1,...$ (see section 1.1).
If X and Y are independent, then, for all $i,j=0,1,...$

$$E(X|Y=y_j)=E(X) \quad \text{und} \quad E(Y|X=x_i)=E(Y).$$

The expected value of the product XY is

$$E(XY)=\sum_{i=0}^{\infty}\sum_{j=0}^{\infty} x_i y_j r_{ij}.$$

If X and Y are independent, then

$$E(XY)=E(X)E(Y). \tag{1.24}$$

Continuous components Let the components X and Y of the random vector (X, Y) be continuous random variables with respective distribution functions and densities

$$F_X(x) = P(X \leq x) \text{ and } F_Y(y) = P(Y \leq y),$$

$$f_X(x) = \frac{dF_X(x)}{dx} \quad \text{and} \quad f_Y(y) = \frac{dF_Y(y)}{dy}.$$

The *joint distribution function* of the random vector (X, Y) is defined by the probability

$$F_{X,Y}(x,y) = P(X \leq x \cap Y \leq y)$$

as a function of x and y; $x, y \in (-\infty, +\infty)$. The joint distribution function of the random vector (X, Y) characterizes its *joint* or *two-dimensional probability distribution*. (In case of discrete components the joint distribution function is, of course, defined in the same way.) $F_{X,Y}(x,y)$ has properties

1) $F_{X,Y}(-\infty, -\infty) = 0, \quad F_{X,Y}(+\infty, +\infty) = 1,$

2) $0 \leq F_{X,Y}(x,y) \leq 1,$

3) $F_{X,Y}(x, +\infty) = F_X(x); \quad F_{X,Y}(+\infty, y) = F_Y(y),$ (1.25)

4) For $x_1 \leq x_2$ and $y_1 \leq y_2$,

$$F_{X,Y}(x_1, y_1) \leq F_{X,Y}(x_2, y_1) \leq F_{X,Y}(x_2, y_2),$$

$$F_{X,Y}(x_1, y_1) \leq F_{X,Y}(x_1, y_2) \leq F_{X,Y}(x_2, y_2).$$

Conversely, any function of two variable which has these properties is the joint distribution function of a random vector (X, Y).

Assuming its existence, the second partial derivative of $F(x,y)$ with respect to x and y,

$$f_{X,Y}(x,y) = \frac{\partial^2 F_{X,Y}(x,y)}{\partial x \, \partial y},$$

is called the *joint probability density* of (X, Y). The joint density can equivalently be defined by

$$F_{X,Y}(x,y) = \int_{-\infty}^{x} \int_{-\infty}^{y} f_{X,Y}(u, v) du \, dv. \tag{1.26}$$

Every joint (probability) density has property

$$\int_{-\infty}^{+\infty} \int_{-\infty}^{+\infty} f_{X,Y}(x,y) \, dx \, dy = 1.$$

Conversely, any nonnegative function of two variables x and y satisfying this condition can be considered to be the joint density of a random vector (X, Y).

The probability that the random vector (X, Y) assumes a realization in the region B of the (x,y)-plane is given by the area integral

$$P((X, Y) \in B) = \iint_B f_{X,Y}(x,y)\, dx\, dy \tag{1.27}$$

Putting $y = +\infty$ and $x = +\infty$, respectively, one obtains the *marginal distribution functions* of $F_{X,Y}(x,y)$

$$F_{X,Y}(x,\infty) = P(X \le x, Y \le \infty) = P(X \le x) = F_X(x)$$

$$F_{X,Y}(\infty,y) = P(X \le \infty, Y \le y) = P(Y \le y) = F_Y(x)$$

Thus, the marginal distribution functions belonging to $F_{X,Y}(x,y)$ are simply the distribution functions of X and Y, respectively. Similarly, the densities of X and Y are the *marginal densities* belonging to $f_{X,Y}(x,y)$. In view of (1.26),

$$f_X(x) = \int_{-\infty}^{+\infty} f_{X,Y}(x,y)\, dy, \quad f_Y(y) = \int_{-\infty}^{+\infty} f_{X,Y}(x,y)\, dx \tag{1.28}$$

Two random variables X and Y with joint distribution function $F_{X,Y}(x,y)$ are *independent* if, for all x and y,

$$P(X \le x,\ Y \le y) = P(X \le x)\, P(Y \le y),$$

or, equivalently,

$$F_{X,Y}(x,y) = F_X(x)\, F_Y(y)$$

If the joint density $f_{X,Y}(x,y)$ of (X, Y) exists, then the independence of X and Y is equivalent to $f_{X,Y}(x,y)$ being the product of the marginal densities:

$$f_{X,Y}(x,y) = f_X(x) f_Y(y)$$

If X and Y are not independent, then the conditional distribution function of X given $Y = y$,

$$F_X(x|y) = P(X \le x | Y = y),$$

and the corresponding conditional density of X given $Y = y$,

$$f_X(x|y) = \frac{dF_X(x|y)}{dx},$$

are of interest. It can be shown that

$$f_X(x|y) = \frac{f_{X,Y}(x,y)}{f_Y(y)} \tag{1.29}$$

Hence,

$$F_X(x) = \int_{-\infty}^{+\infty} F_X(x|y) f_Y(y)\, dy.$$

Thus, $F_X(x)$ can be interpreted as the expected value of the conditional distribution function of X given Y:

$$F_X(x) = E(F_X(x|Y)). \tag{1.30}$$

The *conditional expected value of X given $Y = y$* is

$$E(X|Y=y) = \int_{-\infty}^{+\infty} x f_X(x|y)\, dx.$$

Therefore, the expected value of X given Y is

$$E(E(X|Y)) = \int_{-\infty}^{+\infty} \int_{-\infty}^{+\infty} x f_X(x|y)\, dx\, f_Y(y)\, dy.$$

In view of (1.29),

$$E(E(X|Y)) = E(X) \quad \text{and} \quad E(E(Y|X)) = E(Y). \tag{1.31}$$

The expected values of the sum and of the product of X and Y are given by

$$E(X+Y) = \int_{-\infty}^{+\infty} \int_{-\infty}^{+\infty} (x+y) f_{X,Y}(x,y)\, dx\, dy,$$

$$E(XY) = \int_{-\infty}^{+\infty} \int_{-\infty}^{+\infty} xy f_{X,Y}(x,y)\, dx\, dy.$$

Taking into account (1.28) one obtains the same formulas as in case of discrete components:

$$E(X+Y) = E(X) + E(Y) \tag{1.32}$$

and for independent X and Y

$$E(XY) = E(X)\, E(Y). \tag{1.33}$$

The *covariance* $Cov(X,Y)$ between the random variables X and Y is defined by

$$Cov(X,Y) = E\{[X - E(X)][Y - E(Y)]\}.$$

An equivalent representation for the covariance is

$$Cov(X,Y) = E(XY) - E(X)\, E(Y).$$

In particular,

$$Cov(X,X) = Var(X).$$

From (1.33) it follows that if X and Y are independent, then $Cov(X,Y) = 0$. But if $Cov(X,Y) = 0$, then X and Y are not necessarily independent.

The *correlation coefficient* between X and Y is defined by

$$\rho(X, Y) = \frac{Cov(X, Y)}{\sqrt{Var(X)}\,\sqrt{Var(Y)}} \tag{1.34}$$

The correlation coefficient has the following properties:

1) If X and Y are independent, then $\rho(X, Y) = 0$.

2) If $Y = aX + b$ for any constants a and b, then $\rho(X, Y) = \pm 1$.

3) For any random variables X and Y, $-1 \le \rho(X, Y) \le 1$.

The correlation coefficient is, therefore, a measure of the linear relationship between random variables. X and Y are said to be *uncorrelated* if $\rho(X, Y) = 0$. Otherwise they are called *positively* or *negatively correlated* depending on the sign of $\rho(X, Y)$. Obviously, X and Y are uncorrelated if and only if

$$E(XY) = E(X)\,E(Y).$$

Thus, if X and Y are independent, then they are uncorrelated. But if X and Y are uncorrelated, they need not be independent.

Example 1.4 (*bivariate normal distribution*) The random vector (X, Y) is *bivariate normal distributed* with parameters

$$\mu_x, \mu_y, \sigma_x, \sigma_y \text{ and } \rho; \quad -\infty < \mu_x, \mu_y < \infty, \; \sigma_x > 0, \sigma_y > 0, \; -1 < \rho < 1,$$

if it has joint density (Figure 1.4)

$$f_{X,Y}(x, y) = \frac{1}{2\pi\sigma_x\sigma_y\sqrt{1-\rho^2}} \exp\left(-\frac{1}{2(1-\rho^2)}\left[\frac{(x-\mu_x)^2}{\sigma_x^2} - 2\rho\frac{(x-\mu_x)(y-\mu_y)}{\sigma_x\sigma_y} + \frac{(y-\mu_y)^2}{\sigma_y^2}\right]\right),$$

$$-\infty < x, y < +\infty$$

The corresponding marginal densities are obtained by using (1.28):

$$f_X(x) = \frac{1}{\sqrt{2\pi}\,\sigma_x} \exp\left(-\frac{(x-\mu_x)^2}{2\sigma_x^2}\right), \quad -\infty < x < +\infty$$

$$f_Y(y) = \frac{1}{\sqrt{2\pi}\,\sigma_y} \exp\left(-\frac{(y-\mu_y)^2}{2\sigma_y^2}\right), \quad -\infty < y < +\infty$$

Corollary If (X, Y) is bivariate normally distributed with parameters μ_x, σ_x, μ_y, σ_y, and ρ, then X and Y are normally distributed with parameters μ_x, σ_x and μ_y, σ_y; respectively. X and Y are independent if and only if $\rho = 0$. (Note that the independence of X and Y is equivalent to $f_{X,Y}(x, y) = f_X(x) f_Y(y)$.)

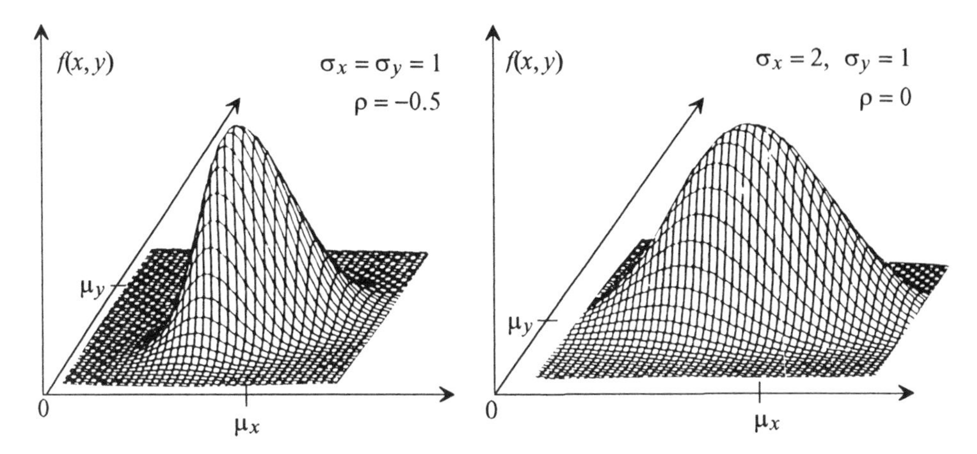

Figure 1.4 Probability densities of the bivariate normal distribution

It can be shown that the parameter ρ is equal to the correlation coefficient between X and Y ($\rho = \rho(X, Y)$)). Therefore,

> *If the random vector (X, Y) has a bivariate normal distribution, then X and Y are independent if and only if they are uncorrelated.*

The conditional density of X given $Y = y$ is obtained from $f_{X,Y}(x,y)$ and formula (1.29):

$$f_X(x|y) = \frac{1}{\sqrt{2\pi}\,\sigma_x\sqrt{1-\rho^2}} \exp\left\{-\frac{1}{2\sigma_x^2(1-\rho^2)}\left[x - \rho\frac{\sigma_x}{\sigma_y}(y-\mu_y) - \mu_x\right]^2\right\}$$

Thus, the random variable X given $Y = y$ is normally distributed as

$$N\left(\rho\frac{\sigma_x}{\sigma_y}(y-\mu_y) + \mu_x,\ \sigma_x^2(1-\rho^2)\right)$$

The parameters of this distribution are the conditional expected value of X given that $Y = y$ and the conditional variance of X given that $Y = y$:

$$E(X|Y=y) = \rho\frac{\sigma_x}{\sigma_y}(y-\mu_y) + \mu_x$$

$$Var(X|Y=y) = \sigma_x^2(1-\rho^2)$$

Of course, in these formulas the roles of X and Y can be changed. Sums of normally distributed random variables are considered in section 1.4. □

1.3.2 *n*-Dimensional Random Vectors

Let $X_1, X_2, \ldots, X_n$ be continuous random variables with distribution functions $F_{X_1}(x_1), F_{X_2}(x_2), \ldots, F_{X_n}(x_n)$ and densities $f_{X_1}(x_1), f_{X_2}(x_2), \ldots, f_{X_n}(x_n)$. The *joint distribution function* of the random vector $\mathbf{X} = (X_1, X_2, \ldots, X_n)$ is

$$F_{\mathbf{X}}(x_1, x_2, \ldots, x_n) = P(X_1 \le x_1, X_2 \le x_2, \ldots, X_n \le x_n)$$

Provided its existence, the n th mixed partial derivative of the joint distribution function with respect to the $x_1, x_2, \ldots, x_n$:

$$f_{\mathbf{X}}(x_1, x_2, \ldots, x_n) = \frac{\partial^n F_{\mathbf{X}}(x_1, x_2, \ldots, x_n)}{\partial x_1 \partial x_2 \cdots \partial x_n}$$

is called the *joint* (*probability*) *density* of the random vector **X**. The functions $F_{\mathbf{X}}(x_1, x_2, \ldots, x_n)$ and $f_{\mathbf{X}}(x_1, x_2, \ldots, x_n)$ are also called the *n- dimen- sional distribution function* and *n-dimensional* (*probability*) *density*, respectively, of **X**. They determine the *n-dimensional probability distribution* of **X**. The characteristic properties of two-dimensional distribution functions and probability densities can be extended in a straightforward way to *n*-dimensional distribution functions and densities. Hence they will not be given here.

The distribution functions and the densities of the X_i can be obtained from the joint distribution function and density, respectively, analogously to the two-dimensional case:

$$F_{X_i}(x_i) = F_{\mathbf{X}}(\infty, \ldots, \infty, x_i, \infty, \ldots, \infty)$$

$$f_{X_i}(x_i) = \int_{-\infty}^{+\infty} \cdots \int_{-\infty}^{+\infty} f_{\mathbf{X}}(x_1, \ldots, x_i, \ldots, x_n)\, dx_1 \cdots dx_{i-1} dx_{i+1} \cdots dx_n \qquad (1.35)$$

The joint distribution function of (X_i, X_j), $i < j$, is given by

$$F_{X_i, X_j}(x_i, x_j) = F_{\mathbf{X}}(\infty, \ldots, \infty, x_i, \infty, \ldots, \infty, x_j, \infty, \ldots, \infty)$$

The random variables $X_1, X_2, \ldots, X_n$ are said to be *independent* if

$$F_{\mathbf{X}}(x_1, x_2, \ldots, x_n) = F_{X_1}(x_1) F_{X_2}(x_2) \cdots F_{X_n}(x_n)$$

Assuming the existence of the densities $f_{X_i}(x_i)$, independence of the X_i is equivalent to

$$f_{\mathbf{X}}(x_1, x_2, \ldots, x_n) = f_{X_1}(x_1) f_{X_2}(x_2) \cdots f_{X_n}(x_n) \qquad (1.36)$$

Conditional densities can be obtained analogously to the two-dimensional case: For instance, the conditional density of $\boldsymbol{X}$ given $X_i = x_i$ is

$$f_{\mathbf{X}}(x_1,...,x_{i-1},x_{i+1},...,x_n \,|\, x_i) = \frac{f_{\mathbf{X}}(x_1,x_2,...,x_n)}{f_{X_i}(x_i)}, \quad i = 1,2,...,n,$$

whereas the conditional density of $\mathbf{X}$ given $X_1 = x_1 \cap X_2 = x_2$ is

$$f_{\mathbf{X}}(x_3,x_4,...,x_n|x_1,x_2) = \frac{f_{\mathbf{X}}(x_1,x_2,...,x_n)}{f_{(X_1,X_2)}(x_1,x_2)}$$

In this formula,

$$f_{(X_1,X_2)}(x_1,x_2) = \int_{-\infty}^{+\infty}\int_{-\infty}^{+\infty}\cdots\int_{-\infty}^{+\infty} f_{\mathbf{X}}(x_1,x_2,\cdots,x_n)\,dx_3\,dx_4...\,dx_n$$

is the joint density of (X_1,X_2).

The expected value of the product $X_1 X_2 \cdots X_n$ is defined by

$$E(X_1 X_2 \cdots X_n) = \int_{-\infty}^{+\infty}\int_{-\infty}^{+\infty}\cdots\int_{-\infty}^{+\infty} x_1 x_2 \cdots x_n f_{\mathbf{X}}(x_1,x_2,...,x_n)\,dx_1 dx_2 \cdots dx_n$$

In view of (1.35) and (1.36), for independent X_i this n-dimensional integral simplifies to

$$E(X_1 X_2 \cdots X_n) = E(X_1)\,E(X_2)\cdots E(X_n) \tag{1.37}$$

The expected value of the product of independent random variables is equal to the product of the expected values of these random variables.

Starting with (1.33), this fact can be more easily proved by induction. Hence it is also valid for discrete random variables X_i.

Let

$$\sigma_{ij} = Cov(X_i,X_j) \text{ and } \rho_{ij} = \rho(X_i,X_j); \quad i,j = 1,2,...,n,$$

be the covariance and the correlation coefficient between X_i and X_j, respectively. (Note that $\sigma_{ii} = Var(X_i)$ and $\rho_{ii} = 1$.) It is useful to combine these parameters in the *covariance matrix* Σ and in the *correlation matrix* ρ, respectively:

$$\Sigma = ((\sigma_{ij})), \quad \rho = ((\rho_{ij})); \quad i,j = 1,2,...,n$$

Example 1.5 (*n-dimensional normal distribution*) Let $X_1,X_2,\cdots,X_n$ be random variables with vector of expected values $\mu = (\mu_1,\mu_2,\cdots,\mu_n)$ and with covariance matrix Σ. Furthermore, let be $|\Sigma|$ the positive determinant of Σ, Σ^{-1} the inverse of Σ, and $\mathbf{x} = (x_1,x_2,\cdots,x_n)$. Then the random vector $\mathbf{X} = (X_1,X_2,\cdots,X_n)$ is n-dimensionally normally distributed if it has joint density

$$f_{\mathbf{X}}(\mathbf{x}) = \frac{1}{\sqrt{(2\pi)^n |\Sigma|}} \exp\left(-\frac{1}{2}(\mathbf{x}-\boldsymbol{\mu})\Sigma^{-1}(\mathbf{x}-\boldsymbol{\mu})^T\right),$$

where $(\mathbf{x}-\boldsymbol{\mu})^T = \begin{pmatrix} x_1 - \mu_1 \\ x_2 - \mu_2 \\ \cdot \\ x_n - \mu_n \end{pmatrix}$ is the transpose of the vector $\mathbf{x}-\boldsymbol{\mu}$.

In detail, this density is

$$f_{\mathbf{X}}(x_1, x_2, \cdots, x_n) = \frac{1}{\sqrt{(2\pi)^n |\Sigma|}} \exp\left(-\frac{1}{2|\Sigma|} \sum_{i=1}^{n} \sum_{j=1}^{n} \Sigma_{ij}(x_i - \mu_i)(x_j - \mu_j)\right),$$

where Σ_{ij} is the cofactor of σ_{ij}.

As a special case, for $n = 2$, $x_1 = x$ and $x_2 = y$ one obtains the density of the bivariate normal distribution (example 1.4). Generalizing from the bivariate case it can be shown that the X_i are $N(\mu_i, \sigma_i^2)$-distributed random variables; $i = 1, 2, ..., n$. If the X_i are uncorrelated, then $\Sigma = ((\sigma_{ij}))$ is a diagonal matrix ($\sigma_{ij} = 0$ for $i \neq j$) so that the product form (1.36) of the density and, therefore, the independence of the X_i follows:

$$f_{\mathbf{X}}(x_1, x_2, \cdots, x_n) = \prod_{i=1}^{n} \left[\frac{1}{\sqrt{2\pi}\,\sigma_i} \exp\left(-\frac{1}{2}\left(\frac{x_i - \mu_i}{\sigma_i}\right)^2\right)\right] \quad \square$$

Theorem 1.1 Let the random vector $(X_1, X_2, ..., X_n)$ be n-dimensionally normally distributed and let the random variables $Y_1, Y_2, ..., Y_m$ be linear combinations of the X_i:

$$Y_i = \Sigma_{j=1}^{m_i} a_{ij} X_j; \quad i = 1, 2, ..., m$$

Then $(Y_1, Y_2, ..., Y_m)$ is m-dimensionally normally distributed. ■

1.4 Sums and Sequences of Random Variables

Let $X_1, X_2, \cdots, X_n$ be a finite sequence of random variables with distribution functions $F_{X_i}(x_i)$, densities $f_{X_i}(x_i)$, finite expected values $E(X_i)$ and finite variances $Var(X_i)$; $i = 1, 2, ..., n$. Furthermore, let $f_{\mathbf{X}}(x_1, x_2, ..., x_n)$ be the joint density of the random vector $\mathbf{X} = (X_1, X_2, ..., X_n)$. Next expected value and variance of the sum of the random variables $X_1, X_2, \cdots, X_n$ are determined.

Expected value of a sum The expected value of the sum $X_1 + X_2 + \cdots + X_n$ is defined by

$$E\left(\sum_{i=1}^{n} X_i\right) = \int_{-\infty}^{+\infty}\int_{-\infty}^{+\infty}\cdots\int_{-\infty}^{+\infty} (x_1 + x_2 + \cdots + x_n) f_{\mathbf{X}}(x_1, x_2, \ldots, x_n)\, dx_1 dx_2 \cdots dx_n$$

From (1.35),

$$E\left(\sum_{i=1}^{n} X_i\right) = \sum_{i=1}^{n} \int_{-\infty}^{+\infty} x_i f_{X_i}(x_i)\, dx_i$$

Hence,

$$E\left(\sum_{i=1}^{n} X_i\right) = \sum_{i=1}^{n} E(X_i) \tag{1.38}$$

The expected value of the sum of any random variables is equal to the sum of the expected values of these random variables.

Using formula (1.32), this fact can be more easily proved by induction. In view of (1.23), formula (1.38) is also valid for discrete random variables X_i.

Variance of a sum The variance of the sum $X_1 + X_2 + \cdots + X_n$ is

$$Var\left(\sum_{i=1}^{n} X_i\right) = \sum_{i=1}^{n}\sum_{j=1}^{n} Cov(X_i, X_j)$$

Since

$$Cov(X_i, X_i) = Var(X_i) \quad \text{and} \quad Cov(X_i, X_j) = Cov(X_j, X_i),$$

this formula implies

$$Var\left(\sum_{i=1}^{n} X_i\right) = \sum_{i=1}^{n} Var(X_i) + 2 \sum_{\substack{i,j=1 \\ i<j}}^{n} Cov(X_i, X_j)$$

Thus, for uncorrelated X_i,

$$Var\left(\sum_{i=1}^{n} X_i\right) = \sum_{i=1}^{n} Var(X_i) \tag{1.39}$$

The variance of a sum of uncorrelated random variables is equal to the sum of the variances of these random variables.

Let $\alpha_1, \alpha_2, \cdots, \alpha_n$ be any sequence of finite real numbers. Then,

$$E\left(\sum_{i=1}^{n} \alpha_i X_i\right) = \sum_{i=1}^{n} \alpha_i E(X_i)$$

$$Var\left(\sum_{i=1}^{n} \alpha_i X_i\right) = \sum_{i=1}^{n} \alpha_i^2 Var(X_i) + 2 \sum_{\substack{i,j=1 \\ i<j}}^{n} \alpha_i \alpha_j Cov(X_i, X_j)$$

If the X_i are uncorrelated,

$$Var\left(\sum_{i=1}^{n} \alpha_i X_i\right) = \sum_{i=1}^{n} \alpha_i^2 Var(X_i) \tag{1.40}$$

The random variables $X_1, X_2, \ldots$ are said to be *identically distributed as X* if all of them have the same probability distribution as X. From a probabilistic point view, there is no difference between identically distributed random variables. In the case of independent, identically as X distributed random variables, formulas (1.38) and (1.39) simplify to

$$E\left(\sum_{i=1}^{n} X_i\right) = n\,E(X), \qquad Var\left(\sum_{i=1}^{n} X_i\right) = n\,Var(X)$$

Distribution of a sum Let $Z = X + Y$ be the sum of two independent discrete random variables with probability distributions

$$\{p_i = P(X=i);\ i = 0, 1, \ldots\} \text{ and } \{q_j = P(Y=j);\quad j = 0, 1, \ldots\}.$$

Then,

$$P(Z=k) = P(X+Y=k) = \sum_{i=0}^{k} P(X=i)\,P(Y=k-i)$$

Letting $r_k = P(Z=k)$ yields for all $k = 0, 1, \ldots$

$$r_k = \sum_{i=0}^{k} p_i q_{k-i} = p_0 q_k + p_1 q_{k-1} + \ldots + p_k q_0 \tag{1.41}$$

A discrete probability distribution $\{r_k;\ k = 0, 1, \ldots\}$ with this structure is called a *convolution* of the probability distributions $\{p_i; i = 0, 1, \ldots\}$ and $\{q_j; j = 0, 1, \ldots\}$.

Now let X and Y be two independent, continuous random variables with densities $f_X(x)$ and $f_Y(y)$. Then their sum $Z = X + Y$ has density

$$f_Z(z) = \int_{-\infty}^{+\infty} f_Y(z-x) f_X(x)\,dx = \int_{-\infty}^{+\infty} f_X(z-y) f_Y(y)\,dy \tag{1.42}$$

Integration yields the corresponding formula for the distribution function $F_Z(z)$ of Z:

$$F_Z(z) = \int_{-\infty}^{+\infty} F_Y(z-x) f_X(x)\,dx = \int_{-\infty}^{+\infty} F_X(z-y) f_Y(y)\,dy \tag{1.43}$$

Since $f(x) = \dfrac{dF(x)}{dx}$ or $dF(x) = f(x)\,dx$, this relationship can be written in the form

$$F_Z(z) = \int_{-\infty}^{+\infty} F_Y(z-x)\,dF_X(x) = \int_{-\infty}^{+\infty} F_X(z-y)\,dF_Y(y)$$

If X and Y are nonnegative, then (1.42) and (1.43) become

$$f_Z(z) = \int_0^z f_Y(z-x) f_X(x)\,dx = \int_0^z f_X(z-y) f_Y(y)\,dy, \quad z \ge 0 \tag{1.44}$$

$$F_Z(z) = \int_0^z F_Y(z-x) f_X(x)dx = \int_0^z F_X(z-y) f_Y(y)dy, \quad z \ge 0 \tag{1.45}$$

The integrals in (1.42) and (1.43) are called *convolutions* of the densities f_X and f_Y and the distribution functions F_X and F_Y, respectively. Hence, analogously to discrete random variables, the following statement is valid:

> *The probability density (distribution function) of the sum of two independent random variables is given by the convolution of their probability densities (distribution functions).*

The density and distributions function of a sum $Z = X_1 + X_2 + \cdots + X_n$ of n independent, continuous random variables X_i are obtained by repeated application of the formulas (1.42) and (1.43), respectively. The resulting functions are respective the *convolutions* of the probability densities f_{X_i} and of the distribution functions F_{X_i}, $i = 1, 2, ..., n$. They are denoted by

$$f_Z(z) = f_1 * f_2 * \cdots * f_n(z) \quad \text{and} \quad F_Z(z) = F_1 * F_2 * \cdots * F_n(z),$$

respectively. If, in particular, the X_i are independent and identically distributed with density $f(x)$, then the density of the sum $Z = X_1 + X_2 + \cdots + X_n$ is denoted by $f^{*(n)}(z)$. This *n th convolution power* of $f(x)$ can be recursively obtained from

$$f^{*(i)}(z) = \int_{-\infty}^{+\infty} f^{*(i-1)}(z-x) f(x)dx\,, \tag{1.46}$$

$i = 2, 3, ..., n;\ \ f^{*(1)}(x) \equiv f(x)$.

Analogously, the distribution of a sum of n independent, identically distributed random variables X_i, each with distribution function $F(x)$, is given by the *n th convolution power* $F^{*(n)}(z)$ of $F(x)$, which can be recursively obtained from

$$F^{*(i)}(z) = \int_{-\infty}^{+\infty} F^{*(i-1)}(z-x)\,dF(x)\,, \tag{1.47}$$

$i = 2, 3, ..., n;\ \ F^{*(1)}(x) \equiv F(x)$.

For nonnegative X_i the formulas (1.46) and (1.47) become

$$f^{*(i)}(z) = \int_0^z f^{*(i-1)}(z-x) f(x)\, d(x), \quad z \ge 0 \tag{1.48}$$

$$F^{*(i)}(z) = \int_0^z F^{*(i-1)}(z-x)\, dF(x), \quad z \ge 0 \tag{1.49}$$

Example 1.6 (*Erlang distribution*) Let the independent random variables X_1 and X_2 be exponentially distributed with parameters λ_1 and λ_2:

$$f_{X_i}(x) = \lambda_i e^{-\lambda_i x}, \quad F_{X_i}(x) = 1 - e^{-\lambda_i x}; \; x \ge 0, \; i = 1, 2$$

From (1.44), $Z = X_1 + X_2$ has density

$$f_Z(z) = \int_0^z \lambda_2 e^{-\lambda_2 (z-x)} \lambda_1 e^{-\lambda_1 x}\, dx = \lambda_1 \lambda_2 e^{-\lambda_2 z} \int_0^z e^{-(\lambda_1 - \lambda_2)x}\, dx$$

If $\lambda_1 = \lambda_2 = \lambda$, then

$$f_Z(z) = \lambda^2 z e^{-\lambda z}, \quad z \ge 0$$

This is the density of an Erlang distribution with parameters $n = 2$ and λ (see Table 1.2).

If $\lambda_1 \ne \lambda_2$, then

$$f_Z(z) = \frac{\lambda_1 \lambda_2}{\lambda_1 - \lambda_2}\left(e^{-\lambda_2 z} - e^{-\lambda_1 z}\right), \quad z \ge 0$$

Now let $X_1, X_2, ..., X_n$ be independent, identically distributed exponential random variables with parameter λ. Then the sum $Z = X_1 + X_2 + \cdots + X_n$ is Erlang distributed with parameters n and λ. This statement can be easily proved by induction: The assertion has been already proved for $n = 2$. Assuming $X_1 + X_2 + \cdots + X_{n-1}$ is Erlang distributed with parameter n-1 and λ. Then, according to Table 1.2,

$$f^{*(n-1)}(x) = \lambda \frac{(\lambda x)^{n-2}}{(n-2)!} e^{-\lambda x}, \quad x \ge 0, \; n > 2,$$

where $f(x) = \lambda e^{-\lambda x}$, $x > 0$. On condition that this assumption is true, the density of $Z = X_1 + X_2 + \cdots + X_n$ is by (1.44)

$$f^{*(n)}(z) = \int_0^z \lambda \frac{[\lambda(z-x)]^{n-2}}{(n-2)!} e^{-\lambda(z-x)} \lambda e^{-\lambda x}\, dx$$

Thus,

$$f^{*(n)}(z) = \lambda \frac{\lambda^{n-1}}{(n-2)!} e^{-\lambda z} \int_0^z (z-x)^{n-2}\, dx$$

$$= \lambda \frac{\lambda^{n-1}}{(n-2)!} e^{-\lambda z} \left[-\frac{(z-x)^{n-1}}{n-1} \right]_0^z$$

$$= \lambda \frac{(\lambda z)^{n-1}}{(n-1)!} e^{-\lambda z}, \quad z \geq 0$$

But this is the density of an Erlang distributed random variable with parameters n and λ. The corresponding distribution function is

$$F(z) = 1 - e^{-\lambda z} \sum_{i=0}^{n-1} \frac{(\lambda z)^i}{i!}, \quad z \geq 0 \tag{1.50}$$

□

Example 1.7 (*Normal distribution*) Let the random variables X_i be independent and normally distributed with parameters μ_i and σ_i^2; $i = 1, 2$:

$$f_{X_i}(x) = \frac{1}{\sqrt{2\pi}\, \sigma_i} \exp\left(-\frac{1}{2} \frac{(x-\mu_i)^2}{\sigma_i^2} \right); \quad i = 1, 2$$

According to (1.42), the probability density of the sum $Z = X_1 + X_2$ is

$$f_Z(z) = \int_{-\infty}^{+\infty} \frac{1}{\sqrt{2\pi}\, \sigma_2} \exp\left(-\frac{1}{2} \frac{(z-x-\mu_2)^2}{\sigma_2^2} \right) \frac{1}{\sqrt{2\pi}\, \sigma_1} \exp\left(-\frac{1}{2} \frac{(x-\mu_1)^2}{\sigma_1^2} \right) dx\,.$$

Substituting $u = x - \mu_1$ and $v = z - \mu_1 - \mu_2$ yields

$$f_Z(z) = \frac{1}{2\pi\sigma_1\sigma_2} \int_{-\infty}^{+\infty} \exp\left[-\frac{1}{2} \left(\frac{(v-u)^2}{\sigma_2^2} + \frac{u^2}{\sigma_1^2} \right) \right] du$$

Replacing the term $(v-u)^2/\sigma_2^2 + u^2/\sigma_1^2$ in the exponent of the integrand by

$$\left(\frac{\sqrt{\sigma_1^2 + \sigma_2^2}}{\sigma_1\sigma_2} u - \frac{\sigma_1}{\sigma_2\sqrt{\sigma_1^2 + \sigma_2^2}} v \right)^2 + \frac{1}{\sigma_1^2 + \sigma_2^2} v^2$$

and substituting

$$t = \frac{\sqrt{\sigma_1^2 + \sigma_2^2}}{\sigma_1\sigma_2} u - \frac{\sigma_1}{\sigma_2\sqrt{\sigma_1^2 + \sigma_2^2}} v$$

yields

$$f_Z(z) = \frac{1}{2\pi\sqrt{\sigma_1^2+\sigma_2^2}} \exp\left(-\frac{v^2}{2\left(\sigma_1^2+\sigma_2^2\right)}\right) \int_{-\infty}^{+\infty} e^{-t^2/2}\, dt$$

Since $\int_{-\infty}^{+\infty} e^{-t^2/2}\, dt = \sqrt{2\pi}$, the density of Z becomes

$$f_Z(z) = \frac{1}{\sqrt{2\pi(\sigma_1^2+\sigma_2^2)}} \exp\left(-\frac{(z-\mu_1-\mu_2)^2}{2\left(\sigma_1^2+\sigma_2^2\right)}\right); \quad -\infty < z < +\infty \quad \square$$

By theorem 1.1, a sum of normally distributed random variables is also normally distributed. For independent random variables, example 1.7 yields, by induction, a sharpening of this statement:

Corollary Let $Z = X_1 + X_2 + ... + X_n$ be the sum of independent random variables $X_i = N(\mu_i, \sigma_i^2);\ i = 1, 2, ..., n$. Then,

$$Z = N\left(\mu_1 + \mu_2 + ... + \mu_n,\ \sigma_1^2 + \sigma_2^2 + ... + \sigma_n^2\right)$$

From theorem 1.1 we know that if $(X_1, X_2, ..., X_n)$ has a joint normal distribution, then the sum $X_1 + X_2 + \cdots + X_n$ has a normal distribution. In particular, if (X, Y) has a bivariate normal distribution with parameters μ_x, μ_y, σ_x, σ_y and ρ, then $X+Y$ has a normal distribution with

$$E(X+Y) = \mu_x + \mu_y, \quad Var(X+Y) = \sigma_x^2 + 2\rho\sigma_x\sigma_y + \sigma_y^2$$

Standardized random variables A random variable X is called *standardized* if $E(X) = 0$ and $Var(X) = 1$. If X is any random variable with finite expected value and finite variance, then

$$Y = \frac{X - E(X)}{\sqrt{Var(X)}},$$

is a standardized random variable, since obviously $E(Y) = 0$ and $Var(Y) = 1$. The random variable Y is called the *standardized version* of X.

If X is normally distributed, then the standardized version of X is also normally distributed: $Y = N(0, 1)$. The density and distribution function of an $N(0, 1)$-distributed random variable, which is referred to as a *standard normal random variable,* are denoted by $\phi(x)$ and $\Phi(x)$, respectively:

$$\phi(x) = \frac{1}{\sqrt{2\pi}} e^{-x^2/2}, \quad \Phi(x) = \frac{1}{\sqrt{2\pi}} \int_{-\infty}^{x} e^{-u^2/2}\, du \tag{1.51}$$

Theorem 1.2 (*Central limit theorem*) Let $X_1, X_2, \ldots$ be an infinite sequence of independent random variables with finite expected values $E(X_i) = \mu$ and finite variances $Var(X_i) = \sigma^2$; $i = 1, 2, \ldots$. Furthermore, let $Z_n = X_1 + X_2 + \cdots + X_n$ and

$$Y_n = \frac{Z_n - n\mu}{\sigma\sqrt{n}}$$

Then,

$$\lim_{n\to\infty} P(Y_n \leq x) = \Phi(x)$$ ■

Corollary For n sufficiently large, Z_n is approximately normally distributed with expected value $n\mu$ and variance $n\sigma^2$:

$$Z_n \approx N(n\mu, n\sigma^2)$$

Theorem 1.3 deals with the sum of a random number of random variables. To state this theorem, another important concept has to be introduced:

Definition 1.2 (*Stopping time*) An integer-valued, positive random variable N is said to be a *stopping time* for the sequence of independent random variables $X_1, X_2, \ldots$ if the random event "$N = n$" is independent of all $X_{n+1}, X_{n+2}, \ldots$ for $n = 1, 2, \ldots$ ●

Comment A random event A is *independent of a random variable* X if for all x the random events A and "$X \leq x$" are independent.

Intuitively, the notation "stopping time" can be motivated by assuming that the random variables X_i are observed in turn. For $N = n$, this process is stopped after having observed $X_1, X_2, \ldots, X_n$, i.e., before observing $X_{n+1}, X_{n+2}, \ldots$

Example 1.8 Let $X_i = 1$ if after the ith tossing of a coin a "head" is observed and $X_i = 0$ otherwise. Further, let $P(X_i = 1) = P(X_i = 0) = 0.5$ for all $i = 0, 1, \ldots$ Then

$$N = \min\{n;\ X_1 + X_2 + \cdots + X_n = 8\}$$

is a stopping time for the sequence $X_1, X_2, \ldots$. □

Theorem 1.3 (*Wald's equation*) Let $X_1, X_2, \ldots$ be a sequence of independent random variables and let N be a stopping time for this sequence. Assuming the X_i to be identically distributed as X and the expected values $E(X)$ and $E(N)$ to be finite, then

$$E\left(\sum_{i=1}^{N} X_i\right) = E(X)\,E(N) \tag{1.52}$$

Proof Let the binary random variables Y_i be defined by

$$Y_i = \begin{cases} 1, & \text{if } N \geq i, \\ 0, & \text{if } N < i, \end{cases} \quad i = 1, 2, \ldots .$$

$Y_i = 1$ holds if and only if no stopping has occured after observing the random variables $X_1, X_2, \ldots, X_{i-1}$. Since N is a stopping time, Y_i is independent of the random variables $X_i, X_{i+1}, \ldots$ In particular, Y_i is independent of X_i so that

$$E(X_i Y_i) = E(X_i) E(Y_i)$$

Since $E(Y_i) = P(N \geq i)$, formula (1.11) yields

$$\begin{aligned} E\left(\sum_{i=1}^{N} X_i\right) &= E\left(\sum_{i=1}^{\infty} X_i Y_i\right) \\ &= \sum_{i=1}^{\infty} E(X_i) E(Y_i) \\ &= E(X) \sum_{i=1}^{\infty} E(Y_i) \\ &= E(X) \sum_{i=1}^{\infty} P(N \geq i) \\ &= E(X) E(N) \end{aligned}$$

■

Obviously, formula (1.52) holds if N is independent of all $X_1, X_2, \ldots$

Since $E(X_i) = 1/2$, applying Wald's equation to example 1.8 yields

$$E(X_1 + X_2 + \cdots + X_n) = \frac{1}{2} E(N) .$$

According to the definition of N, $X_1 + X_2 + \cdots + X_n = 8$. Hence, $E(N) = 16$.

Now, let X_i denote the random gain (loss) of a player when tossing the i th coin (depending on whether "head" or "tail" is observed. If all X_i are identically distributed according to $P(X_i = -1) = P(X_i = +1) = 1/2$; $i = 1, 2, \ldots$; then

$$N = \min\{n, \; X_1 + X_2 + \cdots + X_n = 3\}$$

is a stopping time with respect to $X_1, X_2, \ldots$ A formal application of Wald's equation would yield

$$E(X_1 + X_2 + \cdots + X_N) = E(X) E(N).$$

The left hand side of this equation is equal to 3. The right hand side contains the factor $E(X) = 0$. This situation requires the assumption $E(N) = \infty$. Therefore, in this case Wald's equation is not applicable.

1.5 Transformations of Probability Distributions

The analytical treatment of stochastic models can frequently be facilitated by going over to transformations of the probability distributions of the underlying random variables. This section deals with the *z-transformation* for discrete random variables and with the *Laplace transformation* for continuous random variables. Taking into account the applications dealt with in this book, the discussion of these transformations is restricted to nonnegative random variables. As usual, the probability distribution of a discrete random variable is identified with the set of the probabilities of its realizations, and the probability distribution of a continuous random variable is identified with its distribution function or its probability density.

The set of all transformable probability distributions is called the *pre-image space*, whereas the set of all images of probability distributions is called the *image space*. The following fact is the basis for the successful application of the transformations mentioned:

There is a one-to-one-correspondence between image space and pre-image space.

This fact allows the application of the following algorithm for tackling stochastic models: The problem, originally given in the pre-image space, is transformed into the image space. There it is solved. The solution obtained in the image space is retransformed into the pre-image space to obtain the solution of the original problem. Hence the application of the transformations is only efficient if the problem in the image space can a priori be more easily solved than the original problem. For instance, if a system of linear differential equations of the first order with constant coefficients has to be solved for the time-dependent probability distribution of a discrete random variable, then the z-transformation reduces this problem to solving a linear algebraic equation system with constant coefficients.

Retransforming the solution obtained in the image space is usually the most difficult step when applying the algorithm. But even without retransforming, the image in many cases provides important or even sufficient information on the pre-image. For instance, knowledge of the z-transformation and the Laplace transformation of probability distributions makes it possible to compute moments of all orders of probability distributions. Therefore, these transformations are also called *moment generating functions*. In particular, expected values and variances of random variables can be immediately obtained from the images of their probability distributions.

1.5.1 z-Transformation

Let X be a discrete random variable with range $\{0, 1, 2, ...\}$ and probability distribution

$$\{p_i = P(X_i = i);\ i = 0, 1, 2, ...\}$$

Definition 1.3 The *z-transform* of the random variable X and its probability distribution $\{p_0, p_1, p_2, ...\}$, respectively, is defined as the infinite series

$$M(z) = \sum_{i=0}^{\infty} p_i z^i$$ •

Clearly, $M(z)$ is the expected value of z^X:

$$M(z) = E(z^X) \tag{1.53}$$

$M(z)$ converges absolutely for $|z| \le 1$:

$$|M(z)| \le \sum_{i=0}^{\infty} p_i \left|z^i\right| \le \sum_{i=0}^{\infty} p_i = 1$$

Therefore, $M(z)$ can be differentiated (and integrated) term by term:

$$M'(z) = \sum_{i=0}^{\infty} i p_i z^{i-1}$$

Letting $z = 1$ yields

$$M'(1) = \sum_{i=0}^{\infty} i p_i = E(X)$$

Taking the second derivative of $M(z)$ gives

$$M''(z) = \sum_{i=0}^{\infty} (i-1) i p_i z^{i-2}$$

Letting $z = 1$ yields

$$M''(1) = \sum_{i=0}^{\infty} (i-1) i p_i = \sum_{i=0}^{\infty} i^2 p_i - \sum_{i=0}^{\infty} i p_i = E(X^2) - E(X)$$

Thus, the first two moments of X are

$$E(X) = M'(1), \quad E(X^2) = M''(1) + M'(1) \tag{1.54}$$

Continuing in this way, all moments of X can be expressed by means of $M(z)$. Since (1.10) is also valid for discrete random variables, the variance of X is

$$Var(X) = M''(1) + M'(1) - [M'(1)]^2$$

Now let X and Y be two independent random variables with probability distributions $\{p_i = P(X=i);\ i = 0, 1, ...\}$ and $\{q_j = P(Y=j);\ j = 0, 1, ...\}$, respectively. As it has been shown in section 1.4, the probability distribution of the sum $Z = X+Y$ is given by the convolution $\{r_k;\ k = 0, 1, ...\}$ of the probability distributions of X and Y. Let $M_Z(z)$, $M_X(z)$ and $M_Y(z)$ denote the z-transforms of X, Y and Z, respectively. Then, according to (1.41),

$$M_Z(z) = \sum_{k=0}^{\infty} r_k z^k = \sum_{k=0}^{\infty} \left(\sum_{i=0}^{k} p_i q_{k-i} \right) z^k$$

Making use of the well-known relationship

$$\sum_{k=0}^{\infty} \sum_{i=0}^{k} a_{ik} = \sum_{i=0}^{\infty} \sum_{k=i}^{\infty} a_{ik} \tag{1.55}$$

yields

$$M_Z(z) = \sum_{i=0}^{\infty} \sum_{k=i}^{\infty} p_i q_{k-i} z^k = \sum_{i=0}^{\infty} p_i z^i \left(\sum_{k=i}^{\infty} q_{k-i} z^{k-i} \right)$$

$$= \left(\sum_{i=0}^{\infty} p_i z^i \right) \left(\sum_{j=0}^{\infty} q_j z^j \right)$$

Hence,

$$M_Z(z) = M_X(z)\, M_Y(z)$$

The z-transform of the sum of two independent discrete random variables is equal to the product of the z-transforms of these random variables.

From this result, a more general statement is easily deduced by induction:

If $X_1, X_2, \dots, X_n$ are independent discrete random variables with the z-transforms $M_1(z), M_2(z), \dots, M_n(z)$, then the z-transform of the sum $Z = X_1 + X_2 + \dots + X_n$ is given by the product

$$M_Z(z) = M_1(z)\, M_2(z) \cdots M_n(z)$$

Example 1.9 (*Poisson distribution*) Let X be Poisson distributed with parameter λ. Then,

$$p_i = \frac{\lambda^i}{i!} e^{-\lambda}; \quad i = 0, 1, ...$$

By making use of $e^x = \sum_{i=0}^{\infty} \frac{x^i}{i!}$, the corresponding z-transform is obtained as follows:

$$M(z) = \sum_{i=0}^{\infty} \frac{\lambda^i}{i!} e^{-\lambda} z^i = e^{-\lambda} \sum_{i=0}^{\infty} \frac{\lambda^i}{i!} z^i = e^{-\lambda} \sum_{i=0}^{\infty} \frac{(\lambda z)^i}{i!} = e^{-\lambda}\, e^{+\lambda z}$$

Hence,

$$M(z) = e^{\lambda(z-1)}$$

The first both derivatives are

$$M'(z) = \lambda\, e^{\lambda(z-1)}, \quad M''(z) = \lambda^2\, e^{\lambda(z-1)}$$

Letting $z = 1$ yields

$$M'(1) = \lambda, \quad M''(1) = \lambda^2$$

Thus, expected value, variance and second moment of X are given by

$$E(X) = \lambda, \quad Var(X) = \lambda, \quad E(X^2) = \lambda(\lambda + 1)$$

Now let $X_1, X_2, \ldots, X_n$ be independent, Poisson distributed random variables with parameters $\lambda_1, \lambda_2, \ldots, \lambda_n$. Then the z-transform of the sum

$$Z = X_1 + X_2 + \ldots + X_n$$

is given by the product

$$M_Z(z) = e^{\lambda_1(z-1)} \cdot e^{\lambda_2(z-1)} \cdots e^{\lambda_n(z-1)}$$

Hence,

$$M_Z(z) = e^{(\lambda_1+\lambda_2+\ldots+\lambda_n)(z-1)}$$

Corollary The sum of independent, Poisson distributed random variables is Poisson distributed. The parameter of this distribution is equal to the sum of the parameters of the individual terms. (Remember the one-to-one-correspondence between probability distributions and their z-transforms.) □

1.5.2 Laplace-Transformation

Let $f(x)$ be a real-valued function having the following properties:

1) $f(x)$ is a piecewise continuous function on $[0, \infty)$.

2) There exist constants a and b such that $f(x) \le b\, e^{ax}$ for all $x > 0$.

Definition 1.4 The *Laplace-transform* $\hat{f}(s)$ of $f(x)$ is the parameter integral

$$\hat{f}(s) = \int_0^{\infty} e^{-sx} f(x)\, dx$$

as a function of the complex parameter s with $Re(s) > a$. •

In view of the assumptions on $f(x)$, the integral exists in the area given by $Re(s) > a$. If in particular $f(x)$ is the density of a nonnegative random variable X, then $\hat{f}(s)$ has a simple probabilistic interpretation:

$$\hat{f}(s) = E\left(e^{-sX}\right) \tag{1.56}$$

If $f(x) = dF(x)/dx$, then the Laplace transform of $f(x)$ is also written in the form

$$\hat{f}(s) = \int_0^\infty e^{-sx}\, dF(x)$$

In what follows, it is sometimes more convenient to denote the Laplace-transform of a function f by $L\{f\}$. Partial integration yields

$$L\left\{\int_0^x f(u)\, du\right\} = \hat{F}(s) = \frac{1}{s}\hat{f}(s)$$

and

$$L\{f'\} = \hat{f}'(s) = s\,\hat{f}(s) - f(0)$$

More generally, if $f^{(n)}(x) = \dfrac{d^n f(x)}{dx^n}$ denotes the $n\,th$ derivative of $f(x)$, then

$$\hat{f}^{(n)}(s) = s^n\,\hat{f}(s) - s^{n-1} f(0) - s^{n-2} f'(0) - \cdots - s^1 f^{(n-2)}(0) - f^{(n-1)}(0)$$

Now let $f_1(x)$ and $f_2(x)$ be two functions satisfying conditions 1) and 2). Then,

$$L\{f_1 + f_2\} = L\{f_1\} + L\{f_2\} = \hat{f}_1(s) + \hat{f}_2(s) \tag{1.57}$$

The *convolution* $(f_1 * f_2)(x)$ of $f_1(x)$ and $f_2(x)$ has already been introduced in section 1.4:

$$(f_1 * f_2)(x) = \int_0^x f_2(x-u) f_1(u)\, du$$

The following formula is of special importance:

$$L\{f_1 * f_2\} = L\{f_1\} L\{f_2\} = \hat{f}_1(s)\,\hat{f}_2(s) \tag{1.58}$$

A proof of this relationship is easily established:

$$\begin{aligned}
L\{f_1 * f_2\} &= \int_0^\infty e^{-sx} \int_0^x f_2(x-u) f_1(u)\, du\, dx \\
&= \int_0^\infty e^{-su} f_1(u) \int_u^\infty e^{-s(x-u)} f_2(x-u)\, dx\, du \\
&= \int_0^\infty e^{-su} f_1(u) \int_0^\infty e^{-sy} f_2(y)\, dy\, du \\
&= \hat{f}_1(s)\hat{f}_2(s).
\end{aligned}$$

In proving this relationship, *Dirichlet's formula* has been applied:

$$\int_0^z \int_0^y f(x,y)\, dx\, dy = \int_0^z \int_x^z f(x,y)\, dy\, dx \tag{1.59}$$

Formula (1.58) is equivalent to the following statement:

The Laplace transform of the convolution of two functions is equal to the product of the Laplace transforms of these functions.

In view of (1.44), the probabilistic meaning of this is

If X and Y are two independent random variables, then the Laplace transform of the probability density of X+Y is equal to the product of the Laplace transforms of the probability densities of X and Y.

By induction, formulas (1.57) and (1.58) can be easily generalized to $n > 2$ terms and factors, respectively. (Mind also the analogy to the corresponding properties of the z-transformation.)

Properties (1.57) and (1.58) of the Laplace transformation suggest that Laplace transforms be decomposed as far as possibles into terms and factors (for instance, decomposing a fraction into partial fractions), because the retransformations of the arising terms and factors can usually be carried out more easily than the retransformation of the original image. The retransformation can, moreover, be facilitated by using *contingency tables.* These tables contain important functions and their Laplace transforms. There exists, moreover, an explicit formula for obtaining the pre-image of a Laplace transform. However, its application requires knowledge of complex calculus.

The k th derivative of $\hat{f}(s)$ with respect to s is

$$\frac{d^k \hat{f}(s)}{ds^k} = (-1)^k \int_0^\infty x^k e^{-sx} f(x)\, dx$$

Hence, if $f(x)$ is the density of a nonnegative random variable X, then its moments can immediately be obtained from $\hat{f}(s)$:

$$E(X^k) = (-1)^k \left. \frac{d^k \hat{f}(s)}{ds^k} \right|_{s=0} \tag{1.60}$$

Example 1.10 Let X be an exponentially distributed random variable with parameter λ, that is,

$$f(x) = \lambda e^{-\lambda x}, \quad x \geq 0$$

Then,

$$\hat{f}(s) = \int_0^\infty e^{-sx} \lambda e^{-\lambda x}\, dx = \lambda \int_0^\infty e^{-(s+\lambda)x}\, dx = \frac{\lambda}{s+\lambda}$$

The k *th* derivative of $\hat{f}(s)$ is

$$\frac{d^k \hat{f}(s)}{ds^k} = (-1)^k \frac{\lambda k!}{(s+\lambda)^{k+1}}$$

Thus, the k *th* moment is

$$E(X^k) = \frac{k!}{\lambda^k}$$

If $Z_n = X_1 + X_2 + \cdots + X_n$ is the sum of n independent, identically distributed exponential random variables with parameter λ, then Z_n is Erlang distributed with parameters n and λ (example 1.6). The Laplace transform of the density $f(x;\, \lambda, n)$ of Z_n is, therefore,

$$\hat{f}(s;\, \lambda, n) = \left(\frac{\lambda}{s+\lambda}\right)^n$$

The k *th* derivative of $\hat{f}(s;\, \lambda, n)$ with respect to s is

$$\frac{d^k \hat{f}(s;\, \lambda, n)}{ds^k} = (-1)^k \frac{\lambda^n n(n+1)\cdots(n+k-1)}{(s+\lambda)^{n+k}}$$

Hence, the k *th* moment of Z_n is given by

$$E(Z_n^k) = \frac{n(n+1)\cdots(n+k-1)}{\lambda^k}$$

□

Exercises

1.1) 4 semiconductors are checked and 3 random events A, B, C are defined as follows:

A = "at least one of the semiconductors is defective"
B = "no semiconductor is defective"
C = "at most three semiconductors are defective"

(1) Characterize the random events $A \cap C$, $A \cup C$ and $A \cap B \cap C$ verbally.

(2) By introducing a suitable sample space, determine the sets of elementary events belonging to the random events specified under (1).

1.2) Castings are produced weighing either 1, 5, 10 or 20 *kg*. Let A_1, A_2, A_3, and A_4 be the events that a casting does not weigh more than 1, 5, 10 and 20 *kg*, respectively.

Characterize the events $\bigcup_{i=1}^{4} A_i$, $A_1 \cap A_3$ and $A_2 \cap \overline{A}_4$ verbally.

1.3) Let $P(A) = 0.3$; $P(B) = 0.5$ and $P(A \cap B) = 0.2$

Determine the probabilities $P(A \cup B)$, $P(\overline{A} \cap B)$ and $P(\overline{A} \cup \overline{B})$.

1.4) 200 plates from a supplier are checked for surface finish (acceptable, non acceptable) and for satisfying given tolerance limits of the diameter (yes, no). The results are:

		surface finish	
		acceptable	*non acceptable*
diameter	*yes*	170	15
	no	8	7

A plate is selected at random from these 200. Let A be the event that its diameter is within the tolerance limits and let B the event that its surface finish is acceptable.

1) What are the probabilities of the random events $A,\ B,\ A \cap B,\ A \cup B$ and $\bar{A} \cup \bar{B}$?
2) Are A and B independent?

1.5) A computer producer optionally equips his newly developed PC *Ibson* with one and two hard disk drives and with or without extra software. He analyzes the first 1000 orders:

		hard disk drives	
		two	*one*
extra software	*yes*	520	90
	no	70	320

An order is selected at random from the first 1000. Let A be the event that this PC has two hard disk drives and let B be the event this PC has extra software.
Find the probabilities $P(A \cup B)$, $P(A \cap B)$, $P(A|B)$, $P(B|A)$, $P(A \cup B|\bar{B})$ and $P(\bar{A}|\bar{B})$.

1.6) A random experiment has been repeated many times and it has been recorded how often the events A and B occurred.
Suggest how to check whether A and B are independent.

1.7) 1000 bits are independently transmitted from a source to a sink. The probability of a faulty transmission of a bit is 0.0005. What is the probability that the transmission of at least two bits is not successful?

1.8) To construct a circuit a student needs, among others, 12 chips of a certain type. He knows that 4% of these chips are defective. How many chips have to be provided so that with a probability of at least 0.9 the student has a sufficient number of nondefective chips to be able to construct the circuit?

1.9) It costs *$50* to find out whether a spare part required for the repair of a failed device is faulty or not. Installing a faulty spare part causes damage of *$* 1000. Is it on average more profitable to use a spare part without checking if
(1) 1% of all spare parts of that type
(2) 3% of all spare parts of that type
(3) 10% of all spare parts of that type are faulty?

1.10) A test procedure for diagnosing faults in circuits indicates no fault with probability 0.99 if the circuit is faultless. It indicates a fault with probability 0.90 if the circuit is faulty. Let the probability that a circuit is faulty be 0.02.
(1) What is the probability that the test indicates that a circuit is faulty?
(2) What is the probability that a circuit is faulty if the test procedure indicates a fault?
(3) What is the probability that a circuit is faultless if the test procedure indicates that it is faultless?

1.11) A telephone connection between **s** and **t** (see figure) can be established if there is at least one closed path between **s** and **t**. The figure indicates the possible interruption of an edge (connection between two nodes of the transmission graph) by a switch. Such an interruption can in practice arise from a cable break or if the transmission capacity of a channel is exceeded. The switches 1 to 5 operate independently. Each one is closed with probability p and open with probability $1 - p$.

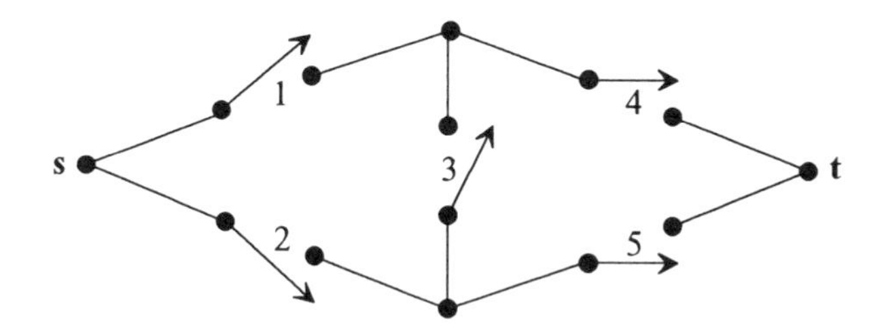

(1) What is the probability $w(p)$ that there is a phone connection between **s** and **t**?
(2) Draw the graph of $w(p)$ as a function of p, $0 \leq p \leq 1$.

1.12) The symbols 0 and 1 are transmitted independently from each other in proportion 1 : 4. Random noise may cause transmission failures: If a 0 was transmitted, then the sink will receive a 0 with probability 0.9. If a 1 was transmitted, then the sink will receive a 1 with probability 0.95.

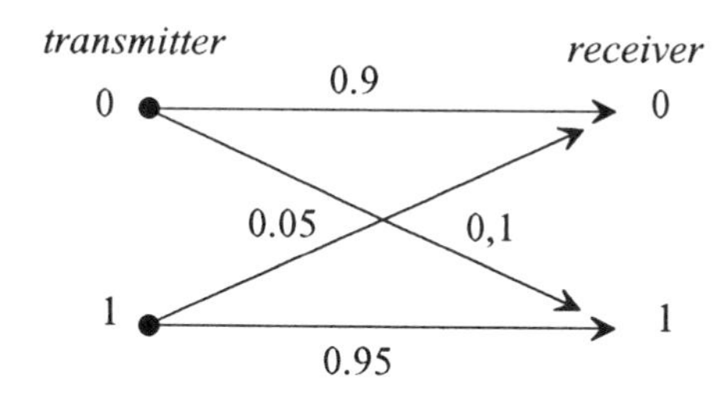

(1) What is the probability that a received symbol is a "1" ?
(2) A "1" has been received. What is the probability that a "1" had actually been transmitted?
(3) A "0" has been received. What is the probability that a "1" had actually been transmitted?

1.13) The random number of crackle sounds produced per hour by an old radio has a Poissondistribution with parameter $\lambda = 12$.
What is the probability that there is no crackle sound during the 4 minutes of transmission a listener's favourite hit?

1.14) According to the timetable, a lecture begins at 8:15. The arrival time X of professor *Unpopular* in the venue has a uniform distribution over the time interval [8:13, 8:20], whereas the arrival time Y of student *Sluggish* has a uniform distribution over the time interval [8:05, 8:30].
What is the probability that *Sluggish* arrives after *Unpopular* in the venue.
Hint Consider the problem in the (x, y)-plane.

1.15) A road traffic light is switched on every day at 5:00 a.m. It always begins with "red" and holds this colour for 2 minutes. Then it changes to "green" and holds this colour 4 minutes. This cycle continues till midnight. A car driver arrives at this traffic light at a time point which has uniform distribution over [9:00, 9:10].
(1) What is the probability that the driver has to wait in front of the traffic light?
(2) Determine the same probability on condition that his arrival time point has a uniform distribution over [8:58, 9:08].

1.16) The lifetimes of bulbs of a particular type are exponential with parameter λ $[h^{-1}]$. Five bulbs of this type are switched on at time 0.
What is the probability that after $1/\lambda$ hours
(1) all 5 bulbs,
(2) at least 3 bulbs have failed?

1.17) The density function of the annual energy consumption of an enterprise $[10^8 kwh]$ is

$$f(x) = 30(x-2)^2\left[1 - 2(x-2) + (x-2)^2\right], \quad 2 \le x \le 3.$$

(1) What is the mean annual energy consumption?
(2) What is the probability that the annual energy consumption exceeds 2.8?

1.18) The times between the arrivals of taxis at a rank are independent and exponential with parameter $\lambda = 4$ $[h^{-1}]$. Assuming, a customer arriving at the taxi rank does not find an available taxi and the last one left 3 minutes earlier.
What is the probability that the customer has to wait at least 5 minutes for the next available taxi?

1.19) The response time of an average male car driver is normal with expected value 0.5 and standard deviation 0.06 (in seconds).
(1) What is the probability that the response time is greater than 0.6 seconds?
(2) What is the probability that the response time is between 0.5 and 0.55 seconds?

1.20) The annual consumptions of energy X and Y of two consumers have a joint normal distribution with parameters [in $10^4 kwh$]

$$\mu_x = 7.2, \quad \sigma_x = 0.8, \quad \mu_y = 10.8, \quad \sigma_y = 1.2, \text{ and } \rho = 0.5.$$

(1) What is the probability that the annual total consumption of the two consumers is between 17 and 19 $\left[10^4 kwh\right]$?

(2) Find the conditional probability that X is greater than 8.0 given $Y = 12$ $[10^4 kwh]$.

1.21) A supermarket employes 24 shop-assistants. 20 of them achieve an average daily turnover of $\$ 8000$, whereas 4 achieve an average daily turnover of $\$ 10\,000$. The corresponding standard deviations are $\$2400$ and $\$3000$, respectively. The daily turnovers of all shopassistants are independent and normally distributed. Let Z be the daily total turnover of all shop-assistants.

(1) Determine $E(Z)$ and $\sqrt{Var(Z)}$.

(2) What is the probability that Z is greater than $\$ 190\,000$?

1.22) Every day a car dealer sells X cars of type I and Y cars of type II. The table shows the joint distribution $\left\{r_{ij} = P(X=i,\ Y=j);\ i,j = 0,1,2\right\}$ of (X, Y):

		Y		
		0	1	2
X	0	0.1	0.1	0
	1	0.1	0.3	0.1
	2	0	0.2	0.1

(1) Determine the marginal distributions of $\left\{r_{ij};\ i,j = 0,1,2\right\}$.

(2) Are X and Y independent?

(3) Determine $E(X)$, $E(Y)$, $Var(X)$ and $Var(Y)$.

(4) Determine the conditional expected values $E(X|Y=1)$ and $E(Y|X=0)$.

1.23) The joint probability density of the random vector (X, Y) is

$$f_{X,Y}(x,y) = x + y;\quad 0 \le x, y \le 1.$$

(1) Verify that $f_{X,Y}(x,y)$ is the joint probability density of a random vector (X, Y).

(2) Determine the conditional probability density $f_X(x|y)$ and the conditional expected values $E(X|Y=0,5)$ and $E(Y|X=1)$.

(3) Are X and Y independent?

1.24) The random vector (X, Y) has joint probability density

$$f_{X,Y}(x,y) = \frac{1}{16}xy;\quad 0 \le x \le 2,\ 0 \le y \le 4.$$

Check whether X and Y are independent.

1.25) The times between successive orders for individual spare parts of a particular type from a store are independent exponential random variables with parameter $\lambda = 0{,}01\ [h^{-1}]$. There are only 3 spare parts of this type left in the store.
What is the probability that the demand can be satisfied within the following 200 hours if there is no delivery during this time period?

1.26) A helicopter is allowed to carry at most 8 persons provided that their total weight does not exceed 620 *kg*. The weights of the passengers are independent normal random variables with expected value 76 *kg* and standard deviation 18 *kg*.
(1) What are the probabilities of exceeding the permissible load with 7 and 8 passengers, respectively?
(2) What would the maximum total allowable load have to be to ensure with probability 0.99 that with 8 passengers the heliocopter will be allowed to fly?

1.27) In a country, the height X of married women has a normal distribution with parameters $\mu_x = 168\,cm$ and $\sigma_x = 8\,cm$, whereas the height Y of married men has a normal distribution with parameters $\mu_y = 175\,cm$ and $\sigma_y = 10\,cm$. Assuming independence of X and Y, determine the percentage of married couples with women being taller than their husbands.
Hint You have to determine the probability $P(X \geq Y) = P(X + (-Y) \geq 0)$.

1.28) Let X have a uniform distribution according to

$$P(X = i) = \frac{1}{n}; \quad i = 1, 2, \ldots, n.$$

(1) Determine the corresponding z-transform and from it $E(X)$.
(2) Let X_1 and X_2 be independent random variables which are identically distributed as X. Check whether $X_1 + X_2$ has also a uniform distribution.

1.29) Let X have a geometric distribution with parameter p (Table 1.1).
Determine the corresponding z-transform and by means of it $E(X)$.

1.30) (1) Determine the z-transform of the binary random variable

$$X = \begin{cases} 1 & \text{with probability } p \\ 0 & \text{with probability } 1-p \end{cases}.$$

(2) Let $X_1, X_2, \ldots, X_n$ be independent random variables which are identically distributed as X.
Determine the z-transform of the sum $X_1 + X_2 + \ldots + X_n$.

1.31) Let X have a binomial distribution with parameters p and n.
(1) Determine the z-transform and by means of it expected value and variance of X.
(2) Compare this z-transform with the one obtained in exercise 1.30, (2) and interpret the result.
(3) Let $X_1, X_2, \ldots, X_m$ be independent, binomially distributed random variables with parameters (n_i, p_i); $i = 1, 2, \ldots, m$.
Under which condition has the sum $X_1 + X_2 + \ldots + X_m$ also a binomial distribution?

1.32) Let X be uniformly distributed over the interval $[0, T]$.
(1) Determine the Laplace transform of the probability density of this distribution and by means of it $E(X)$.
(2) Let X_1 and X_2 be independent random variables which are identically distributed as X. Has the sum $X_1 + X_2$ a uniform distribution over the interval $[0, 2T]$?

2 Stochastic Processes

2.1 Introduction

A random variable X is the outcome of a random experiment under fixed conditions. A change in these conditions will influence the outcome of the random experiment, i.e. the probability distribution of X will change. Varying conditions can be taken into account by considering random variables which depend on a deterministic parameter t: $X = X(t)$. This approach leads to more general random experiments than the ones considered in chapter 1. Such generalized random experiments are illustrated by two simple examples.

Example 2.1 a) At one and the same geographical point the temperature is measured each day at 7:00. Apart from the random fluctuations of the temperature, it is quite evident that the measured values also depend on a deterministic parameter, namely on the time, or, more precisely, on the day of the year. Let x_i be the temperature measured at 7:00 on the *i th* day of a year. This x_i will vary from year to year and hence it can be considered to be a realization of a random variable X_i. Thus, X_i is the (random) temperature measured on the *i th* day of any year at 7:00. Obviously, the temperatures on the first of January and on the first of July, i.e. X_1 and X_{182}, are random variables which generally have different probability distributions. However, if one is only interested in the temperatures at the first 5 days of the year, then the $X_1, X_2, ..., X_5$ are at least approximately identically distributed as a random variable X. Nevertheless, indexing the daily temperatures is necessary, because modeling the obvious statistical dependence between the daily temperatures requires information on the joint probability distribution of the random vector $\{X_1, X_2, ..., X_5\}$. This situation and the problems connected with it motivate the introduction of the generalized random experiment "daily measurement of the temperature at a given geographical point at 7:00". The random outcomes of this generalized random experiment are sequences of random variables X_i which are generally neither independent nor identically distributed: $\{X_1, X_2, ..., X_{365}\}$. The temperatures which are actually measured on each day constitute the deterministic outcome $\{x_1, x_2, ..., x_{365}\}$. This vector is a realization of $\{X_1, X_2, ..., X_{365}\}$.

b) If the temperature is continuously recorded by a sensor over the whole year, then the outcome of the measurement during a year is a continuous function of the time: $x = x(t),\ 0 \le t \le 1$. In view of random temperature fluctuations, this function will vary from year to year. Therefore, $x(t)$ has to be interpreted as a realization of a random variable $X(t)$. Analogously to example 2.1, it makes sense to introduce the generalized random experiment "continuous measurement of the temperature during a year". It will be denoted by $\{X(t),\ 0 \le t \le 1\}$. The outcomes of this generalized random experiment are real functions $x = x(t)$ defined on $0 \le t \le 1$. A complete probabilistic characterization of this generalized random experiment requires knowledge of the joint probability distributions of all possible random vectors ($X(t_1), X(t_2), ..., X(t_n)), 0 \le t_1 < t_2 < ... < t_n \le 1;\ n = 1, 2, ...$ Otherwise, the statistical dependence between the X_{t_i} in any particular sequence of random variables $X(t_1), X(t_2), ..., X(t_n)$ could not be quantified. It is quite obvious that for small differences between t_i and t_{i+1} there is a strong statistical dependence between $X(t_i)$ and $X(t_{i+1})$. (If X_i is defined as in example 2.1, the dependence between X_j and X_k, $j \ne k$, is mainly due to the inertia of the weather over an area.) □

The following simple, but rather instructive example has been in principle already considered by *Cramer/Leadbetter* (1967). It illustrates that the parameter t need not denote time.

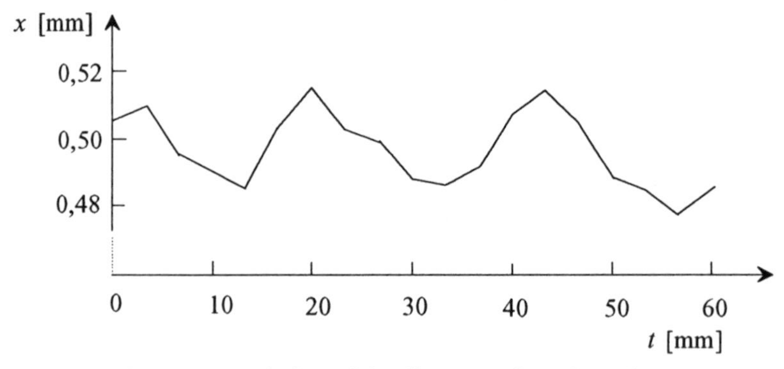

Figure 2.1 Variation of the diameter of a nylon wire

Example 2.2 A machine continuously produces nylon wires of length $L = 60$ *mm* with a nominal diameter of 0.5 *mm*, always under the same conditions Despite constant production conditions being maintained, minor variations of the wire diameter can technologically not be avoided. Thus, when measuring the actual diameter x of a single wire at a distance t from the origin, one gets a function $x = x(t)$, $0 \le t \le L$. Measurements have shown that the variations in $x(t)$ are bounded by $\pm 0{,}02$ *mm* (Figure 2.1). The functions $x(t)$ will vary randomly from wire to wire. This suggests the introduction of a generalized random experiment "continuous measurement of the wire diameter, depending on the distance t from the origin". If $X(t)$ denotes the diameter of a randomly selected wire at a distance t from the origin, then it makes sense to denote this generalized random experiment by $\{X(t),\ 0 \le t \le L\}$. The outcomes of this experiment are real functions $x = x(t)$ defined on $0 \le t \le L$. □

In contrast to the random experiments (random variables) considered in chapter 1 with outcomes (realizations) being real numbers, the outcomes of the generalized random experiments introduced in the examples 2.1 and 2.2 are real functions. Hence, such generalized random experiments should be called *random functions.* However, the terminology *stochastic processes* is more common and will be used throughout the book. In order to characterize them more precisely, some further notation is required: Let the random variable of interest X be depend on a parameter t, which assumes values from a set $\mathbf{T}$: $X = X(t),\ t \in \mathbf{T}$. To simplify the treatment and recognizing the overwhelming majority of applications, in what follows the parameter t is considered to be time. Thus, $X(t)$ is the random variable X at time t, whereas $\mathbf{T}$ denotes the whole observation time span. Further, let $\mathbf{Z}$ denote the set of all states (realizations) which the $X(t),\ t \in \mathbf{T}$, can assume.

Definition 2.1 (*stochastic process*) A family of random variables $\{X(t),\ t \in \mathbf{T}\}$ is called a *stochastic process* with parameter space $\mathbf{T}$ and *state space* $\mathbf{Z}$. ●

If $\mathbf{T}$ is a finite or countably infinite set, then $\{X(t),\ t \in \mathbf{T}\}$ is called a *stochastic process in discrete time* or a *discrete-time stochastic process.* Such processes can be written as a sequences of random variables $\{X_1, X_2, ...\}$ (example 2.1 a). Conversely, every sequence of random variables can be interpreted as stochastic process in discrete time. If $\mathbf{T}$ is an interval, then $\{X(t), t \in \mathbf{T}\}$ is a *stochastic process in continuous time* or a *continuous-time stochastic process.* The stochastic process $\{X(t),\ t \in \mathbf{T}\}$ is said to be *discrete* if its state space $\mathbf{Z}$ is a finite or a countably infinite set. The stochastic process $\{X(t),\ t \in \mathbf{T}\}$ is said to be *continuous* if $\mathbf{Z}$ is an interval. Thus, there are discrete stochastic processes in discrete time, discrete stochastic processes in continuous time, continuous stochastic processes in discrete time, and continuous stochastic processes in continuous time. (Unfortunately, the terminology is not unique.) With the exception of chapter 8, throughout this book $\mathbf{Z}$ will be assumed to be a subset of the real axis.

If the stochastic process $\{X(t),\ t \in \mathbf{T}\}$ is observed over the whole time period $\mathbf{T}$, i.e. the realizations of $X(t)$ are registered for all $t \in \mathbf{T}$, then one obtains a real function $x = x(t),\ t \in \mathbf{T}$. Such a function is a *realization* of the stochastic process. Realizations of stochastic processes are usually called *sample paths* or *trajectories*. In this book the concept *sample path* is used. The sample paths of a stochastic process in discrete time are, therefore, sequences of real numbers, whereas the sample paths of stochastic processes in continuous time are generally continuous functions of time (Figure 2.1). The sample paths of a discrete stochastic process in continuous time are piece-wise constant functions (step functions). The set of all sample paths of a stochastic process with parameter space $\mathbf{T}$ is therefore a subset of all functions over the domain $\mathbf{T}$.

In engineering and science there are many parameter-dependent and in particular time-dependent phenomena which can be modeled by stochastic processes. For instance, if a sensor continuously registers the oil pressure in a pipeline at a fixed distance from its origin, then random fluctuations will be observed. The outcome of this experiment can be considered to be a sample path of a stochastic process. In this case it also makes sense to measure the oil pressure over the whole length of the pipeline at a fixed time point. An outcome of this experiment would be a sample path of a stochastic process in which the parameter is the distance of the measuring point from the origin of the pipeline.

Comment The random oil pressure can also be modeled with respect to its dependence both on time and distance. This leads to a stochastic process with parameter space being a vector space. The sample paths of such stochastic processes are surfaces. However, this book deals only with one-dimensional parameter spaces.

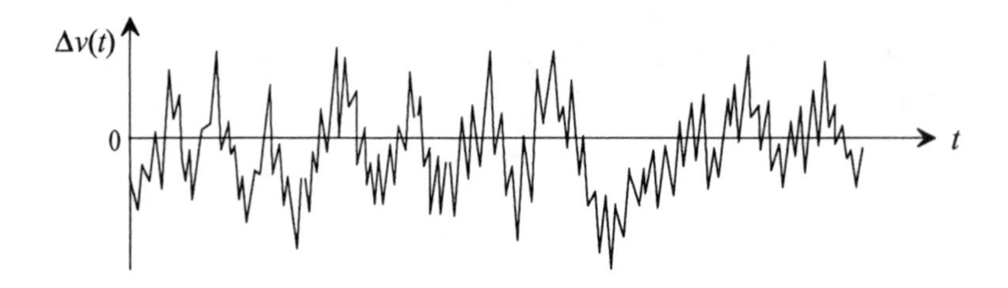

Figure 2.2 Voltage fluctuations caused by thermal noise under high temperature

In an electrical circuit it is not possible to keep the voltage strictly constant. Random fluctuations of the voltage are, for instance, caused by *thermal noise.* If $v(t)$ denotes the voltage measured at time t, then $v = v(t)$ is a sample path of a stochastic process $\{V(t),\ t \geq 0\}$, where $V(t)$ is the random voltage at time t. Figure 2.2 shows a typical graph of the voltage fluctuations $\Delta v(t)$ in an electrical circuit around a nominal value caused by thermal noise.

Producers of radar and satellite supported communication systems have to take into account a phenomenon called *fading.* This is characterized by random fluctuations in the energy of received signals caused by the dispersion of radio waves as a result of inhomogeinities in the atmosphere and by *metereological* and *industrial noise.* (Both metereological and industrial noise cause electrical discharges in the atmosphere which occur at random time points with randomly varying power.)

Stochastic processes also play an important role in the economy and in the economic sciences. "Classic" applications of stochastic processes in this field are modeling share prices, rendits, and prices of precious metals over time. A well-known example of a sample path of a stochastic process is the development of wheat prices (corrected for currency fluctuations) in Central and Western Europe from 1500 to 1869 (Beveridge (1921), Figure 2.3). (However, since the underlying generalized random experiment cannot be repeated, there is some doubt as to whether it is justified to consider Figure 2.3 a sample path of a stochastic process.) A number of applications of stochastic processes in the economy, particularly in the field of operations research, will be discussed in chapters 3 to 7.

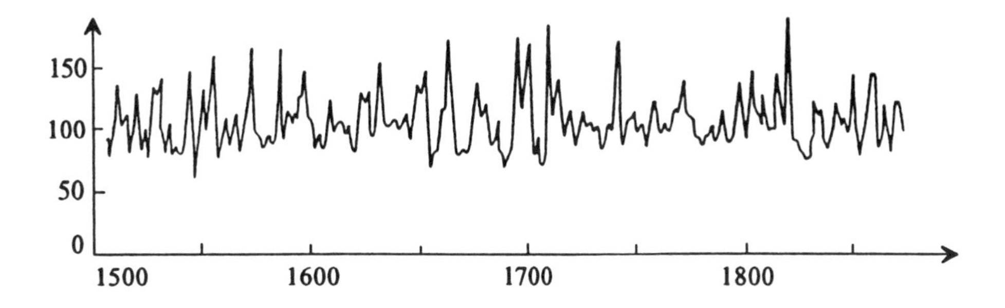

Figure 2.3 Fluctuation of wheat prices from 1500 to 1869 (Beveridge (1921))

2.2 Characteristics of Stochastic Processes

Let $F_t(x)$ be the distribution function of $X(t)$, $t \in \mathbf{T}$:

$$F_t(x) = P(X(t) \leq x) \tag{2.1}$$

The family of one-dimensional distribution functions $\{F_t(x),\ t \in \mathbf{T}\}$ characterizes the *one-dimensional probability distribution* of $\{X(t),\ t \in \mathbf{T}\}$. In view of the statistical dependence which generally exists between $X(t_1), X(t_2), ..., X(t_n)$ for any $t_1, t_2, ..., t_n$, the specification of $\{F_t(x),\ t \in \mathbf{T}\}$ does not completely characterize a stochastic process. This fact has been already illustrated in the examples 2.1 and 2.2. A stochastic process $\{X(t),\ t \in \mathbf{T}\}$ is only then completely determined if for all integers $n = 1, 2, ...$ and for all n-tuples $\{t_1, t_2, ..., t_n\}$ with $t_i \in \mathbf{T}$, the joint probability distribution functions of the random vectors $(X(t_1), X(t_2), ..., X(t_n))$ are known:

$$F_{t_1, t_2, ..., t_n}(x_1, x_2, ..., x_n) = P(X(t_1) \leq x_1,\ X(t_2) \leq x_2, ...,\ X(t_n) \leq x_n) \tag{2.2}$$

The set of all these joint distribution functions defines the *probability distribution* of the stochastic process. For discrete stochastic processes it is generally simpler to characterize their probability distributions by the probabilities

$$P(X(t_1) \in A_1,\ X(t_2) \in A_2, ... ,\ X(t_n) \in A_n) \tag{2.3}$$

for all $t_1, t_2, ..., t_n$ with $t_i \in \mathbf{T}$ and $A_i \subseteq \mathbf{Z}$; $i = 1, 2, ..., n$; $n = 1, 2, ...$

Assuming the existence of $E(X(t))$ for all $t \in \mathbf{T}$, the *trend* or the *trend function* of the stochastic process $\{X(t), t \in \mathbf{T}\}$ is the expected value of $X(t)$ as a function of t (average process development):

$$m(t) = E(X(t)), \quad t \in \mathbf{T} \tag{2.4}$$

Thus, the trend function describes the average development of a stochastic process. If the densities

$$f_t(x) = \frac{dF_t(x)}{dx}, \quad t \in \mathbf{T},$$

exist, then

$$m(t) = \int_{-\infty}^{\infty} x f_t(x)\, dx, \quad t \in \mathbf{T} \tag{2.5}$$

The *covariance function* of a stochastic process $\{X(t), t \in \mathbf{T}\}$ is the covariance between the random variables $X(s)$ and $X(t)$ as a function of s and t:

$$C(s,t) = Cov(X(s), X(t))$$

$$= E([X(s) - m(s)]\,[X(t) - m(t)])\,; \quad s,t \in \mathbf{T}$$

or

$$C(s,t) = E(X(s)\,X(t)) - m(s)m(t); \quad s,t \in \mathbf{T} \tag{2.6}$$

In particular,

$$C(t,t) = Var(X(t))$$

The covariance function is a symmetric function of s and t:

$$C(s,t) = C(t,s).$$

Analogously, the *correlation function* of $\{X(t), t \in \mathbf{T}\}$ is the correlation coefficient $\rho(s,t) = \rho(X(s), X(t))$ between $X(s)$ and $X(t)$ as a function of s and t. Thus, according to (1.34),

$$\rho(s,t) = \frac{Cov(X(s), X(t))}{\sqrt{Var(X(s)}\,\sqrt{Var(X(t)}} \tag{2.7}$$

Since $X(s)$ and $X(t)$ are expected to be almost independent if the time difference $|t-s|$ is sufficiently large, one anticipates that

$$\lim_{|t-s|\to\infty} C(s,t) = \lim_{|t-s|\to\infty} \rho(s,t) = 0 \tag{2.8}$$

However, already simple examples considered in section 2.4 prove that (2.8) does not hold for all stochastic processes. The covariance function of a stochastic process is also called *autocovariance function* and the correlation function *autocorrelation function*. This is useful when considering covariances and correlations between two different stochastic processes.

2.3 Properties of Stochastic Processes

Special importance attaches to those stochastic processes for which the joint distribution functions (2.2) only depend on the distances between neighbouring t_i, i.e. where only the relative positions of $t_1, t_2, ..., t_n$ to each other have an impact on the joint distribution of the random variables $X(t_1), X(t_2), ..., X(t_n)$.

Definition 2.2 (***strict stationarity***) A stochastic process $\{X(t),\ t \in \mathbf{T}\}$ is said to be *strictly stationary*) if for all $n = 1,2,...$, for any h and for all n-tuples $(t_1, t_2, ..., t_n)$ with $t_i \in \mathbf{T}$ and $t_i + h \in \mathbf{T}$, $i = 1,2,..., n$; as well as for all n-tuples $(x_1, x_2, ..., x_n)$,

$$F_{t_1, t_2, ..., t_n}(x_1, x_2, ..., x_n) = F_{t_1+h, t_2+h, ..., t_n+h}(x_1, x_2, ..., x_n) \tag{2.9}$$

●

Roughly speaking: The probability distribution of a stochastic process which is strictly stationary is invariant against absolute time shifts. In particular, the one-dimensional distribution functions $F_t(x)$ do not depend on t. (This statement results from (2.9) by choosing $n = 1$.):

$$F_t(x) \equiv F(x) \tag{2.10}$$

Thus,

$$m(t) = E(X(t)) \equiv m \tag{2.11}$$

$$Var(X(t)) \equiv \text{constant}$$

The trend function of strictly stationary processes is, therefore, a parallel to the time axis and the fluctuations of their sample paths around the trend function will experience no systematic changes with increasing t. The sample paths shown in Figures 2.1 to 2.3 have approximately these properties.

Substituting $n = 2$, $t_1 = 0$, $t_2 = t - s$ and $h = s$ in (2.9) yields for all $s < t$ and any x_1, x_2:

$$F_{0, t-s}(x_1, x_2) = F_{s,t}(x_1, x_2), \tag{2.12}$$

i.e. the joint distribution function of the random vector (X_s, X_t), and, therefore, the expected value of the product $X_s X_t$, depend only on the difference $\tau = t - s$, and not on the absolute values of s and t. Since the covariance function is given by

$$C(s,t) = E[X(s)X(t)] - m^2 \quad \text{for any } s, t \in \mathbf{T},$$

it must have the same property:

$$C(s,t) = C(s, s+\tau) = C(0, \tau)$$

$$= C(\tau)$$

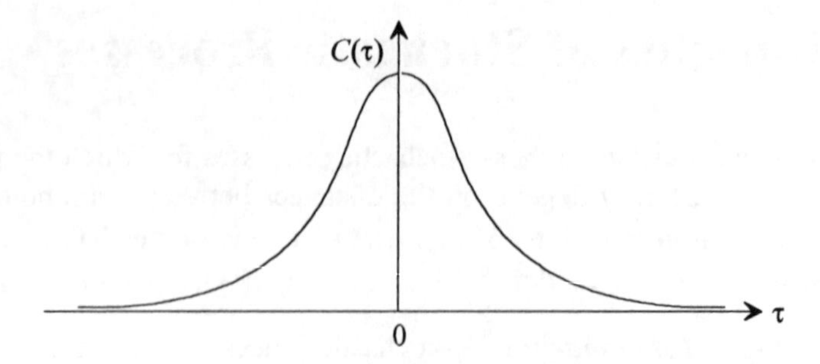

Figure 2.4 Typical graph of the covariance function of a strictly stationary process

The covariance function of strictly stationary processes is, therefore, given by

$$C(\tau) = Cov(X(s), X(s+\tau)) \tag{2.13}$$

for all $s \in \mathbf{T}$ and τ with $s+\tau \in \mathbf{T}$.

Since the covariance function $C(s,t)$ of any stochastic process is symmetric in s and t, the covariance function of a strictly stationary process is symmetric with respect to $\tau = 0$, i.e. $C(\tau) = C(-\tau)$ or, equivalently,

$$C(\tau) = C(|\tau|)$$

For strictly stationary processes, property (2.8) simplifies to

$$\lim_{|\tau| \to \infty} K(\tau) = 0 \tag{2.14}$$

Figure 2.4 shows a typical graph of the covariance function of a strictly stationary process.

In practical situations it is generally not possible to determine all n-dimensional probability distributions of a stochastic process in order to check whether it is strictly stationary or not. The user of stochastic processes is, therefore, frequently satisfied with the validity of properties (2.11) and (2.13). Hence, based on these two properties, another concept of stationarity has been defined. It is, however, only defined for second order processes. A stochastic process $\{X(t), t \in \mathbf{T}\}$ is said to be a *second order process* if

$$E(X^2(t)) < \infty \text{ for all } t \in \mathbf{T},$$

i.e. the existence of the second moments of all $X(t)$ is assumed. This assumption implies the existence of the expected values $E(X(t))$, of the variances $Var(X(t))$, and of the covariance function $C(s,t)$ for all s and t. (Up till now the existence of moments has been an implied condition if required.)

Definition 2.3 (*wide-sense stationarity*) A second order process is said to be *wide-sense stationary* (*stationary in wide sense*) if it has properties (2.11) and (2.13). ●

A strictly stationary process is not necessarily a wide-sense stationary process, since there are strictly stationary processes which are not second order processes. But, if a second order process is strictly stationary, then it is, as shown above, also wide-sense stationary. Wide-sense stationary processes are also called *weakly stationary, covariance stationary* or *second-order stationary processes.*

Further important properties of stochastic processes are based on properties of their increments. The *increment* of a stochastic process $\{X(t), t \in \mathbf{T}\}$ with respect to the interval $[t_1, t_2]$ is the difference $X(t_2) - X(t_1)$. Increments can, of course, be negative or positive.

Definition 2.4 (*homogeneous increments*) A stochastic process $\{X(t), t \in \mathbf{T}\}$ is said to have *homogeneous* or *stationary increments* if its increments

$$X(t_2 + \tau) - X(t_1 + \tau)$$

have the same probability distribution for all τ with $t_1 + \tau \in \mathbf{T}$, $t_2 + \tau \in \mathbf{T}$; t_1, t_2 fixed, but arbitrary. ●

An equivalent definition of processes with homogeneous increments is the following one: $\{X(t), t \in \mathbf{T}\}$ has homogeneous increments if the probability distribution of $X(t+\tau)$ - $X(t)$ does not depend on t for any fixed τ; $t, t+\tau \in \mathbf{T}$.

A stochastic process with homogeneous (stationary) increments need not be stationary (in any sense).

Definition 2.5 (*independent increments*) A stochastic process $\{X(t), t \in \mathbf{T}\}$ has *independent increments* if for all $n = 3, 4, ...$ and for all n-tuples $\{t_1, t_2, ..., t_n\}$ with $t_1 < t_2 < ... < t_n$ and $t_i \in \mathbf{T}$, the increments

$$X(t_2) - X(t_1),\ X(t_3) - X(t_2), \dots,\ X(t_n) - X(t_{n-1})$$

are independent random variables. ●

Thus, the increments over disjoint time intervals of a stochastic process with independent increments are independent random variables.

Definition 2.6 (*Gaussian process*) A stochastic process $\{X(t), t \in \mathbf{T}\}$ is a *Gaussian process* if the random vectors $(X(t_1), X(t_2), ..., X(t_n))$ have a joint Gaussian (Normal) distribution for all n-tuples $(t_1, t_2, ..., t_n)$ with $t_i \in \mathbf{T}$ and $t_1 < t_2 < ... < t_n$; $n = 1, 2, ...$ ●

Important examples of Gaussian processes are considered in chapter 7.

Definition 2.7 (*Markovian property*) A stochastic process $\{X(t), t \in \mathbf{T}\}$ has the *Markov(ian)-property* if for all $n = 2, 3, \ldots$, for all $(n+1)$-tuples $\{t_1, t_2, \ldots, t_{n+1}$ with $t_i \in \mathbf{T}$ and $t_1 < t_2 < \ldots < t_{n+1}$, and for any $A_i \subseteq \mathbf{Z}$; $i = 1, 2, \ldots, n+1$;

$$P(X(t_{n+1}) \in A_{n+1} | X(t_n) \in A_n, X(t_{n-1}) \in A_{n-1}, \ldots, X(t_1) \in A_1)$$
$$= P(X(t_{n+1}) \in A_{n+1} | X(t_n) \in A_n)$$ ●

The Markovian property has the following implication: If t_{n+1} is a time point in the future, t_n the present time and, correspondingly, $t_1, t_2, \ldots, t_{n-1}$ time points in the past, then the future development of a process having the Markovian property does not depend on its evolvement in the past, but only on its present state.

Stochastic processes having the Markovian property are called *Markov processes.* They always have independent increments. Markov processes have enormous practical implications mainly for three reasons: 1) Many practical phenomena can be modeled by Markov processes. 2) The input necessary for their practical application is generally much more easily to provide than that for other classes of stochastic processes. 3) Computer algorithms are available for numerical evaluations. The practical importance of Markov processes is illustrated by many examples in the chapters 5 and 6.

A Markov process with finite or countably infinite parameter space **T** is called a *discrete-time Markov process.* Otherwise it is called a *continuous-time Markov process.* Markov processes with finite or countably infinite state spaces **Z** are called *Markov chains.* Thus, a discrete-time Markov chain has both a discrete state space and a discrete parameter space. However, deviations from this notation are found in the literature.

Theorem 2.1 A Markov process is strictly stationary if and only if its one-dimensional probability distributions do not depend on time, i.e. if there is a distribution function $F(x)$ with

$$F_t(x) = P(X(t) \le x) = F(x) \quad \text{for all } t \in \mathbf{T}$$ ■

Definition 2.8 A second order process $\{X(t), t \in \mathbf{T}\}$ is said to be *mean-square continuous at point* $t = t_0 \in \mathbf{T}$ if

$$\lim_{h \to 0} E([X(t_0 + h) - X(t_0)]^2) = 0$$

The process $\{X(t), t \in \mathbf{T}\}$ is said to be *mean-square continuous in the region* $\mathbf{T}_0$, $\mathbf{T}_0 \subseteq \mathbf{T}$, if it is mean-square continuous at all points $t \in \mathbf{T}_0$. ●

(The convergence criterium used in definition 2.8 is called *convergence in mean square.*) There is a simple criterion for a second order stochastic process to be mean-square continuous at t_0:

Theorem 2.2 A second order process $\{X(t), t \in \mathbf{T}\}$ is mean-square continuous at t_0 if and only if its covariance function $C(s,t)$ is continuous at $(s,t) = (t_0, t_0)$. ■

Corollary A wide-sense stationary process $\{X(t), t \in (-\infty, +\infty)\}$ is mean-square continuous in $(-\infty, +\infty)$ if and only if it has this property at $t = 0$.

2.4 Special Stochastic Processes

2.4.1 Continuous-Time Stochastic Processes

This and the following section present examples of rather simple stochastic processes, which are, however, of considerable importance in many fields of application, e.g. in electronics, communication (noise modeling), and in time series analysis. Most of them can be found, for example, in *Anděl* (1984), *Gardner* (1989), *Hellstrom* (1984)), and *Jaglom* (1962).

All references to stationarity refer to wide-sense stationarity.

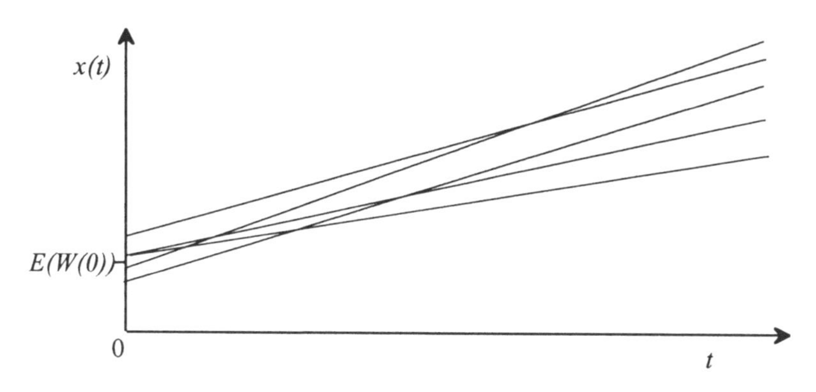

Figure 2.5 Possible sample paths of the stochastic process in example 2.3

Example 2.3 (***process with linear sample paths***) For describing the development of maintenance costs and of the degree of mechanical wear over time the following model is sometimes used (*Družinin* (1977), *Beichelt* and *Franken* (1984)):

$$X(t) = Vt + W .$$

V and W are assumed to be random variables with finite expected values and variances.

The trend function of the stochastic process $\{X(t), t \geq 0\}$ is

$$m(t) = E(V)t + E(W)$$

According to (2.6), its covariance function is

$$C(s,t) = E([Vs + W][Vt + W]) - m(s)m(t)$$

Thus,

$$C(s,t) = stVar(V) + (s + t)Cov(V, W) + Var(W)$$

Figure 2.5 shows some possible sample paths of the process $\{X(t), t \geq 0\}$. □

Example 2.4 ***(Cosine wave with random amplitude)*** Let

$$X(t) = A\cos\omega t\,,$$

where A is a nonnegative random variable with $E(A) < \infty$.

The process $\{X(t), t \geq 0\}$ may be interpreted as the output of an oscillator which is selected from a set of identical ones. (Random deviations of the amplitudes from a nominal value are technologically unavoidable.) Its trend function is

$$m(t) = E(A)\cos\omega t$$

The covariance function of this process is

$$C(s,t) = E([A\cos\omega s]\,[A\cos\omega t]) - m(s)\,m(t)$$

$$= (E(A^2) - (E(A))^2)\,(\cos\omega s)\,(\cos\omega t)$$

Hence,

$$C(s,t) = Var(A)\,(\cos\omega s)\,(\cos\omega t)$$

□

The stochastic processes considered in the examples 2.3 and 2.4 are nonstationary. In particular, their covariance functions do not have the property (2.8). In both cases this is due to the deterministic dependence between $X(s)$ and $X(t)$ for any s and t. The correlation functions of both processes are identical to 1: $\rho(s,t) \equiv 1$

For analyzing examples 2.5 and 2.6 two addition formulas in trigonometry are needed:

$$\cos\alpha\,\cos\beta = \tfrac{1}{2}[\cos(\beta - \alpha) + \cos(\alpha + \beta)]$$

$$\cos(\beta - \alpha) = \cos\alpha\,\cos\beta + \sin\alpha\,\sin\beta$$

Example 2.5 ***(Cosine wave with random amplitude and random phase)*** Let

$$X(t) = A\cos(\omega t + \Phi),$$

where A is a nonnegative random variable with finite expected value and finite variance. Φ is assumed to be uniformly distributed in $[0, 2\pi]$ and independent of A.

The stochastic process $\{X(t),\ t \in (-\infty, +\infty)\}$ might be the output of an oscillator which is selected from a set of ocillators of the same kind which have been turned on at different times. Since

$$\begin{aligned} E(\cos(\omega t + \Phi)) &= \frac{1}{2\pi}\int_0^{2\pi} \cos(\omega t + \phi)\, d\phi \\ &= \frac{1}{2\pi}[\sin(\omega t + \phi)]_0^{2\pi} \\ &= 0, \end{aligned}$$

the trend function of this process is identically zero:

$$m(t) \equiv 0$$

Hence its covariance function is

$$\begin{aligned} C(s,t) &= E\{[A\cos(\omega s + \Phi)][A\cos(\omega t + \Phi)]\} \\ &= E(A^2)\frac{1}{2\pi}\int_0^{2\pi} \cos(\omega s + \phi)\cos(\omega t + \phi)\, d\phi \\ &= E(A^2)\frac{1}{2\pi}\int_0^{2\pi} \frac{1}{2}\{\cos\omega(t-s) + \cos[\omega(s+t) + 2\phi]\}\, d\phi \end{aligned}$$

The first integrand is a constant with respect to integration. Since the integral of the second term is zero, $C(s,t)$ depends only on the difference $\tau = t - s$:

$$C(\tau) = \frac{1}{2}E(A^2)\cos w\tau$$

Thus, the process is stationary. □

Example 2.6 Let A and B be two uncorrelated random variables satisfying

$$E(A) = E(B) = 0 \quad \text{and} \quad Var(A) = Var(B) = \sigma^2 < \infty$$

Define the stochastic process $\{X(t),\ t \in (-\infty, +\infty)\}$ by

$$X(t) = A\cos\omega t + B\sin\omega t$$

Since $Var(X(t)) = \sigma^2 < \infty$ for all t, it is a second order process. Its trend function is identically zero:

$$\begin{aligned} m(t) &= E(A)\cos\omega t + E(B)\sin\omega t = 0\cos\omega t + 0\sin\omega t \\ &= 0 \end{aligned}$$

Thus, from (2.6),

$$C(s,t) = E(X(s)\,X(t))$$

Since A and B are assumed to be uncorrelated, $E(A\,B) = E(A)\,E(B) = 0$. Hence,

$$
\begin{aligned}
C(s,t) = &\ E(A^2 \cos\omega s \cos\omega t + B^2 \sin\omega s \sin\omega t) \\
&+ E(AB \cos\omega s \sin\omega t + AB \sin\omega s \cos\omega t) \\
= &\ \sigma^2 (\cos\omega s \cos\omega t + \sin\omega s \sin\omega t) \\
&+ E(AB)(\cos\omega s \sin\omega t + \sin\omega s \cos\omega t) \\
= &\ \sigma^2 \cos\omega(t-s)
\end{aligned}
$$

Therefore, the covariance function depends only on the difference $\tau = t - s$:

$$C(\tau) = \sigma^2 \cos\omega\tau$$

Thus, the process is stationary. □

The stochastic processes considered in the examples 2.3 to 2.6 have an important property in common: Once the realizations of the random variables V, A, Φ, and B are known, then the process develops in a strictly deterministic way. That means, by observing a sample path of such a process over an arbitrarily small time interval one can predict the further development of the sample path with absolute certainty. Under such a condition it is not possible for $X(s)$ and $X(t)$ to become independent as $|\tau| = |t-s| \to \infty$. Hence it is obvious that none of these processes has property (2.8). (In the examples 2.5 and 2.6 the covariance functions are equal to 0 if and only if $\tau = \frac{\pi}{2\omega}n$; $n = \pm 1, \pm 2, ...$) More complicated problems arise when random influences continuously, or at least repeatedly, affect the development of the stochastic process. All the examples considered in section 2.1 belong to this category. The stochastic processes analyzed in the following two examples also have rather simple structures, but random influences affect the processes repeatedly and, moreover, at random time points. These processes are of importance in physics, electrical engineering, and communication.

Example 2.7 ***(pulse code modulation)*** A source generates the symbols 1 or 0 independently with probabilities $1-p$ and p, respectively. The symbol 1 is transmitted by sending a pulse with constant amplitude A and duration T. The symbol 0 is transmitted by sending nothing during an interval of length T. A signal modulated in this way is represented by the stochastic process $\{X(t), t \in (-\infty, +\infty)\}$ with structure

$$X(t) = \sum_{n=-\infty}^{\infty} A_n h(t - nT), \quad nT \le t < (n+1)T,$$

where the A_n; $n = 0, \pm 1, ...$ are independent binary random variables defined by

$$A_n = \begin{cases} 0 & \text{with probability} \quad p \\ A & \text{with probability} \quad 1-p \end{cases}$$

and $h(t)$ is given by

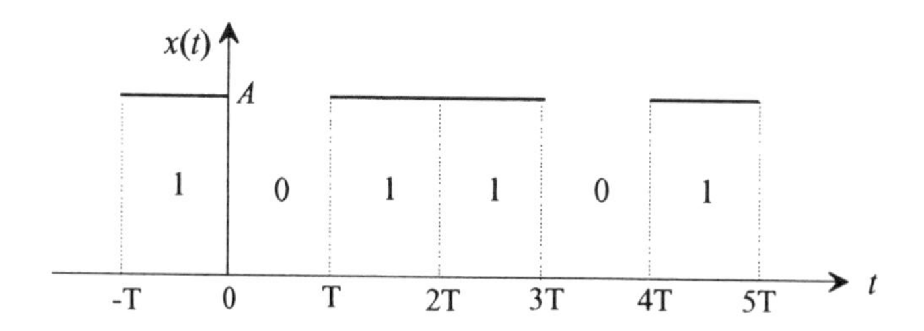

Figure 2.6 Pulse code modulation-binary case

$$h(t) = \begin{cases} 1 & \text{for} \quad 0 \le t < T \\ 0 & \quad \text{elsewhere} \end{cases}$$

In particular, for any t

$$X(t) = \begin{cases} 0 & \text{with probability} \quad p \\ A & \text{with probability} \quad 1-p \end{cases}$$

For instance, the section of a sample path $x = x(t)$ depicted in Figure 2.6 is generated by the following partial sequence of a signal:

... 1 0 1 1 0 1 ...

Note that the time point $t = 0$ has been chosen in such a way that it coincides with the beginning of a new transmission period.

The trend function of the process is constant:

$$m(t) = A\,P(X(t) = A) + 0\,P(X(t) = 0) = A(1-p)$$

For $nT \le s, t < (n+1)T;\ n = 0, \pm 1, \pm 2, \ldots,$

$$\begin{aligned} E(X(s)X(t)) = &\ E(X(s)X(t)|X(s) = A)P(X(s) = A) \\ &+E(X(s)X(t)|X(s) = 0)P(X(s) = 0) \\ = &\ A^2(1-p) \end{aligned}$$

If $mT \le s < (m+1)T$ and $nT \le t < (n+1)T$ with $m \ne n$, then $X(s)$ and $X(t)$ are independent. Hence the covariance function of the process is

$$C(s,t) = \begin{cases} A^2 p(1-p) & \text{for} \quad nT \le s, t < (n+1)T;\ n = 0, \pm 1, \pm 2, \ldots \\ 0 & \qquad\qquad \text{elsewhere} \end{cases}$$

Notwithstanding its constant trend function, the process $\{X(t),\ t \in (-\infty, +\infty)\}$ is not stationary. □

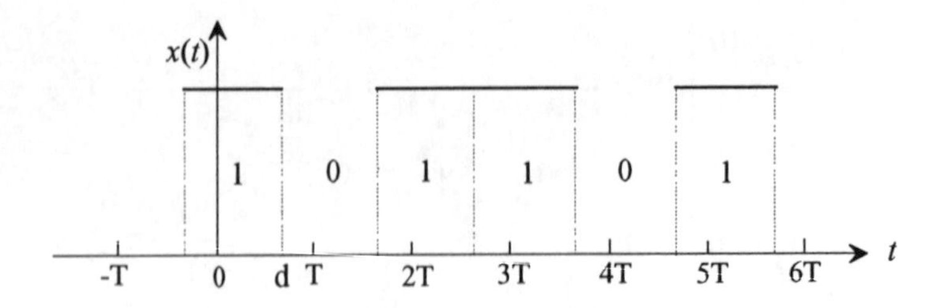

Figure 2.7 Randomly delayed pulse code modulation - binary case

Example 2.8 (*randomly delayed pulse code modulation*) Using the stochastic process $\{X(t), t \in (-\infty, +\infty)$ defined in example 2.7, the process

$$\{Y(t), t \in (-\infty, +\infty) \quad \text{with} \quad Y(t) = X(t - D)$$

is introduced, where D is uniformly distributed over $[0, T]$. Thus, when shifting the sample paths of the process $\{X(t), t \in (-\infty, +\infty)\}$ D time units to the right, one obtains the corresponding sample paths of the process $\{Y(t), t \in (-\infty, +\infty)\}$. For instance, shifting the section of the sample path shown in Figure 2.6 exactly $D = d$ time units to the right yields the corresponding section of the sample path of the process $\{Y(t), t \in (-\infty, +\infty)\}$ shown in Figure 2.7.

As in example 2.7, the trend function of the process $\{Y(t), t \in (-\infty, +\infty)\}$ is

$$m(t) = A(1-p)$$

$X(s)$ and $X(t)$ are independent if $|t-s| > T$ and/or s and t are separated by a switching point $nT+D$; $n = 0, \pm 1, \pm 2, \ldots$. In this case, $C(s,t) = 0$. If $|t-s| \le T$ the random variables $X(s)$ and $X(t)$ are only independent if s and t are separated by a switching point $nT+D$; $n = 0, \pm 1, \pm 2, \ldots$ This random event is denoted by B. The probabilities of B and $\overline{B}$ are

$$P(B) = \frac{|t-s|}{T}, \quad P(\overline{B}) = 1 - \frac{|t-s|}{T}$$

Therefore, the covariance function of the stochastic process $\{Y(t), t \ge 0\}$, given $|t-s| \le T$, is

$$\begin{aligned} C(s,t) &= E(X(s)X(t)|B)P(B) + E(X(s)X(t)|\overline{B})P(\overline{B}) - m(s)\,m(t) \\ &= E(X(s))E(X(t))P(B) + E([X(s)]^2)P(\overline{B}) - m(s)\,m(t) \\ &= [A(1-p)]^2 \frac{|t-s|}{T} + A^2(1-p)\left(1 - \frac{|t-s|}{T}\right) - [A(1-p)]^2 \end{aligned}$$

Finally, with $\tau = t - s$, the covariance function becomes

$$C(\tau) = \begin{cases} A^2 p(1-p)\left(1 - \frac{|\tau|}{T}\right) & \text{for } |\tau| \le T \\ 0 & \text{elsewhere} \end{cases}$$

Thus, the process $\{Y(t), t \in (-\infty, +\infty)\}$ is stationary. Figure 2.8 shows the graph of its covariance function.
Analogously to the transition from example 2.5 to example 2.6, stationarity is achieved by introducing a uniformly distributed phase shift. □

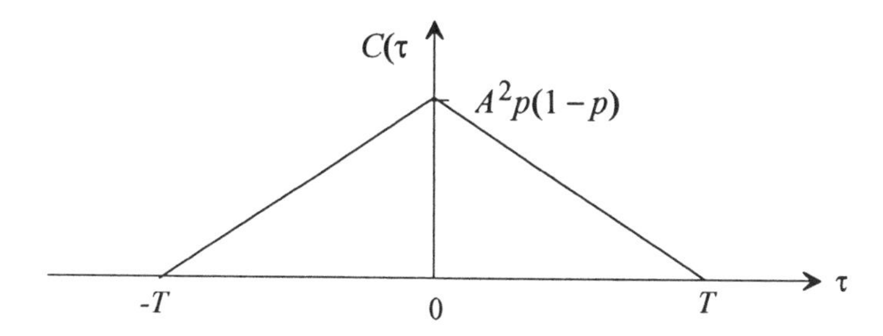

Figure 2.8 Covariance function in case of randomly delayed pulse code modulation

Example 2.9 (*shot noise*) At random time points T_n pulses of a random intensity A_n are induced. The sequence of the T_n, $n = 0, \pm 1, \pm 2, \ldots$ is assumed to be increasing unboundedly and the A_n are assumed to be independent, identically distributed random variables with finite expected value. The sequence

$$\{(T_n, A_n);\ n = 0, \pm 1, \pm 2, \ldots\}$$

is said to be a *pulse process*. If the A_n are identically constant, then the pulse process reduces to the sequence $\{T_n;\ n = 0, \pm 1, \pm 2, \ldots\}$. Let $h(t)$ be the *response* of a system to a pulse. $h(t)$ is assumed to be a real function satisfying

$$h(t) = 0 \text{ for } t < 0 \quad \text{and} \quad \lim_{t \to \infty} h(t) = 0$$

Then the stochastic process $\{X(t), t \in (-\infty, +\infty)\}$ defined by

$$X(t) = \sum_{n=-\infty}^{\infty} A_n h(t - T_n) \tag{2.15}$$

is called a *shot noise process* or just *shot noise*. It quantifies the additive superposition of the responses of the system to the pulses. The factors A_n are sometimes called *amplitudes* of the shot noise process.

A well-known physical phenomenon which can be modeled by a shot noise process is the fluctuation of the anode current in vacuum tubes ("tube noise"). This fluctuation is caused by random current impulses which are initiated by emissions of electrons from the anode at random time points T_n (*Schottky-effect*). The pulse code modulation process (example 2.7) is obviously also a shot noise process.
The term *shot noise* is due to the fact that the effect of firing small shot at a metal slab can also be modeled by a stochastic process with structure (2.15). □

Shot noise processes are discussed in more detail in section 3.1 (example 3.4), where special assumptions about the underlying pulse process are made.

2.4.2 Stationary Discrete-Time Stochastic Processes

Stochastic processes in discrete time are simply sequences of random variables. Hence they are also called *random sequences.* This section considers only stationary random sequences. They play an important role in communication theory as stochastic models of random signals, and in time series analysis for predicting the development of stationary random phenomena over time.

Example 2.10 (***purely random sequence***) Let $\{..., X_{-2}, X_{-1}, X_0, X_1, X_2, ...\}$ be a sequence of uncorrelated random variables distributed identically as X with

$$E(X_t) = 0 \quad \text{and} \quad Var(X_t) = \sigma^2 < \infty; \quad t = 0, \pm 1, \pm 2 \;... \tag{2.16}$$

(Following common practice in denoting random sequences, the parameter appears as an index.) In view of the assumed finite variance of the X_i, the purely random sequence is a second order process. Its trend function is identically zero:

$$m(t) = 0\,; \qquad t = 0, \pm 1, \pm 2, \ldots$$

The covariance function is given by $C(s,t) = E(X_s\, X_t)$, where s and t are integers. Hence,

$$C(s,t) = E(X_s)E(X_t) = 0 \text{ for } s \neq t \text{ and } C(s,t) = E(X^2) = \sigma^2 \text{ for } s = t.$$

Thus, letting $C(s,t) = C(\tau)$ with $\tau = t - s$ yields

$$C(\tau) = \begin{cases} \sigma^2 & \text{for} \quad \tau = 0 \\ 0 & \text{for} \quad \tau \neq 0 \end{cases} \tag{2.17}$$

The purely random sequence is stationary and has independent increments. The notation *purely random sequence* for $\{..., X_{-2}, X_{-1}, X_0, X_1, X_2, ...\}$ does not need a motivation. It is also called *discrete white noise.* The motivation for this is given in chapter 8. □

Example 2.11 ***(sequence of moving averages of order n - MA(n))*** Let a random signal Y_t be given by

$$Y_t = \sum_{i=0}^{n} c_i X_{t-i}\,; \quad t = 0, \pm 1, \pm 2, \ldots;$$

where n is a positive integer, $c_0, c_1, \ldots, c_n$ are finite real numbers, and $\{X_t\}$ is the purely random sequence with parameters (2.16). Thus, Y_t is constructed from the "present" X_t as well as from the n "preceeding" random variables $X_{t-1}, X_{t-2}, \ldots, X_{t-n}$. This construction principle is known as the *principle of moving averages.* (The term *principle of moving sums* would possibly be more accurate.) In view of (1.39),

$$Var(Y_t) = \sigma^2 \sum_{i=0}^{n} c_i^2 < \infty; \quad t = 0, \pm 1, \pm 2, \ldots;$$

so that $\{Y_t;\ t = 0, \pm 1, \pm 2, \ldots\}$ is a second order process. Its trend function is identically zero:

$$m(t) = E(Y_t) = 0 \quad \text{for } t = 0, \pm 1, \pm 2, \ldots$$

For integer-valued s and t,

$$C(s,t) = E(Y_s\, Y_t) = E\left(\left[\sum_{i=0}^{n} c_i X_{s-i}\right]\left[\sum_{j=0}^{n} c_j X_{t-j}\right]\right)$$

$$= E\left(\sum_{i=0}^{n}\sum_{j=0}^{n} c_i c_j\, X_{s-i} X_{t-j}\right)$$

Since $E(X_{s-i} X_{t-j}) = 0$ for $s - i \neq t - j$, the double sum is zero when $|t-s| > n$. Otherwise there exist i and j for which $s - i = t - j$. In this case $C(s,t)$ becomes

$$C(s,t) = E\left(\sum_{\substack{0 \le i \le n \\ 0 \le |t-s|+i \le n}} c_i\, c_{|t-s|+i}\, X_{s-i}^2\right)$$

$$= \sigma^2 \sum_{i=0}^{n-|t-s|} c_i\, c_{|t-s|+i}$$

Letting $\tau = t - s$, the covariance function $C(s,t) = C(\tau)$ is therefore given by

$$C(\tau) = \begin{cases} \sigma^2 (c_0 c_{|\tau|} + c_1 c_{|\tau|+1} + \cdots + c_{n-|\tau|}\, c_n) & \text{for } 0 \le |\tau| \le n \\ 0 & \text{for } \quad |\tau| > n \end{cases} \tag{2.18}$$

Thus, the sequence of moving averages $\{Y_t;\ t = 0, \pm 1, \pm 2, \ldots\}$ is stationary. □

Example 2.12 (*sequence of moving averages of order n - MA(n), equal weights*) By letting in example 2.11

$$c_i = \frac{1}{n+1}; \quad i = 0, 1, \ldots, n;$$

one obtains the *sequence of moving averages*

$$Y_t = \frac{1}{n+1} \sum_{i=0}^{n} X_{t-i}; \quad t = 0, \pm 1, \pm 2, \ldots$$

From (2.18),

$$C(\tau) = \begin{cases} \frac{\sigma^2}{n+1}\left(1 - \frac{|\tau|}{n+1}\right) & \text{for } 0 \le |\tau| \le n \\ 0 & \text{elsewhere} \end{cases}; \quad \tau = 0, \pm 1, \pm 2, \ldots$$

□

Example 2.13 (*sequence of moving averages of unbounded order*) Let

$$Y_t = \sum_{i=0}^{\infty} c_i X_{t-i}; \quad t = 0, \pm 1, \pm 2, \ldots, \tag{2.19}$$

where $\{X_t\}$ is a purely random sequence with parameters (2.16) and the c_i are real numbers. (The random sequence $\{Y_t;\ t = 0,\ \pm 1,\ \pm 2, \ldots\}$ defined in this way is sometimes called a *linear stochastic process.*)

To guarantee the convergence of the infinite series (2.19) in mean square, the c_i have to satisfy

$$\sum_{i=0}^{\infty} c_i^2 < \infty . \tag{2.20}$$

The covariance function is

$$C(\tau) = \sigma^2 \sum_{i=0}^{\infty} c_i c_{|\tau|+i}; \quad \tau = 0, \pm 1, \pm 2, \ldots \tag{2.21}$$

In particular,

$$Var(Y_t) = C(0) = \sigma^2 \sum_{i=0}^{\infty} c_i^2; \quad t = 0, \pm 1, \pm 2, \ldots$$

If the doubly infinite sequence of real numbers $\{\ldots c_{-2}, c_{-1}, c_0, c_1, c_2, \ldots\}$ satisfies the condition

$$\sum_{i=-\infty}^{\infty} c_i^2 < \infty,$$

then the doubly infinite sequence of random variables $\{\ldots Y_{-2}, Y_{-1}, Y_0, Y_1, Y_2, \ldots\}$ defined by

$$Y_t = \sum_{i=-\infty}^{\infty} c_i X_{t-i}; \quad t = 0, \pm 1, \pm 2, ... \tag{2.22}$$

is also stationary and has covariance function

$$C(\tau) = \sigma^2 \sum_{i=-\infty}^{\infty} c_i c_{|\tau|+i}; \quad \tau = 0, \pm 1, \pm 2, ... \tag{2.23}$$

In order to distinguish between random sequences of structure (2.19) and (2.22) they are called one- and two-sided sequences of moving averages, respectively.

Example 2.14 ***(first order autoregressive sequence - AR(1))*** Let a and b be finite real numbers, $|a| < 1$. Then a double infinite sequence $\{Y_t\}$ is defined by

$$Y_t = aY_{t-1} + bX_t; \quad t = 0, \pm 1, \pm 2, ..., \tag{2.24}$$

where $\{X_t\}$ is the purely random sequence with parameters (2.16). Thus, the "present" state Y_t of the process depends on the directly preceeding one Y_{t-1} and on a random disturbance term bX_t with expected value 0 and with variance $b^2\sigma^2$. The n-fold application of (2.24) yields

$$Y_t = a^n Y_{t-n} + b \sum_{i=0}^{n-1} a^i X_{t-i} \tag{2.25}$$

This formula shows that the influence of a past state Y_{t-n} on the present state Y_t on an average decreases as the distance n between Y_{t-n} and Y_t increases. Hence it can be anticipated that the solution of the recurrence formula (2.24) is a stationary process. This stationary solution is obtained taking the limit as $n \to \infty$ in (2.25): Since $\lim_{n\to\infty} a^n = 0$,

$$Y_t = b \sum_{i=0}^{\infty} a^i X_{t-i}; \quad t = 0, \pm 1, \pm 2, ... \tag{2.26}$$

The random sequence $\{Y_t;\ t = 0, \pm 1, \pm 2, ...\}$ defined in this way is called a *first-order autoregressive sequence* (shortly: $AR(1)$). This sequence is obviously a special case of the random sequence generated by (2.19): letting $c_i = b a^i$ makes these sequences formally identical and (2.20) then implies that

$$b^2 \sum_{i=0}^{\infty} (a^i)^2 = b^2 \sum_{i=0}^{\infty} a^{2i} = \frac{b^2}{1-a^2} < \infty$$

Thus, the first order autoregressive sequence is a stationary random sequence the covariance function of which is given by (2.21) with $c_i = b a^i$:

$$C(\tau) = (b\sigma)^2 \sum_{i=0}^{\infty} a^i a^{|\tau|+i} = a^{|\tau|} (b\sigma)^2 \sum_{i=0}^{\infty} a^{2i}$$

Hence,

$$C(\tau) = \frac{(b\sigma)^2}{1-a^2} a^{|\tau|}; \quad \tau = 0, \pm 1, \pm 2, ...$$

□

Example 2.15 (*autoregressive sequence of order r - AR(r)*) In generalizing (2.24), let for a given finite sequence of real numbers $a_1, a_2, ..., a_r$ random variables Y_t be defined by

$$Y_t + a_1 Y_{t-1} + a_2 Y_{t-2} + ... + a_r Y_{t-r} = b X_t, \tag{2.27}$$

where $\{X_t\}$ is a purely random sequence with parameters (2.16). Then the random sequence $\{Y_t;\, t = 0, \pm 1, \pm 2, ...\}$ is called an *autoregressive sequence of the order r*.

Given

$$\sum_{i=0}^{\infty} c_i^2 < \infty,$$

it is interesting to investigate whether analogously to the previous example a stationary sequence $\{Y_t;\; t = 0, \pm 1, \pm 2, ...\}$ of structure

$$Y_t = \sum_{i=0}^{\infty} c_i X_{t-i} \tag{2.28}$$

exists which is solution of (2.27). Substituting (2.28) into (2.27) yields a linear algebraic system of equations in the unknowns c_i:

$$\begin{aligned} c_0 &= b \\ c_1 + a_1 c_0 &= 0 \\ c_2 + a_1 c_1 + a_2 c_0 &= 0 \\ &\cdots\cdots\cdots \\ c_r + a_1 c_{r-1} + ... + a_r c_0 &= 0 \\ c_i + a_1 c_{i-1} + ... + a_r c_{i-r} &= 0; \quad i = r+1, r+2, ... \end{aligned}$$

It can be shown (*Anděl* (1984)) that a nontrivial solution of this system exists if the absolute values of the solutions $y_1, y_2, ..., y_r$ of the algebraic equation

$$y^r + a_1 y^{r-1} + ... + a_{r-1} y + a_r = 0 \tag{2.29}$$

are all less than 1. (This is a property of the sequence $a_1, a_2, ..., a_r$.) In this case, the random sequence $\{Y_t;\; t = 0, \pm 1, \pm 2, ...\}$ generated by (2.28) is a stationary solution of (2.27).

In particular, let $r = 2$. If y_1 and y_2 are the solutions of

$$y^2 + a_1 y + a_2 = 0, \tag{2.30}$$

then the covariance function of the autoregressive sequence of order 2 can, for $y_1 \neq y_2$, be shown to be

$$C(\tau) = C(0)\frac{(1-y_1^2)y_2^{|\tau|+1} - (1-y_2^2)y_1^{|\tau|+1}}{(y_2-y_1)(1+y_1y_2)}; \quad \tau = 0, \pm 1, \pm 2, \ldots \tag{2.31}$$

and for $y_1 = y_2 = y_0$,

$$C(\tau) = C(0)\left(1 + \frac{1-y_0^2}{1+y_0^2}|\tau|\right)y_0^{|\tau|}; \quad \tau = 0, \pm 1, \pm 2, \ldots \tag{2.32}$$

In this formula, the variance $C(0) = Var(Y_t)$ is given by

$$C(0) = \frac{1+a_2}{(1-a_2)\left[(1+a_2)^2 - a_1^2\right]}(b\sigma)^2$$

If y_1 and y_2 are complex, then there exist real numbers λ and ω such that

$$y_1 = \lambda e^{i\omega} \text{ and } y_2 = \lambda e^{-i\omega}$$

and the covariance function assumes a more convenient form than (2.31):

$$C(\tau) = C(0)\,\alpha\,\lambda^{|\tau|}\sin(\omega|\tau| + \beta); \quad \tau = 0, \pm 1, \pm 2, \ldots,$$

where

$$\alpha = \frac{1}{\sin\beta}, \qquad \beta = \arctan\left(\frac{1+\lambda^2}{1-\lambda^2}\tan\omega\right)$$

If $y_1 = y_2 = \lambda$, then this representation is identical to (2.32).

As a numerical example, consider an autoregressive sequence of the form

$$Y_t - 0.6\,Y_{t-1} + 0.05\,Y_{t-2} = 2X_t; \quad t = 0, \pm 1, \pm 2, \ldots \tag{2.33}$$

with $Var(X_t) = 1$. It is obvious that the influence of Y_{t-2} on Y_t is almost negligible compared to the influence of Y_{t-1} on Y_t. The corresponding algebraic equation (2.30) is

$$y^2 - 0.6y + 0.05 = 0$$

The solutions are $y_1 = 0.1$ and $y_2 = 0.5$. The absolute values of y_1 and y_2 are less than 1 so that the random sequence generated by (2.33) is stationary. Its covariance function is obtained from (2.31):

$$C(\tau) = 7.017\,(0.5)^{|\tau|} - 1.063\,(0.1)^{|\tau|}; \quad \tau = 0, \pm 1, \pm 2, \ldots$$

In particular,

$$C(0) = Var(Y_t) = 5.954$$

□

ARMA(r,s)-models Let $\{Y_t;\ t = 0, \pm 1, \pm 2, ...\}$ be a random sequence which is generated by

$$Y_t + a_1 Y_{t-1} + a_2 Y_{t-2} + ... + a_r Y_{t-r} = b_0 X_t + b_1 X_{t-1} + ... + b_s X_{t-s}, \qquad (2.34)$$

where $\{X_i\}$ is a purely random sequence with parameters (2.16). This recursive formula for generating random sequences obviously combines the principle of moving averages and the autoregressive approach. Random sequences generated by (2.34) are therefore called *autoregressive moving averages of order* (r,s) (abbreviation: $ARMA(r,s)$). It can be shown that formula (2.34) also yields stationary sequences if the absolute values of the solutions of the algebraic equation (2.29) are less than 1.

In practice, *ARMA*-sequences are mainly used for modeling time series. A *time series* is a sequence of real numbers which are obtained by observing a process at discrete time points. Thus, in the notation used up till now, a time series may be considered to be a sample path of a discrete-time stochastic process. Continuous -time stochastic processes also lead to time series when these processes are scanned only at discrete time points. Given a finite number of data, the purpos e of a time series analysis essentially consists in identifying the characteristic properties of the underlying stochastic process, for example, whether there is an autoregressive dependence and/or whether the principle of moving averages is in effect. If these or other suitable properties can be confirmed, then reliable predictions on the future development of time series are possible. Moreover, in this case the application of computer simulation is possible to generate time series so that the direct observation of the underlying stochastic process is not necessary.

Time series which are not sample paths of stationary random sequences

$$\{Y_t,\ t = 0, \pm 1, \pm 2, ...\}$$

with constant trend function, can be sometimes transformed into at least approximately stationary sequences $\{Z_t,\ t = 0, \pm 1, \pm 2, ...\}$ by applying the simple transformation

$$Z_t = Y_t - m(t),$$

where $m(t) = E(Y_t)$. The trend function of the sequence $\{Z_t,\ t = 0, \pm 1, \pm 2, ...\}$ is identically zero:

$$E(Z_t) = E[Y_t - m(t)] = m(t) - m(t) = 0; \quad t = 0, \pm 1, \pm 2, ...$$

The numerical work with *ARMA*-models is facilitated by the use of statistical software packages. Important problems are: estimation of the parameters a_i and b_i in the equations (2.27) and (2.34), estimation of trend functions, detection and quantification of possible periodic fluctuations, prediction.

Exercises

2.1) Let the one-dimensional distribution of a process $\{X(t),\ t > 0)\}$ be given by

$$F_t(x) = P(X(t) \le x) = 1 - e^{-(x/t)^2},\ x \ge 0$$

1) Determine its trend function.
2) Can such a process be stationary?

2.2) Let the one-dimensional distribution of a process $\{X(t),\ t > 0)\}$ be given by

$$F_t(x) = P(X(t) \le x) = \frac{1}{\sqrt{2\pi t}\,\sigma} \int_{-\infty}^{x} e^{-\frac{(u-\mu t)^2}{2\sigma^2 t}}\, du;\quad \mu > 0,\ \sigma > 0;\ x \in (-\infty + \infty)$$

1) Determine its trend function.
2) For $\mu = 2$ and $\sigma = 0.5$ sketch the functions

$$y_1(t) = m(t) + \sqrt{Var(X(t))} \quad \text{and} \quad y_2(t) = m(t) - Var(X(t))$$

2.3) Let $X(t) = A\,\sin(\omega\, t + \Phi)$, where A and Φ are independent, non-negative random variables, $E(A) < \infty$, and let Φ be uniformly distributed over $[0, 2\pi]$.
Determine the covariance function and the correlation function of the stochastic process $\{X(t),\ t \in (-\infty, +\infty)\}$.

2.4) Let $X(t) = A(t)\,\sin(\omega\, t + \Phi)$, where $A(t)$ and Φ are independent, non-negative random variables for all t, and let Φ be uniformly distributed over $[0, 2\pi]$.
Prove: If $\{A(t),\ t \in (-\infty, +\infty)\}$ is a wide-sense stationary process, then $\{X(t),\ t \in (-\infty, +\infty)\}$ is also wide-sense stationary.

2.5) Let $\{a_1, a_2, ..., a_n\}$ be a sequence of real numbers and $\{\Phi_1, \Phi_2, ..., \Phi_n\}$ a sequence of independent random variables which are uniformly distributed over $[0, 2\pi]$.
Determine the covariance function and the correlation function of the stochastic process $\{X(t),\ t \in (-\infty, +\infty)\}$ defined by

$$X(t) = \sum_{i=1}^{n} a_i\, \sin(\omega\, t + \Phi_i).$$

2.6) A modulated signal (pulse code modulation) $\{X(t),\ t \in (-\infty, +\infty)\}$ is given by

$$X(t) = \sum_{n=-\infty}^{\infty} A_n h(t - nT),$$

where the A_n are independent and identically distributed random variables with the expected value 0, which can only take on the realizations -1 and +1. Further, let

$$h(t) = \begin{cases} 1 & \text{for } 0 \le t < T/2 \\ 0 & \text{elsewhere} \end{cases}$$

1) Sketch a possible sample path of the stochastic process $\{X(t),\ t \in (-\infty, +\infty)\}$.
2) Determine the covariance function of this process.
3) Let the stochastic process $\{Y(t),\ t \in (-\infty, +\infty)\}$ be defined by $Y(t) = X(t - D)$, where the random variable D has a uniform distribution over $[0, T]$.
Is the stochastic process $\{Y(t),\ t \in (-\infty, +\infty)\}$ wide-sense stationary?

2.7) Let $\{X(t),\ t \in (-\infty, +\infty)\}$ and $\{Y(t),\ t \in (-\infty, +\infty)\}$ be two independent, wide-sense stationary stochastic processes with the same covariance function $C(\tau)$ and with trend functions being both identically zero.
Show that the stochastic process $\{Z(t),\ t \in (-\infty, +\infty)\}$ defined by

$$Z(t) = X(t)\cos\omega t - Y(t)\sin\omega t\,,$$

is wide-sense stationary.

2.8) Let $X(t) = \sin\Phi t$, where Φ is uniformly distributed over the interval $[0, 2\pi]$.
Verify: (1) The random sequence $\{X(t),\ t = 1, 2, \ldots\}$ is wide-sense stationary, but not strictly stationary.
(2) The stochastic process $\{X(t),\ t \geq 0\}$ is neither wide-sense nor strictly stationary.

2.9) Let $\{X(t),\ t \in (-\infty, +\infty)\}$ and $\{Y(t),\ t \in (-\infty, +\infty)\}$ be two independent stochastic processes with trend- and covariance functions $m_X(t)$, $m_Y(t)$ and $C_X(s,t), C_Y(s,t)$, respectively. Further, let $U(t) = X(t) + Y(t)$ and $V(t) = X(t) - Y(t)$, $t \in (-\infty, +\infty)$.
Determine the covariance functions $C_U(s,t)$ and $C_V(s,t)$ of the stochastic processes

$$\{U(t),\ t \in (-\infty, +\infty)\} \text{ and } \{V(t),\ t \in (-\infty, +\infty)\}.$$

2.10) Let $Y_t - 0.8\,Y_{t-1} = X_t;\ t = 0, \pm 1, \pm 2, \ldots,$ where $\{X_t;\ t = 0, \pm 1, \pm 2, \ldots\}$ is a purely random sequence with parameters $E(X_t) = 0$ and $Var(X_t) = 1$.
Determine the covariance function and sketch the correlation function of the autoregressive sequence of order 1 $\{Y_t;\ t = 0, \pm 1, \pm 2, \ldots\}$.

2.11) Let an autoregressive sequence of the order 2 $\{Y_t;\ t = 0, \pm 1, \pm 2, \ldots\}$ be given by

$$Y_t - 0{,}8Y_{t-1} - 0.09Y_{t-1} = X_t;\ t = 0, \pm 1, \pm 2, \ldots,$$

where $\{X_t;\ t = 0, \pm 1, \pm 2, \ldots\}$ is the same purely random sequence as in problem 2.10.
1) Prove that the sequence $\{Y_t;\ t = 0, \pm 1, \pm 2, \ldots\}$ is wide-sense stationary.
2) Determine its covariance and correlation functions.
3) Sketch its correlation function and compare its graph with the one obtained in problem 2.10. Comment the result.

2.12) Let an autoregressive sequence of order 2 $\{Y_t;\ t = 0, \pm 1, \pm 2, \ldots\}$ be given by

$$Y_t - 1.6\,Y_{t-1} + 0.68\,Y_t = 2X_t;\ t = 0, \pm 1, \pm 2, \ldots,$$

where $\{X_t;\ t = 0, \pm 1, \pm 2, \ldots\}$ is the same purely random sequence as in exercise 2.10.
Prove that $\{Y_t;\ t = 0, \pm 1, \pm 2, \ldots\}$ is a wide sense stationary sequence and determine its covariance function.

2.13) Show that the system of difference equations (2.27) has a stationary solution $\{Y_t;\ t = 0, \pm 1, \pm 2, \ldots\}$ if the absolute values of the solutions of the algebraic equation

$$a_r y^r + a_{r-1} y^{r-1} + \ldots + a_1 y + 1 = 0$$

are greater than 1.

3 Poisson Processes

3.1 Homogeneous Poisson Process

3.1.1 Definition and Properties

In many practical situations engineers and scientists are interested in the frequency and time points at which events of a certain type occur; either in a fixed or variable time interval or over a region in line, plane or space. Examples with respect to time intervals are: 1) the arrival of customers at a service facility, 2) failures of a technical system, and 3) frequency of emission of α-particels by a radioactive source. Examples with respect to lines and spatial regions are: 4) frequencies of cracks over specified sections of a railway track, 5) the numbers of traffic accidents over different road sections within a fixed time period, 6) the number of foreign bodies in castings, relative to the volume of the castings. The general stochastic model for describing such situations is the point process (counting process).

Definition 3.1 (***point process***) A stochastic process $\{N(t), t \geq 0\}$ with state space $\mathbf{Z} = \{0, 1, ...\}$ is called a *point process* if it has the following properties:

1) $N(s) \leq N(t)$ for $s \leq t$,
2) For any s, t with $s < t$, the increment $N(t)-N(s)$ is equal to the number of events of a particular type which occur in $(s, t]$. ●

The sample paths of point processes are non-decreasing step functions. The heights of their jumps are equal to 1 provided that more than one event cannot occur at the same time. Point processes having this property are called *simple*. Figure 3.1 illustrates this situation. (The time point at which the i-th event occurs is denoted by t_i, $i = 0, 1, ...$)

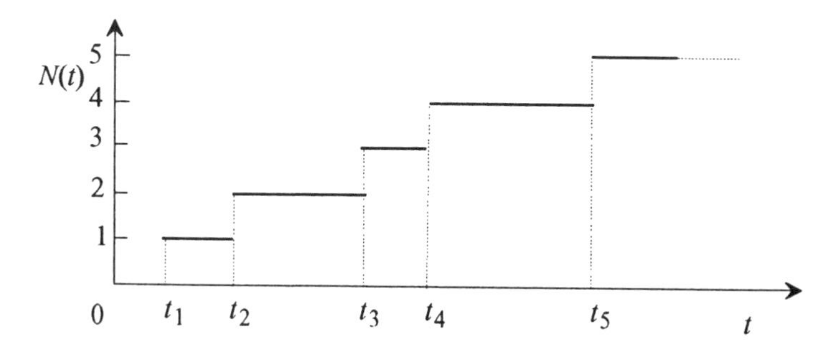

Figure 3.1 Sample path of a point process

Of course, point processes are neither strictly nor wide-sense stationary according to definitions 2.2 and 2.3, respectively. However, the stationarity of point processes is defined as follows:

Definition 3.2 A point process is said to be *homogeneous* or *stationary* if it has homogeneous increments. ●

Definition 3.3 (*homogeneous Poisson process*) A point process $\{N(t), t \geq 0$ is a *homogeneous Poisson process with intensity* λ, $\lambda > 0$, if it has the following properties:

1) $N(0) = 0$,

2) $\{N(t), t \geq 0\}$ is a stochastic process with independent increments.

3) The increments of the process in any interval $[s, t]$, $s < t$, are Poisson distributed with parameter $\lambda(t - s)$:

$$P(N(t) - N(s) = i) = \frac{[\lambda(t-s)]^i}{i!} e^{-\lambda(t-s)}; \quad i = 0, 1, \ldots, \tag{3.1}$$

or, equivalently, introducing the length $\tau = t - s$ of the interval $[s, t]$,

$$P(N(s+\tau) - N(s) = i) = \frac{(\lambda\tau)^i}{i!} e^{-\lambda\tau}; \quad i = 0, 1, \ldots \tag{3.2}$$

●

From (3.2) it follows that the homogeneous Poisson process has homogeneous (stationary) increments, i.e. it is stationary in the sense of definition 3.2.

Theorem 3.1 A point process $\{N(t), t \geq 0\}$ with $N(0) = 0$ is a homogeneous Poisson process with intensity λ if and only if it has the following properties:

(1) $\{N(t), t \geq 0\}$ has homogeneous and independent increments.

(2) The process is *simple,* i.e. $P(N(t+h) - N(t) \geq 2) = o(h)$.

(3) $P(N(t+h) - N(t) = 1) = \lambda h + o(h)$.

(For a definition and properties of *Landau's order symbol* $o(x)$ see appendix 1.)

Proof To prove that definition 3.2 implies properties (1) to (3) it is only necessary to show that a homogeneous Poisson process satisfies properties (2) and (3).

(2) From (3.2),

$$\begin{aligned} P(N(t+h) - N(t) \geq 2) &= e^{-\lambda h} \sum_{i=2}^{\infty} \frac{(\lambda h)^i}{i!} \\ &= \lambda^2 h^2 e^{-\lambda h} \sum_{i=0}^{\infty} \frac{(\lambda h)^i}{(i+2)!} = o(h), \end{aligned}$$

since

$$\lim_{h \to 0} \frac{P(N(t+h)-N(t) \geq 2)}{h} = \lim_{h \to 0} \lambda^2 h e^{-\lambda h} \sum_{i=0}^{\infty} \frac{(\lambda h)^i}{(i+2)!} = 0$$

This result proves that the process is simple.

(3) From (3.2) and the simplicity,

$$\begin{aligned} P(N(t+h)-N(t)=1) &= 1 - P(N(t+h)-N(t)=0) - P(N(t+h)-N(t) \geq 2) \\ &= 1 - e^{-\lambda h} + o(h) \\ &= 1 - (1 - \lambda h) + o(h) \\ &= \lambda h + o(h) \end{aligned}$$

Conversely, it needs to be shown that a process which has properties (1) to (3) is a homogeneous Poisson process. In view of the assumed homogeneity of the increments it is sufficient to prove the validity of (3.1) for $s = 0$. Thus, using the notation

$$p_i(t) = P(N(t)-N(0)=i) = P(N(t)=i); \;\; i = 0, 1, ...$$

it is to show that

$$p_i(t) = \frac{(\lambda t)^i}{i!} e^{-\lambda t}; \;\; i = 0, 1, ... \tag{3.3}$$

From (1),

$$\begin{aligned} p_0(t+h) &= P(N(t+h)=0) \\ &= P(N(t)=0, N(t+h)-N(t)=0) \\ &= P(N(t)=0)\, P(N(t+h)-N(t)=0) \\ &= p_0(t)\, p_0(h) \end{aligned}$$

In view of (2) and (3), this result implies $p_0(t+h) = p_0(t)(1-\lambda h) + o(h)$ or, equivalently,

$$\frac{p_0(t+h)-p_0(t)}{h} = -\lambda p_o(t) + o(h)$$

Taking the limit as $h \to 0$ yields

$$p_0'(t) = -\lambda p_0(t)$$

Since $p_0(0) = 1$, the solution of this differential equation is

$$p_0(t) = e^{-\lambda t}, \;\; t \geq 0,$$

so that (3) holds for $i = 0$. Analogously, for $i \geq 1$

$$p_i(t+h) = P(N(t+h) = i)$$

$$= P(N(t) = i,\ P(N(t+h) - N(t) = 0) + P(N(t) = i-1,\ P(N(t+h) - N(t) = 1)$$

$$+ \sum_{k=2}^{i} P(N(t) = k,\ P(N(t+h) - N(t) = i-k)$$

Because of (3), the sum in the last row is $o(h)$. Using properties (1) and (2), this equation becomes

$$p_i(t+h) = p_i(t)p_0(h) + p_{i-1}(t)p_1(h) + o(h)$$

$$= p_i(t)(1 - \lambda h) + p_{i-1}(t)\lambda h + o(h),$$

or, equivalently,

$$\frac{p_i(t+h) - p_i(t)}{h} = -\lambda[p_i(t) - p_{i-1}(t)] + o(h)$$

Taking the limit as $h \to 0$ yields a system of linear differential equations in the $p_i(t)$:

$$p_i'(t) = -\lambda[p_i(t) - p_{i-1}(t)]; \quad i = 1, 2, ... \tag{3.4}$$

Starting with $p_0(t) = e^{-\lambda t}$, the solution (3.3) is easily obtained by induction. ■

The practical importance of theorem 3.1 is that properties (1) to (3) can generally be verified without any quantitative investigations, only by qualitative reasoning based on the physical nature of the process. In particular, the simplicity implies the practical impossibility of the occurrence of more than one event during a sufficiently small time interval. The possibility of the occurence of two or more events at the same time point can, therefore, be excluded.

Notation In what follows, in order to distinguish the events which are counted by a Poisson process $\{N(t),\ t \geq 0\}$ from any other events, they are called *Poisson events*.

Let T_n; $n = 1,2,...$; be the random time point at which the n-th Poisson event occurs. Since the event $T_n \leq t$ occurs if and only if $N(t) \geq n$, the following equation is obvious:

$$P(T_n \leq t) = P(N(t) \geq n).$$

Therefore, T_n has distribution function

$$F_{T_n}(t) = P(N(t) \geq n) = \sum_{i=n}^{\infty} \frac{(\lambda t)^i}{i!} e^{-\lambda t}; \quad n = 1, 2, ... \tag{3.5}$$

Differentiation with respect to t yields the probability density function of T_n:

$$f_{T_n}(t) = \frac{dF_{T_n}(t)}{dt} = \lambda e^{-\lambda t} \sum_{i=n}^{\infty} \frac{(\lambda t)^{i-1}}{(i-1)!} - \lambda e^{-\lambda t} \sum_{i=n}^{\infty} \frac{(\lambda t)^i}{i!}$$

On the right-hand side of this equation all terms but one cancel:

$$f_{T_n}(t) = \lambda \frac{(\lambda t)^{n-1}}{(n-1)!} e^{-\lambda t}; \quad t \geq 0, \quad n = 1, 2, \dots \tag{3.6}$$

Thus, T_n is Erlang-distributed with parameters n and λ. In particular, T_1 is an exponentially distributed random variable with parameter λ and the *interarrival times* $Y_i = T_i - T_{i-1}$; $i = 1, 2, \dots$; $T_0 = 0$, are independent and identically distributed as T_1. Moreover,

$$T_n = \Sigma_{i=1}^n Y_i$$

From this it follows easily another characterization of the homogeneous Poisson process:

Theorem 3.2 A simple point process $\{N(t), t \geq 0\}$ with $N(0) = 0$ is a homogeneous Poisson process if and only if its times between subsequent jumps are independent, identically distributed exponential random variables. ■

Since a Poisson process $\{N(t), t \geq 0\}$ is statistically equivalent to its corresponding random sequences $\{T_1, T_2, \dots\}$ and $\{Y_1, Y_2, \dots\}$, these sequences are sometimes also called *Poisson processes*. Thus, a Poisson process is a special pulse process (see example 2.9).

Example 3.1 The number of failures $N(t)$, which occur in a computer network over the time interval $[0, t)$, can be described by a homogeneous Poisson process $\{N(t), t \geq 0\}$. On an average, there is a failure after every 4 th hours, i.e. the intensity of the process is equal to $\lambda = 0.25\ [h^{-1}]$.
(1) What is the probability of at most 1 failure in $[0, 8)$, at least 2 failures in $[8, 16)$, and at most 1 failure in $[16, 24)$ (time unit: hour)?
(2) What is the probability that the third failure occurs after 8 hours?

(1) The probability

$$p = P(N(8) - N(0) \leq 1,\ N(16) - N(8) \geq 2,\ N(24) - N(16) \leq 1)$$

is required. In view of the independence and the homogeneity of the increments of a homogeneous Poisson process, it can be determined as follows:

$$\begin{aligned} p &= P(N(8) - N(0) \leq 1)\, P(N(16) - N(8) \geq 2)\, P(N(24) - N(16) \leq 1) \\ &= P(N(8) \leq 1)\, P(N(8) \geq 2)\, P(N(8) \leq 1) \end{aligned}$$

Since

$$\begin{aligned} P(N(8) \leq 1) &= P(N(8) = 0) + P(N(8) = 1) \\ &= e^{-0.25 \cdot 8} + 0.25 \cdot 8 \cdot e^{-0.25 \cdot 8} \\ &= 0.406 \end{aligned}$$

and

$$P(N(8) \geq 2) = 1 - P(N(8) \leq 1) = 0.594$$

the desired probability is

$$p = 0.098$$

(2): Since T_3 is Erlang-distributed with parameters $n = 3$ and $\lambda = 0.25$,

$$P(T_3 > 8) = 1 - F_{T_3}(8)$$

$$= e^{-0.25 \cdot 8}\left(\sum_{i=0}^{2} \frac{(0.25 \cdot 8)^i}{i!}\right)$$

$$= e^{-2}\left(1 + \frac{2^1}{1!} + \frac{2^2}{2!}\right) = 5\,e^{-2}$$

Thus, $P(T_3 > 8) = 0.677$. □

The following examples make use of the hyperbolic sine and cosine functions:

$$\sinh x = \frac{e^x - e^{-x}}{2}, \quad \cosh x = \frac{e^x + e^{-x}}{2}, \quad x \in (-\infty, +\infty)$$

Example 3.2 (*random telegraph signal*) Let a random signal $X(t)$ have structure

$$X(t) = Y(-1)^{N(t)}, \ t \geq 0,$$

where $\{N(t),\ t \geq 0\}$ is a homogeneous Poisson process with intensity λ and Y is a binary random variable with $P(Y = 1) = P(Y = -1) = 1/2$ which is independent of $N(t)$ for all t. Signals of this structure are called *random telegraph signals* (see, e.g., *Kannan* (1979)). Random telegraph signals are basic modules for generating signals with a more complicated structure. Obviously, $X(t) = 1$ or $X(t) = -1$ and Y determines the sign of $X(0)$. Figure 3.2 shows a sample path $x = x(t)$ of the process $\{X(t),\ t \geq 0\}$ given $T_n = t_n;\ n = 1, 2, \ldots$ and $Y = 1$. It will now be shown that $\{X(t),\ t \geq 0\}$ is wide-sense stationary: Since $|X(t)|^2 = 1 < \infty$ for all $t \geq 0$, the stochastic process $\{X(t),\ t \geq 0\}$ is a second-order process. Letting

$$I(t) = (-1)^{N(t)},$$

its trend function is $m(t) = E(X(t)) = E(Y)\,E(I(t))$. Since $E(Y) = 0$, the trend function is identically zero:

$$m(t) \equiv 0$$

It remains to show that the covariance function $C(s,t)$ of this process depends only on $|t - s|$. This requires the determination of the probability distribution of $I(t)$:
A transition from $I(t) = -1$ to $I(t) = +1$ or, conversely, from $I(t) = +1$ to $I(t) = -1$ occurs at those time points where Poisson events occur, i.e. where $N(t)$ jumps:

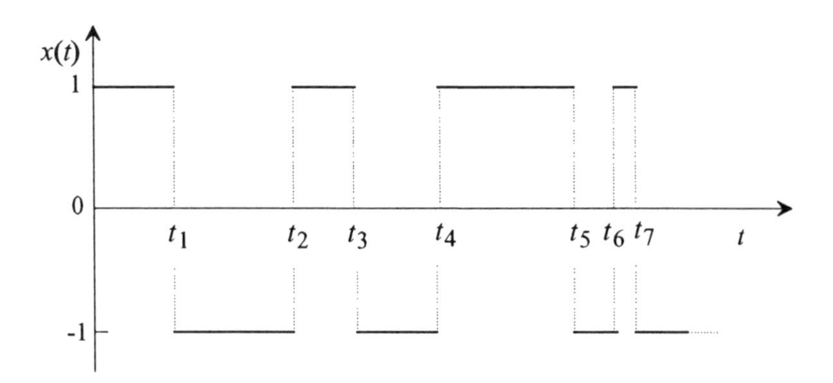

Figure 3.2 Sample path of the random telegraph signal

$$P(I(t)=1) = P(\text{even number of jumps in } [0,t])$$

$$= e^{-\lambda t} \sum_{i=0}^{\infty} \frac{(\lambda t)^{2i}}{(2i)!} = e^{-\lambda t} \cosh \lambda t\,,$$

$$P(I(t)=-1) = P(\text{odd number of jumps } [0,t])$$

$$= e^{-\lambda t} \sum_{i=0}^{\infty} \frac{(\lambda t)^{2i+1}}{(2i+1)!} = e^{-\lambda t} \sinh \lambda t\,.$$

Hence the expected value of $I(t)$ is

$$E[I(t)] = 1 \cdot P(I(t)=1) + (-1) \cdot P(I(t)=-1)$$

$$= e^{-\lambda t}[\cosh \lambda t \; - \; \sinh \lambda t]$$

$$= e^{-2\lambda t}.$$

Since

$$C(s,t) = Cov\,[X(s), X(t)] = E[(X(s)\,X(t))] = E[Y I(s)\, Y I(t)]$$

$$= E\left[\, Y^2\, I(s)\, I(t)\right] = E(Y^2)\, E[I(s)\, I(t)]$$

and $E(Y^2) = 1$,

$$C(s,t) = E[I(s)\, I(t)]$$

Thus, in order to evaluate $C(s,t)$, the joint distribution of the random vector $(I(s), I(t))$ must be determined: From (1.6) and in view of the homogeneity of the increments of $\{N(t),\, t \geq 0\}$, for $s < t$,

$$\begin{aligned} p_{1,1} &= P(I(s)=1,\, I(t)=1) = P(I(s)=1)P(I(t)=1 \mid I(s)=1) \\ &= e^{-\lambda s}\cosh \lambda s\, P(\text{even number of jumps in } (s,t]) \\ &= e^{-\lambda s}\cosh \lambda s\, e^{-\lambda(t-s)}\cosh \lambda(t-s) \\ &= e^{-\lambda t}\cosh \lambda s\, \cosh \lambda(t-s) \end{aligned}$$

Analogously,

$$\begin{aligned} p_{1,-1} &= P(I(s)=1, I(t)=-1) &&= e^{-\lambda t}\cosh \lambda s\, \sinh \lambda(t-s) \\ p_{-1,1} &= P(I(s)=-1, I(t)=1) &&= e^{-\lambda t}\sinh \lambda s\, \sinh \lambda(t-s) \\ p_{-1,-1} &= P(I(s)=-1, I(t)=-1) &&= e^{-\lambda t}\sinh \lambda s\, \cosh \lambda(t-s) \end{aligned}$$

Since $E[I(s)I(t)] = p_{1,1} + p_{-1,-1} - p_{1,-1} - p_{-1,1}$,

$$C(s,t) = e^{-2\lambda(t-s)}, \quad s < t$$

Since the order of s and t can be changed,

$$C(s,t) = e^{-2\lambda|t-s|}$$

The stochastic process $\{X(t),\, t \geq 0\}$ is, therefore, wide-sense stationary. □

Example 3.3 Let $\{N(t),\, t \geq 0\}$ be a Poisson process with the intensity λ. What is the probability that exactly i events occur in the interval $[0, s]$ given that exactly n events occur in the interval $[0, t]$; $s < t,\ i = 0, 1, \ldots, n$?
In view of (1.6) and the homogeneity and independence of the increments of the process $\{N(t),\, t \geq 0\}$

$$\begin{aligned} P(N(s)=i \mid N(t)=n) &= \frac{P(N(s)=i,\, N(t)=n)}{P(N(t)=n)} \\ &= \frac{P(N(s)=i,\, N(t)-N(s)=n-i)}{P(N(t)=n)} \\ &= \frac{P(N(s)=i)\, P(N(t)-N(s)=n-i)}{P(N(t)=n)} \\ &= \frac{\frac{(\lambda s)^i}{i!}e^{-\lambda s}\, \frac{[\lambda(t-s)]^{n-i}}{(n-i)!}e^{-\lambda(t-s)}}{\frac{(\lambda s)^n}{n!}e^{-\lambda t}} \\ &= \binom{n}{i}\left(\frac{s}{t}\right)^i\left(1-\frac{s}{t}\right)^{n-i}; \quad i = 0, 1, \ldots, n. \end{aligned}$$

This is a binomial distribution with parameters $p = s/t$ and n.

For $n = i = 1$, this result implies that given "$N(t) = 1$" the random time T_1 to the first and only Poisson event occuring in $[0, t]$ is uniformly distributed over this interval:

$$P(T_1 \le s | T_1 \le t) = P(N(s) = 1 | N(t) = 1) = \frac{s}{t}$$ □

3.1.2 Poisson Process and Uniform Distribution

The relationship between the homogeneous Poisson process and the uniform distribution pointed out in example 3.3 is a special case of a more general result. To prove it the joint probability density of the random vector $(T_1, T_2, ..., T_n)$ is needed. (As in the previous section, T_i denotes the time point at which the i-th Poisson event occurs.)

Theorem 3.3 The joint probability density of the random vector $(T_1, T_2, ..., T_n)$ is

$$f(t_1, t_2, ..., t_n) = \begin{cases} \lambda^n e^{-\lambda t_n} & \text{for } 0 \le t_1 < t_2 < ... < t_n \\ 0 & \text{elsewhere} \end{cases} \tag{3.7}$$

Proof For $0 \le t_1 < t_2$, the joint distribution function of (T_1, T_2) is given by

$$P(T_1 \le t_1, T_2 \le t_2) = \int_0^{t_1} P(T_2 \le t_2 | T_1 = t) f_{T_1}(t) dt$$

According to theorem 3.2, the interarrival times $Y_i = T_i - T_{i-1}$; $i = 1, 2, ...$; are independent, identically distributed exponential random variables with parameter λ. Therefore, since $T_1 = Y_1$,

$$P(T_1 \le t_1, T_2 \le t_2) = \int_0^{t_1} P(T_2 \le t_2 | T_1 = t) \lambda e^{-\lambda t} dt$$

Given "$T_1 = t$" the random event "$T_2 \le t_2$" is equivalent to "$Y_2 \le t_2 - t$". Thus, the desired two-dimensional distribution function is

$$P(T_1 \le t_1, T_2 \le t_2) = \int_0^{t_1} (1 - e^{-\lambda(t_2 - t)}) \lambda e^{-\lambda t} dt$$
$$= 1 - e^{-\lambda t_1} - \lambda t_1 e^{-\lambda t_2}, \ t_1 < t_2$$

Partial differentiation yields the corresponding two-dimensional probability density

$$f(t_1, t_2) = \begin{cases} \lambda^2 e^{-\lambda t_2} & \text{for } 0 \le t_1 < t_2 \\ 0 & \text{elsewhere} \end{cases}$$

The proof of the theorem is now easily completed by induction. ■

The formulation of the following theorem requires a result from the theory of ordered samples: Let $\{X_1, X_2, ..., X_n\}$ be a random sample, i.e. a sequence of independent, identically distributed random variables, and let the X_i be uniformly distributed over $[0, x]$. Then the joint probability density of the corresponding ordered sample $\{X_1^*, X_2^*, ..., X_n^*\}$, $0 \le X_1^* < X_2^* < ... < X_n^* \le x$, is given by

$$f^*(x_1^*, x_2^*, ..., x_n^*) = \begin{cases} n!/x^n, & 0 \le x_1^* < x_2^* < ... < x_n^* \le x, \\ 0, & \text{elsewhere} \end{cases} \tag{3.8}$$

For the sake of comparision: The joint probability density of the original (unordered) sample $\{X_1, X_2, ..., X_n\}$ is given by (see (1.36))

$$f(x_1, x_2, ..., x_n) = \begin{cases} 1/x^n, & 0 \le x_i \le x \\ 0, & \text{elsewhere} \end{cases} \tag{3.9}$$

With that the prerequisites have been provided to tackle the main problem of this section.

Theorem 3.4 Let $\{N(t), t \ge 0\}$ be a homogeneous Poisson process with intensity λ and let T_i be the time point at which the i-th Poisson event occurs; $i = 1, 2, ...$; $T_0 = 0$. Given that $N(t) = n$, $t > 0$, $n = 1, 2, ...$, the random vector $\{T_1, T_2, ..., T_n\}$ has the same joint probability density as an ordered random sample taken from a uniform distribution over $[0, t]$.

Proof By definition, for disjoint, but otherwise arbitrary subintervals $[t_i, t_i + h_i]$ of $[0, t]$; $i = 1, 2, ..., n$; the joint probability density of $\{T_1, T_2, ..., T_n\}$, given $N(t) = n$, is

$$f(t_1, t_2, ..., t_n | N(t) = n)$$

$$= \lim_{\substack{h_i \to 0 \\ i=1,2,...,n}} \frac{P(t_1 \le T_1 < t_1+h_1, t_2 \le T_2 < t_2+h_2, ..., t_n \le T_n < t_n+h_n | N(t)=n)}{h_1 h_2 \cdots h_n}. \tag{3.10}$$

Since the event "$N(t) = n$" is equivalent to "$T_n \le t < T_{n+1}$",

$$P(t_1 \le T_1 < t_1 + h_1, t_2 \le T_2 < t_2 + h_2, ..., t_n \le T_n < t_n + h_n | N(t) = n)$$

$$= \frac{P(t_i \le T_i < t_i + h_i; i = 1, 2, ..., n; T_n \le t < T_{n+1})}{P(N(t) = n)}$$

$$= \frac{\int_t^\infty \int_{t_n}^{t_n+h_n} \int_{t_{n-1}}^{t_{n-1}+h_{n-1}} \cdots \int_{t_1}^{t_1+h_1} \lambda^{n+1} e^{-\lambda x_{n+1}} dx_1 \ldots dx_n \, dx_{n+1}}{\frac{(\lambda t)^n}{n!} e^{-\lambda t}}$$

$$= \frac{h_1 h_2 \ldots h_n \lambda^n e^{-\lambda t}}{\frac{(\lambda t)^n}{n!} e^{-\lambda t}} = \frac{h_1 h_2 \ldots h_n n!}{t^n}$$

Hence, (3.10) does not depend on the h_i, so that the desired conditional joint probability density is

$$f(t_1, t_2, \ldots, t_n | N(t) = n) = \begin{cases} n!/t^n, & 0 \le t_1 < t_2 < \ldots < t_n \le t \\ 0, & \text{elsewhere} \end{cases} \tag{3.11}$$

Apart from the notation of the variables, this is the joint density (3.8). ■

The relationship between homogeneous Poisson processes and the uniform distribution proved in this theorem motivates the common phrase that a homogeneous Poisson process is a *purely random process*; since, if $N(t) = n$, then the time points at which the n Poisson events occur, are distributed "at random" over $[0, t]$. (There is, however, no relationship with the purely random sequence of example 2.10.)

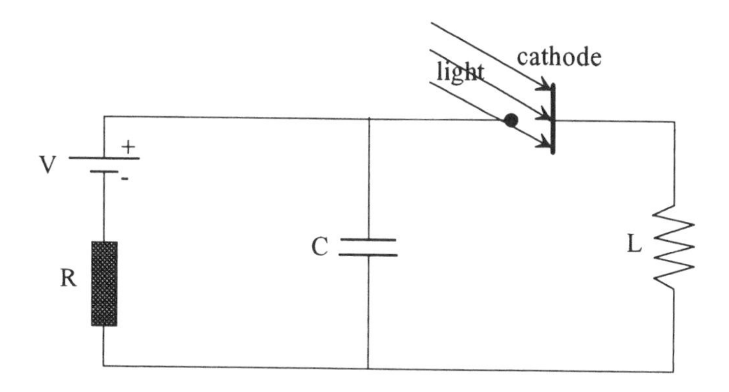

Figure 3.3 Photodetection circuit (Example 3.4)

Example 3.4 (***shot noise***) Besides the practical application of shot noise processes pointed out in example 2.9, another application is now discussed in more detail: In the circuit depicted in Figure 3.3 the light source is switched on at time $t = 0$. A current pulse is initiated in the circuit as soon as a photoelectron is emitted by the cathode due to the light falling on it. Such a current pulse can be quantified by a function $h(t)$ with properties (Figure 3.4)

$$h(t) \ge 0, \quad h(t) = 0 \text{ for } t < 0 \quad \text{and} \quad \int_0^\infty h(t)\,dt < \infty \tag{3.12}$$

Let $T_1, T_2, \ldots$ be the sequence of random time points at which the cathode emits photoelectrons and $N(t) = \max(n;\ T_n \le t)$. Then the total current flowing in the circuit at time t is

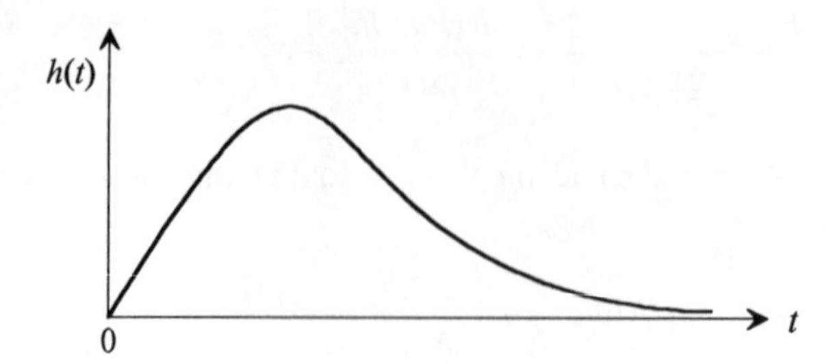

Figure 3.4 Current pulse

$$X(t) = \sum_{i=1}^{N(t)} h(t - T_i) \tag{3.13}$$

In view of properties (3.12), $X(t)$ can also be written in the form

$$X(t) = \sum_{i=1}^{\infty} h(t - T_i)$$

In what follows, the underlying pulse process $\{T_1, T_2, ...\}$, or $\{N(t), t \geq 0$, respectively, is assumed to be a homogeneous Poisson process with parameter λ. (This assumption corresponds to the practical reality.) From (3.11),

$$E(X(t)|N(t) = n) = E\left(\sum_{i=1}^{n} h(t - T_i) \middle| N(t) = n \right)$$

$$= \int_0^t \int_0^{t_{n-1}} \cdots \int_0^{t_3} \int_0^{t_2} \left(\sum_{i=1}^{n} h(t - t_i) \right) \frac{n!}{t^n} dt_1 dt_2 ... dt_n \tag{3.14}$$

The terms in the integrand of (3.14) can be arbitrarily interchanged without influencing the integral. Thus, when determining the conditional expected value, n independent and unordered T_i being uniformly distributed over $[0, t]$, can be assumed. The conditional expected value can therefore also be written in the form

$$E(X(t)|N(t) = n) = \int_0^t \cdots \int_0^t \int_0^t \left(\sum_{i=1}^{n} h(t - t_i) \right) \frac{1}{t^n} dt_1 dt_2 ... dt_n$$

The integration of the single terms $h(t - t_i)$ is equivalent to computing their expected values with respect to the distribution of T_i. Hence,

$$E(X(t)|N(t) = n) = \sum_{i=1}^{n} E(h(t - T_i)) = \sum_{i=1}^{n} \frac{1}{t} \int_0^t h(t - x)\, dx$$

$$= \left(\frac{1}{t} \int_0^t h(x)\, dx \right) n$$

Application of the total probability rule (1.7) yields

$$E(X(t)) = \sum_{n=0}^{\infty} E(X(t)|N(t)=n)P(N(t)=n)$$

$$= \frac{1}{t}\int_0^t h(x)\,dx \sum_{n=1}^{\infty} n\frac{(\lambda t)^n}{n!}e^{-\lambda t}$$

$$= \left(\frac{1}{t}\int_0^t h(x)\,dx\right)E(N(t))$$

$$= \left(\frac{1}{t}\int_0^t h(x)\,dx\right)(\lambda t)$$

Therefore, the trend function of the shot noise process $\{X(t),\ t \geq 0\}$ is

$$m(t) = \lambda \int_0^t h(x)\,dx \tag{3.15}$$

In order to compute the variance function, the expected value of $X(s)X(t)$ is determined:

$$E(X(s)X(t)) = \sum_{i,j=1}^{\infty} E\left[h(s-T_i)\,h(t-T_j)\right]$$

$$= \sum_{i=1}^{\infty} E[h(s-T_i)\,h(t-T_i)]$$

$$+ \sum_{\substack{i,j=1 \\ i \neq j}}^{\infty} E\left[h(s-T_i)\,h(t-T_j)\right]$$

Given $N(t) = n$, an argument analogously to the one which has been used to obtain (3.14) implies that the T_k can be considered independent, unordered random variables which are uniformly distributed over $[0, t]$. Hence, for $s < t$,

$$E(X(s)X(t)|N(t)=n) = \left(\frac{1}{t}\int_0^s h(s-x)h(t-x)dx\right)n$$

$$+\left(\frac{1}{t}\int_0^s h(s-x)dx\right)\left(\frac{1}{t}\int_0^t h(t-x)\,dx\right)(n-1)n$$

Application of the total probability rule yields

$$E(X(s)\,X(t)) = \left(\frac{1}{t}\int_0^s h(x)\,h(t-s+x)\,dx\right)E(N(t))$$

$$+\left(\frac{1}{t}\int_0^s h(x)dx\right)\left(\frac{1}{t}\int_0^t h(x)\,dx\right)\left[E(N^2(t)) - E(N(t))\right]$$

Using this result, the relationships $E(N(t)) = \lambda t$, $E(N^2(t)) = \lambda t(\lambda t + 1)$, and (3.15) yields the covariance function:

$$C(s,t) = \lambda \int_0^s h(x)\,h(t-s+x)\,dx, \quad s < t$$

Note that $C(s,t) = C(t,s)$. Since the roles of s and t can be changed, one obtains for any s and t

$$C(s,t) = \lambda \int_0^{\min(s,t)} h(x)\,h(|t-s|+x)\,dx \tag{3.16}$$

In particular, the variance of $X(t)$ is

$$Var(X(t)) = \lambda \int_0^t h^2(x)\,dx$$

By letting $s \to \infty$, keeping $|\tau| = |t-s|$ constant, the trend and covariance functions become

$$m = \lambda \int_0^\infty h(x)\,dx \tag{3.17}$$

$$C(\tau) = \lambda \int_0^\infty h(x)\,h(|\tau|+x)\,dx \tag{3.18}$$

Thus, for large t the process $\{X(t),\ t \geq 0\}$ is approximately wide-sense stationary. Equation (3.18) is called *Campell's theorem.* (For another proof see exercise 3.7.)

In particular, let $h(t)$ be given by

$$h(t) = \begin{cases} a & \text{for } 0 \leq t \leq d \\ 0, & \text{elsewhere} \end{cases},$$

where a and d are constants. The corresponding trend- and covariance functions are

$$m(t) = \begin{cases} \lambda a t & \text{for } 0 \leq t \leq d \\ \lambda a d & \text{for } \quad t > d \end{cases}$$

$$C(s,t) = \begin{cases} \lambda a^2 s & \text{for } 0 \leq s \leq d - |t-s| \\ \lambda a^2 (d - |t-s|) & \text{for } 0 \leq d - |t-s| < s \\ 0, & \text{elsewhere} \end{cases} \tag{3.19}$$

The formulas (3.17) and (3.18) specialise to

$$m = \lambda a d,$$

$$C(\tau) = \begin{cases} \lambda a^2 (d - |\tau|), & 0 \leq |\tau| < d \\ 0, & \text{elsewhere} \end{cases}$$

Generalization If the current impulses induced by photoelectrons have random intensities, then this can be taken into account by introducing random factors A_i (see also example 2.9):

$$X(t) = \sum_{i=1}^{\infty} A_i\,h(t - T_i) \tag{3.20}$$

Provided that the A_i are identically distributed as A, independent of each other and also independent of all T_k, then computing the trend and covariance functions of the process $\{X(t), t \geq 0\}$ does not give rise to principally new problems. Assuming the existence of the first two moments of A, one obtains

$$m(t) = \lambda E(A) \int_0^t h(x)\, dx$$

$$C(s,t) = \lambda E(A^2) \int_0^{\min(s,t)} h(x)\, h(|t-s|+x)\, dx$$

If the process of inducing current impulses by photoelectrons has already been operating for an unboundedly long time (i.e. the circuit was switched on sufficiently long ago), then the underlying shot noise process $\{X(t), t \in (-\infty, +\infty)\}$ is given by

$$X(t) = \sum_{i=-\infty}^{\infty} A_i\, h(t - T_i)$$

In this case the process is a priori stationary. Its trend- and covariance functions are obtained from (3.17) and (3.18) simply by multiplying the right-hand sides with $E(A)$ and $E(A^2)$, respectively. □

Example 3.5 Customers arrive at a service station (service system, queueing system) according to a homogeneous Poisson process $\{N(t), t \geq 0\}$ with intensity λ. The arrival of a customer is therefore a Poisson event. The number of servers in the system is assumed to be so large that an incoming customer will always find an available server. To cope with this situation the service system must be modeled as having an infinite number of servers. The service times of all customers are assumed to be independent random variables which are identically distributed as B. Let $G(y) = P(B \leq y)$ be the distribution function of B and $X(t)$ the random number of customers which are in the system at time t, $X(0) = 0$. The aim is to determine the *state probabilities* of the system

$$p_i(t) = P(X(t) = i); \quad i = 0, 1, \ldots; \quad t \geq 0$$

According to the total probability rule (1.7),

$$p_i(t) = \sum_{n=i}^{\infty} P(X(t) = i | N(t) = n) \cdot P(N(t) = n)$$

$$= \sum_{n=i}^{\infty} P(X(t) = i | N(t) = n) \cdot \frac{(\lambda t)^n}{n!} e^{-\lambda t} \tag{3.21}$$

A customer arriving at time x is with probability $1 - G(t-x)$ still in the system at time t, $t > x$, i.e. the service has not yet been finished by time t. Given $N(t) = n$, the arrival times $T_1, T_2, \ldots, T_n$ of the n customers in the system are, according to theorem 3.4, independent random variables, uniformly distributed over $[0, t]$. For calculating the state probabilities, the order of the T_i is not important. Hence they can be assumed to be independent, unordered random variables, uniformly distributed over $[0, t]$. Thus, the probability that any one of those n customers is still in the system at time t is

$$p(t) = \int_0^t (1 - G(t-x)) \frac{dx}{t} = \frac{1}{t} \int_0^t (1 - G(x))\, dx$$

Since the service times are independent of each other,

$$P(X(t) = i \,|\, N(t) = n) = \binom{n}{i} [p(t)]^i [1 - p(t)]^{n-i}; \quad i = 0, 1, \ldots, n$$

By (3.21), the desired state probabilities can be obtained as follows:

$$\begin{aligned} p_i(t) &= \sum_{n=i}^{\infty} \binom{n}{i} [p(t)]^i [1 - p(t)]^{n-i} \cdot \frac{(\lambda t)^n}{n!} e^{-\lambda t} \\ &= \frac{[\lambda t p(t)]^i}{i!} e^{-\lambda t} \sum_{k=0}^{\infty} \frac{[\lambda t (1 - p(t))]^k}{k!} \\ &= \frac{[\lambda t p(t)]^i}{i!} e^{-\lambda t} \cdot e^{\lambda t (1 - p(t))} \\ &= \frac{[\lambda t p(t)]^i}{i!} \cdot e^{-\lambda t p(t)}; \quad i = 0, 1, \ldots \end{aligned}$$

Thus, $\{p_i(t);\ i = 0, 1, \ldots\}$ is a Poisson distribution with intensity $E(X(t)) = \lambda t p(t)$. The trend function of the stochastic process $\{X(t), t \geq 0\}$ is, consequently,

$$m(t) = \lambda \int_0^t (1 - G(x))\, dx$$

Let $E(Y) = 1/\lambda$ be the expected interarrival time between two successive customers. Since $E(B) = \int_0^{\infty} (1 - G(x))\, dx$,

$$\lim_{t \to \infty} m(t) = \frac{E(B)}{E(Y)}$$

Thus, for $t \to \infty$ the trend function and the state probabilities of the stochastic process $\{X(t), t \geq 0\}$ become constants. Letting $\rho = E(B)/E(Y)$, the *stationary state probabilities* of the system become

$$p_i = \lim_{t \to \infty} p_i(t) = \frac{\rho^i}{i!} e^{-\rho}; \quad i = 0, 1, \ldots \tag{3.22}$$

In particular, if B is exponentially distributed with parameter μ, then

$$m(t) = \lambda \int_0^t e^{-\mu x}\, dx = \frac{\lambda}{\mu}\left(1 - e^{-\mu t}\right)$$

In this case, $\rho = \lambda/\mu$. □

3.2 Inhomogeneous Poisson Process

3.2.1 Definition and Properties

Now the interest is focussed on the structure of a stochastic process which, except for the homogeneity of its increments, has all the other properties listed in theorem 3.1. This leads to the following definition.

Definition 3.3 A point process $\{N(t),\ t \geq 0\}$ satisfying $N(0)=0$ is called an *inhomogeneous Poisson process with intensity function* $\lambda(t)$ if it has properties

(1) $\{N(t),\ t \geq 0\}$ has independent increments,

(2) $P(N(t+h) - N(t) \geq 2) = o(h)$,

(3) $P(N(t+h) - N(t) = 1) = \lambda(t)h + o(h)$. ●

The following problems are considered:

1) Computation of the probabilities

$$p_i(s,t) = P(N(t) - N(s) = i); \quad s < t,\ i = 0, 1, \ldots \tag{3.23}$$

2) Computation of the probability density of the random time point T_i at which the i-th Poisson event occurs.

3) Computation of the joint probability density of $(T_1, T_2, \ldots, T_n)$, $n = 1, 2, \ldots$

1) In view of the assumed independence of the increments,

$$\begin{aligned} p_0(s, t+h) &= P(N(t+h) - N(s) = 0) \\ &= P(N(t) - N(s) = 0) = 0,\ N(t+h) - N(t) = 0) \\ &= P(N(t) - N(s) = 0) \cdot P(N(t+h) - N(t) = 0) \\ &= p_0(s,t)[1 - \lambda(t)h + o(h)] \end{aligned}$$

Thus,

$$\frac{p_0(s,t+h) - p_0(s,t)}{h} = -\lambda(t)\,p_0(s,t) + \frac{o(h)}{h}$$

Letting $h \to 0$ yields a partial differential equation of the first order:

$$\frac{\partial}{\partial t} p_0(s,t) = -\lambda(t)\,p_0(s,t)$$

Since $p_0(0,0) = 1$, or equivalently, $N(0) = 0$, the solution is

$$p_0(s,t) = e^{-[\Lambda(t)-\Lambda(s)]}, \tag{3.24}$$

where

$$\Lambda(y) = \int_0^y \lambda(x)\,dx$$

Analogously to the proof of theorem 3.1, the probabilities (3.23) are seen to be

$$p_i(s,t) = \frac{[\Lambda(t)-\Lambda(s)]^i}{i!} e^{-[\Lambda(t)-\Lambda(s)]}; \quad i = 0, 1, 2, \ldots \tag{3.25}$$

In particular, the absolute state probabilities

$$p_i(t) = p_i(0,t) = P(N(t) = i)$$

of the inhomogeneous Poisson process at time t are

$$p_i(t) = \frac{[\Lambda(t)]^i}{i!} e^{-\Lambda(t)}; \quad i = 0, 1, 2, \ldots \tag{3.26}$$

2) Let denote $F_{T_1}(t) = P(T_1 \le t)$ the distribution function and $f_{T_1}(t)$ the probability density of the random time T_1 to the occurence of the first Poisson event. Then, from (3.24),

$$p_0(t) = p_0(0,t) = P(T_1 > t) = 1 - F_{T_1}(t)$$

so that

$$p_0(t) = e^{-\Lambda(t)}$$

Thus,

$$F_{T_1}(t) = 1 - e^{-\int_0^t \lambda(x)\,dx}, \quad t \ge 0 \tag{3.27}$$

$$f_{T_1}(t) = \lambda(t) e^{-\int_0^t \lambda(x)\,dx}, \quad t \ge 0$$

A comparision of (3.27) with (1.18) shows that the intensity function $\lambda(t)$ of the inhomogeneous Poisson process is identical to the failure rate belonging to T_1.

More generally, since $F_{T_n}(t) = P(T_n \le t) = P(N(t) \ge n)$,

$$F_{T_n}(t) = \sum_{i=n}^{\infty} \frac{[\Lambda(t)]^i}{i!} e^{-\Lambda(t)}, \quad n = 1, 2, \ldots \tag{3.28}$$

Analogously to the derivation of (3.6), differentiation with respect to t yields the probability density of T_n to be

$$f_{T_n}(t) = \frac{[\Lambda(t)]^{n-1}}{(n-1)!} \lambda(t) e^{-\Lambda(t)}$$

or, equivalently,

$$f_{T_n}(t) = \frac{[\Lambda(t)]^{n-1}}{(n-1)!} f_{T_1}(t), \quad t \geq 0, \quad n = 1, 2, ... \tag{3.29}$$

The expected value of T_n is given by

$$E(T_n) = \int_0^\infty (1 - F_{T_n}(t))\, dt \tag{3.30}$$

$$= \int_0^\infty e^{-\Lambda(t)} \left(\sum_{i=0}^{n-1} \frac{[\Lambda(t)]^i}{i!} \right) dt$$

Hence, the expected time between the $(n-1)$–th and the n-th Poisson event

$$E(Y_n) = E(T_n - T_{n-1}) = E(T_n) - E(T_{n-1})$$

becomes

$$E(Y_n) = \frac{1}{(n-1)!} \int_0^\infty [\Lambda(t)]^{n-1} e^{-\Lambda(t)}\, dt; \quad n = 1, 2, ... \tag{3.31}$$

Letting $\lambda(x) \equiv \lambda$ and $\Lambda(x) \equiv \lambda x$ yields the corresponding characteristics for homogeneous Poisson processes.

3) The conditional distribution function of T_2 given that $T_1 = t_1$ is equal to the probability that at least one Poisson event occurs in $(t_1, t_2]$, $t_1 < t_2$. Thus, using (3.24),

$$F_{T_2}(t_2 | T_1 = t_1) = 1 - p_0(t_1, t_2) = 1 - e^{-[\Lambda(t_2) - \Lambda(t_1)]}$$

Differentiation with respect to t_2 yields the corresponding probability density:

$$f_{T_2}(t_2 | t_1) = \lambda(t_2)\, e^{-[\Lambda(t_2) - \Lambda(t_1)]}$$

By making use of (1.29) and (3.27), the joint probability density of (T_1, T_2) becomes

$$f(t_1, t_2) = \begin{cases} \lambda(t_1) f_{T_1}(t_2) & \text{for } t_1 < t_2 \\ 0, & \text{elsewhere} \end{cases}$$

Proceeding in this way one obtains the joint probability density of $(T_1, T_2, ..., T_n)$:

$$f(t_1, t_2, ..., t_n) = \begin{cases} \lambda(t_1)\lambda(t_2) \cdots \lambda(t_{n-1}) f_{T_1}(t_n) & \text{for } t_1 < t_2 < ... < t_n \\ 0, & \text{elsewhere} \end{cases} \tag{3.32}$$

This formula is a generalization of (3.7).

The inhomogeneous Poisson process $\{N(t),\ t \geq 0\}$ and the corresponding random sequence $\{T_1, T_2, ...\}$ are statistically equivalent, because, if a sample path (realization) of $\{N(t),\ t \geq 0\}$ is given, then there is exactly one realization of the random vector $\{T_1, T_2, ...\}$ belonging to this sample path and conversely. Hence, an inhomogeneous Poisson process with intensity function $\lambda(t)$ is uniquely characterized by the joint probability densities (3.32); $n = 1, 2, ...$

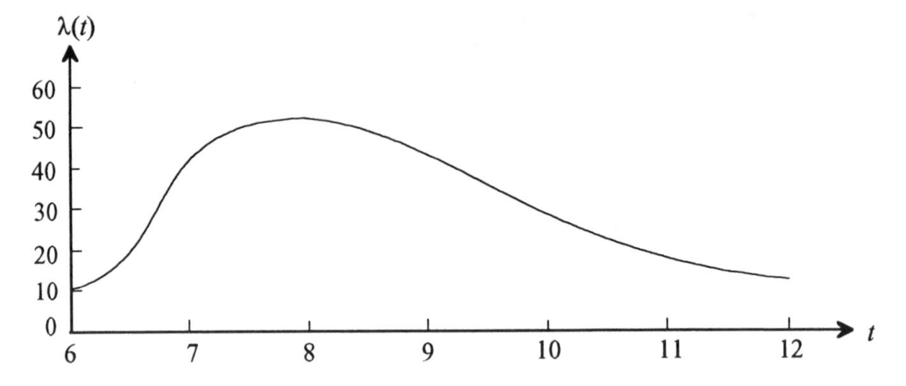

Figure 3.5 Intensity of the arrival of cars at a filling station (example 3.7)

Example 3.7 Based on a large statistical sample it is known that the number of cars which arrive for petrol week-days between 6:00 and 12:00 at a particular filling station can be described by an inhomogeneous Poisson process $\{N(t),\ t \geq 0\}$, the intensity function $\lambda(t)\ [h^{-1}]$ of which is given by (Figure 3.5)

$$\lambda(t) = 10 + 35.4\,(t-6)e^{-\frac{1}{8}(t-6)^2}, \quad 6 \leq t \leq 12$$

1) How many cars on average arrive for petrol week-days between 6:00 and 12:00?

2) What is the probability that at least 90 cars arrive for petrol week-days between 7.00 and 9.00?

1) The average number is

$$\int_6^{12} \lambda(t)\,dt = \int_0^6 \left[10 + 35.4\,t\,e^{-\frac{1}{8}t^2} \right] dt$$

$$= \left[10\,t + 141.6\left(1 - e^{-\frac{1}{8}t^2}\right) \right]_0^6$$

$$= 200$$

2) During the time interval [7:00, 9:00] the random number of arriving cars is Poisson distributed with parameter

$$\int_7^9 \lambda(t)\,dt = \left[10\,t + 141.6\left(1 - e^{-\frac{1}{8}t^2}\right)\right]_1^3$$

$$= 99$$

That is, on average 99 cars arrive for petrol between 7:00 and 9:00. The desired probability is

$$P(N(9) - N(7) \geq 90) = \sum_{n=90}^{\infty} \frac{99^n}{n!}\, e^{-99}$$

For computational reasons, this probability is approximated by making use of the normal approximation to the Poisson distribution:

$$P(N(9) - N(7) \geq 90) \approx 1 - \Phi\left(\frac{90 - 99}{\sqrt{99}}\right)$$

$$\approx 1 - 0.1827$$

$$= 0.8173,$$

where $\Phi(x)$ denotes the distribution function of a standardized normally distributed random variable as defined by (1.51). □

3.2.2 Minimal Repair

The inhomogeneous Poisson process is an important mathematical tool for optimizing maintenance policies with respect to cost and reliability criteria. *Maintenance policies* for technical systems prescribe when principally to carry out (preventive) repairs or renewals of the systems. By definition, after a *renewal* a system is "as good as new", i.e. a renewal restores the state of the system to that at the beginning of its work. *Repairs* after system failures usually only remove the actual causes of failures. A *minimal repair* performed after a failure enables the system to continue its work but does not affect the failure rate of the system. In other words, after finishing a minimal repair the failure rate of the system has the same value as immediately before the failure. For example, if a failure of a complicated electronic system is caused solely by a defective plug and socket connection, then removing this cause of failure is approximately a minimal repair. *Preventive renewals* and *preventive repairs* are carried out while there is no failure at present, however, a possible future one might be prevented or at least postponed. Of course, preventive minimal repairs make no sense.

It is assumed that all renewals and repairs take only negligibly small times and that after finishing a renewal or a repair the system immediately starts working again. The random lifetime T of the system is assumed to have probability density $f(t)$, distribution function $F(t)$, survival probability $\overline{F}(t) = 1 - F(t)$, and failure rate $\lambda(t)$. The following maintenance policy is directly related to the inhomogeneous Poisson process.

Policy 1 Every system failure is removed by a minimal repair.

In policy 1, let T_n be the random time point at which the n th system failure (minimal repair) occurs. Then $Y_n = T_n - T_{n-1}$ is the length of the time span between the $(n$-$1)$ th and the n th system failure, $n = 1, 2, ...;\ T_0 = 0$. The first failure of the system after beginning its work at time $t = 0$ occurs at time $T = T_1$. Given $T_1 = t$ the failure rate of the system immediately after finishing the repair is $\lambda(t)$. Hence, the future failure behaviour of the system is the same as that of a system which has worked up to time point t without failing. Therefore, from (1.13), the random time between the first and the second system failures $Y_2 = T_2 - t$ on condition that $T_1 = t$ has distribution function

$$F_t(y) = P(Y \leq y) = \frac{F(t+y) - F(t)}{\overline{F}(t)}$$

According to (1.19), equivalent representations of $F_t(y)$ are

$$F_t(y) = 1 - e^{-[\Lambda(t+y) - \Lambda(t)]}$$

and

$$F_t(y) = 1 - p_0(t, t+y)$$

These equations also remain valid if t is not the time point of the first failure, but the time of any failure, for instance the n-th failure. Then $F_t(y)$ is the distribution function of the $(n+1)$-th interarrival time $Y_n = T_{n+1} - T_n$ given that $T_n = t$. The occurrence of system failures (minimal repairs) is, therefore, governed by the same probability distribution as the occurrence of Poisson events generated by an inhomogeneous Poisson process with intensity function $\lambda(t)$. Specifically, the random vector $(T_1, T_2, ..., T_n)$ has joint probability density (3.32) for all $n = 1, 2, ...$ Thus, if $N(t)$ denotes the number of system failures (minimal repairs) in $[0, t]$, then $\{N(t), t \geq 0\}$ is an inhomogeneous Poisson process with intensity function $\lambda(t)$. In particular, $N(t)$ is Poisson-distributed with parameter $\Lambda(t)$:

$$E(N(t)) = \Lambda(t) = \int_0^t \lambda(x)\, dx$$

If C_i denotes the random cost of the i-th minimal repair, then the total repair costs over $[0, t]$ amount to

$$K(t) = \sum_{i=1}^{N(t)} C_i \tag{3.33}$$

Assuming the C_i; $i = 1, 2, \ldots$; to be independent of each other and of $N(t)$ and identically distributed as C with $c_m = E(C) < \infty$, the expected repair costs over $[0, t]$ amount to (see theorem 1.3)

$$E(K(t)) = E(C)\, E(N(t)) = c_m\, \Lambda(t) \tag{3.34}$$

Remark The stochastic process $\{K(t), t \geq 0\}$ is called a *compound inhomogeneous Poisson process.* A further example of the occurrence of such processes is the following one: If the random number of customers $N(t)$ which are served in a supermarket within the time interval $[0, t]$ follows an inhomogeneous Poisson process and C_i is the bill of the i-th customer, then the total return to the supermarket in $[0, t]$ is given by (3.33).

In practice policy 1 cannot be maintained over a long period since sooner or later a failure will happen which cannot be removed by a minimal repair. However, policy 1 provides the basis for modeling a number of more sophisticated maintenance policies involving preventive renewals. In what follows, two simple policies of this kind will be considered.

To justify preventive renewals one has to assume that the system which is being maintained is aging (section 1.2.3, definition 1.1), i.e. it has an increasing failure rate $\lambda(t)$. Moreover, the costs c_p and c_m of preventive renewals and minimal repairs, respectively, are assumed to be constant.

Policy 2 A system is preventively renewed at fixed times $\tau, 2\tau, 3\tau, \ldots$ Failures between renewals are removed by minimal repairs.

This policy reflects the frequently applied approach of preventively overhauling complicated systems after fixed time periods and only carrying out repairs which are absolutely necessary in between.

The preventive renewals partition the running time of the system into statistically equivalent time periods (cycles) of length τ, since after every renewal the system is in its initial state both from the failure behaviour and the cost point of view. The total expected maintenance cost per cycle are, according to (3.34), equal to $c_p + c_m\, \Lambda(\tau)$. Thus, the expected total maintenance cost rate per cycle is

$$K(\tau) = \frac{c_p + c_m\, \Lambda(\tau)}{\tau}$$

Remark Formula (4.46) shows that $K(\tau)$ coincides with the long-run maintenance cost rate which arises when the renewal-minimal repair process is indefinetely continued. This remark also refers to policies 3 to 5.

Provided its existence the renewal interval $\tau = \tau^*$ minimizing $K(\tau)$ satisfies the necessary condition $dK(\tau)/d\tau = 0$ or

$$\tau\, \lambda(\tau) - \Lambda(\tau) = c_p / c_m$$

If $\lambda(t)$ tends to ∞ as $t \to \infty$, the there exists a unique solution $\tau = \tau^*$ for this equation. The corresponding minimal cost rate is

$$K(\tau^*) = c_m \lambda(\tau^*)$$

Example 3.8 Let the lifetime of the system have a Rayleigh distribution with distribution function

$$F(t) = 1 - e^{-(t/\theta)^2}, \ t \geq 0$$

The corresponding failure rate is $\lambda(t) = 2t/\theta^2$. The optimal parameters are

$$\tau^* = \theta\sqrt{c_p/c_m}$$

$$K(\tau^*) = \frac{2}{\theta}\sqrt{c_m c_p}$$

For the sake of comparision: If the system is only renewed on failures, then, assuming the same renewal cost c_p, the long-run cost rate is

$$K = \frac{c_p}{\theta\Gamma(1.5)} = 1.13\ c_p/\theta$$

Thus, policy 2 is superior to this policy if $c_m < 0.56\,c_p$. □

Policy 3 A system is renewed at the first failure which occurs after a fixed time τ. Failures which occur between renewals are removed by minimal repairs.

This policy treats the lifetime of the system fully so that from this point of view it is preferable to policy 2. However, the partial uncertainty about the times of renewals generally lead to greater renewal costs than with policy 2. Thus, the cost rate of policy 3 may actually be greater than that of policy 2.

The residual lifetime $T_\tau = T - \tau$ of the system after time τ has distribution function

$$F_\tau(t) = P(T_\tau \leq t) = \frac{F(t+\tau) - F(\tau)}{\bar{F}(\tau)}$$

$$= 1 - e^{-[\Lambda(t+\tau) - \Lambda(\tau)]}$$

(This takes account of the fact that when applying policy 3, the assumption "$T > \tau$" is formally satisfied a priori.) From (1.12), the expected value of T_τ is

$$r(\tau) = E(T_\tau) = e^{\Lambda(\tau)} \int_\tau^\infty e^{-\Lambda(t)} dt$$

The expected maintenance cost per renewal cycle (the interval between two successive renewals) is, from the notational point of view, equal to that of policy 2. Thus, the expected long-run cost rate (= expected total maintenance cost within a cycle per mean cycle length) is

$$K(\tau) = \frac{c_p + c_m \Lambda(\tau)}{\tau + r(\tau)},$$

since $\tau + r(\tau)$ is the mean cycle length. An optimal renewal interval $\tau = \tau^*$ satisfies the necessary condition $dK(\tau)/d\tau = 0$ or

$$\left(\Lambda(\tau) + \frac{c_p}{c_m} - 1\right) r(\tau) = \tau$$

If τ^* exists, then the minimal cost rate is

$$K(\tau^*) = \frac{c_p + c_m[\Lambda(\tau^*) - 1]}{\tau^*}$$

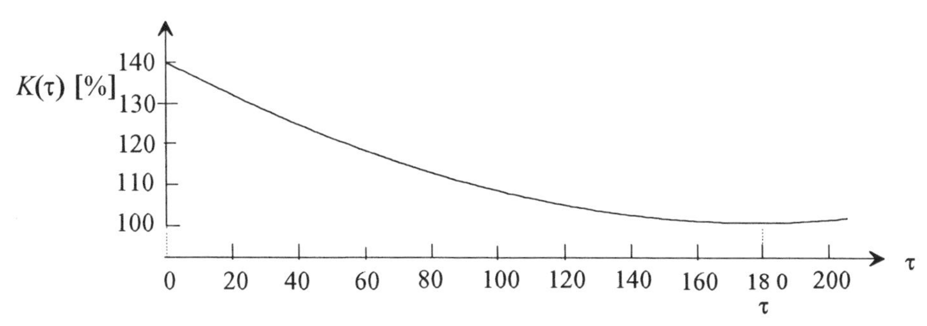

Figure 3.6 Cost rate in percent dependent on the renewal interval

Example 3.9 As in example 3.8, let the system lifetime T have a Rayleigh distribution with failure rate $\lambda(t) = 2t/\theta^2$. $r(\tau)$ is seen to be

$$r(\tau) = \theta\sqrt{\pi}\, e^{(\tau/\theta)^2}\left[1 - \Phi\left(\frac{\sqrt{2}}{\theta}\tau\right)\right]$$

In particular, if $\theta = 100\ [h^{-1}]$, $c_m = 1$, and $c_p = 5$, then the optimal renewal interval is $\tau^* = 180\ [h]$. The corresponding minimal cost rate amounts to $K(\tau^*) = 0.04021$. Assuming identical model parameters this means a reduction of 10% compared to policy 2. Figure 3.6 shows the cost rate dependency on τ, relative to $K(\tau^*) = 100\%$. Clearly, great accuracy is not required when computing τ^*. □

Different failure types It frequently happens that a system failure cannot be removed by a minimal repair, e.g. the repair of a car after a heavy traffic accident is never minimal. Hence it makes sense to introduce two failure types:

Type 1: failures of this type are removed by minimal repairs.
Type 2: failures of this type are removed by renewals.

The successive failure types are assumed to be independent.

Let $1-p$ (p) be the probability that a failure is of type 1 (2) and let the time Y to the first type 2 failure have distribution function $G(t) = P(Y \le t)$. Then

$$P(t < Y \le t+\Delta t \mid Y > t) = p\,\lambda(t)\Delta t + o(\Delta t)$$

Hence,

$$\frac{1}{\bar{G}(t)}\frac{G(t+\Delta t) - G(t)}{\Delta t} = p\,\lambda(t) + \frac{o(\Delta t)}{\Delta t}$$

Letting $\Delta t \to 0$,

$$G'(t)/\bar{G}(t) = p\,\lambda(t)$$

Thus, $p\,\lambda(t)$ is the failure rate belonging to $G(t)$. Therefore,

$$\bar{G}(t) = e^{-p\Lambda(t)} = \left(\bar{F}(t)\right)^{p}, \quad t \ge 0 \tag{3.35}$$

Let $M(t)$ be the random number of minimal repairs in $[0, L_t)$ with

$$L_t = \min(t, Y)$$

Then,

$$P(M(t) = i,\ Y \ge t) = (1-p)^i P(N(t) = i) = (1-p)^i \frac{\Lambda(t)^i}{i!} e^{-\Lambda(t)}; \quad i = 0, 1, \ldots$$

Hence,

$$P(M(t) = i \mid Y \ge t) = \frac{[(1-p)\,\Lambda(t)]^i}{i!} e^{-(1-p)\Lambda(t)}; \quad i = 0, 1, \ldots$$

Consequently, on condition $Y \ge t$, $\{M(t),\ t \ge 0\}$ is an inhomogeneous Poisson process with intensity function $(1-p)\,\Lambda(t)$. (Thus, on condition $Y \ge t$, $\{M(t),\ t \ge 0\}$ is a *thinning* of the inhomogeneous Poisson process $\{N(t),\ t \ge 0\}$.) Note that the condition "$Y \ge t$" can be replaced by "$Y = t$". Therefore,

$$E(M(t) \mid Y \ge t) = E(M(t) \mid Y = t) = (1-p)\,\Lambda(t)$$

and

$$E(M(t) \mid Y < t) = \frac{1}{G(t)} \int_0^t E(M(t) \mid Y = x)\, dG(x)$$

so that

$$E(M(t) \mid Y < t) = \frac{1-p}{p} \frac{1}{G(t)} \left[1 - (p\,\Lambda(t) + 1)\,\bar{G}(t)\right]$$

By the total probability rule,

$$E(M(t)) = \frac{1-p}{p}\, G(t) \tag{3.36}$$

Policy 4 The system is maintained according to the failure type. At time τ after the previous renewal a preventive renewal is carried out.

Let c_m, c_e, and c_p with $c_m < c_p < c_e$ denote the cost of a minimal repair, a renewal after a type 2 failure (emergency renewal) and a preventive renewal, respectively. Then $L_\tau = \min(Y, \tau)$ is the random length of a renewal cycle (time between successive renewals of any type) and $M(\tau)$ is the random number of minimal repairs in a renewal cycle. The expected long-run cost rate has structure

$$K(\tau) = \frac{c_m E(M(\tau)) + c_e G(\tau) + c_p \bar{G}(\tau)}{E(L_\tau)}$$

Since $E(L_\tau) = \int_0^\tau \bar{G}(t)\,dt$ and in view of (3.36), the expected long-run cost rate is

$$K(\tau) = \frac{\left[c_m \frac{1-p}{p} + c_e\right] G(\tau) + c_p \bar{G}(\tau)}{\int_0^\tau \bar{G}(t)\,dt} \tag{3.37}$$

An optimal renewal interval $\tau = \tau*$ satisfies equation

$$p\,\lambda(\tau) \int_0^\tau \bar{G}(t)\,dt - G(\tau) = \frac{p\,c_p}{(c_e - c_p - c_m)p + c_m}$$

A unique solution $\tau*$ exists if $\lambda(t)$ is strictly increasing to infinity and

$$c_e - c_p > c_m(1+p)/p$$

If there is no preventive maintenance, i.e. $\tau = \infty$, then (3.37) simplifies to

$$K = \frac{c_m \frac{1-p}{p} + c_e}{\int_0^\infty \bar{G}(t)\,dt} \tag{3.38}$$

Policy 4 has been proposed in *Beichelt* (1976). This paper also allows for the probability p to depend on time: $p = p(t)$.

Different from the maintenance policies considered so far, the following one explicitely takes into account that repair costs are random variables.

Policy 5 (*Repair cost limit replacement policy*) On failure the system is renewed if the repair cost C exceeds a given limit c. Otherwise a minimal repair is carried out.

The failure type model introduced above applies to this policy when the failure type is generated by the random repair cost C in the following way: A type 1 (type 2) failure occurs if and only if $C \le c$ $(C > c)$. If $R(x) = P(C \le x)$ denotes the distribution function of C, then

$$\bar{p} = R(c), \quad p = \bar{R}(c) = 1 - R(c) \tag{3.39}$$

are the probabilities of type 1 and type 2 failures, respectively. In what follows it is assumed that p does not depend on time, i.e. neither C nor c depend on the system age at failure. As before, the cost c_e of a renewal after a type 2 failure is a constant (mean value). It is reasonable to assume that $0 < c < c_e$ and

$$R(x) = \begin{cases} 1, & \text{if } x \geq c_e \\ 0, & \text{if } x \leq 0 \end{cases}$$

Since p as given by (3.39) does not depend on time, formula (3.38) can be applied for determining the expected long-run maintenance cost rate $K(c)$ under policy 5. But an important peculiarity of policy 5 has to be taken into account: The expected cost of a minimal repair c_m depends now on c and, hence, on p:

$$c_m = E(C|C \leq c) = \frac{1}{R(c)}\left[\int_0^c \bar{R}(x)\,dx - c\bar{R}(c)\right] \tag{3.40}$$

Inserting (3.39) and (3.40) into (3.38) yields the long-run cost rate:

$$K(c) = \frac{\frac{1}{\bar{R}(c)}\int_0^c \bar{R}(x)\,dx + c_e - c}{\int_0^\infty (\bar{F}(t))^{\bar{R}(c)}dt}$$

By applying numerical methods it is in principle easy to obtain a repair cost limit being at least approximately optimal with respect to $K(c)$. However, in case of Weibull-distributed lifetimes, more detailed information on an optimal repair cost limit $c = c*$ can be obtained. Hence, let

$$\bar{F}(t) = \exp(-\alpha t^\beta), \ t \geq 0, \ \beta > 1$$

Then the expected cycle length is

$$E(L) = \int_0^\infty \exp(-\alpha\bar{R}(c)t^\beta)dt$$

$$= \alpha^{-1/\beta}\Gamma(1 + 1/\beta)\left[\bar{R}(c)\right]^{-1/\beta}$$

An optimal $c = c*$ satisfies the equation

$$\frac{1}{\bar{R}(c)}\int_0^c \bar{R}(x)\,dx + \frac{1}{\beta - 1}c = \frac{1}{\beta - 1}c_e$$

For instance, if

$$R(x) = 1 - [(c_e - x)/c_e]^s; \ 0 \leq x \leq c_e, \ s > 0$$

(power distribution), then the optimal repair cost limit is

$$c* = \left[1 - \sqrt[s+1]{\frac{\beta-1}{\beta+s}}\right]c_e$$

For more information on policies 4 and 5 and related ones see *Beichelt, Franken* (1983), *Beichelt* (1993 *a,b*).

Exercises

3.1) The number of catastrophic failures at the chemical works *Lowest, Inc.* can be described by a homogeneous Poisson process with intensity $\lambda = 3$ per year.

(1) What is the probability that at least two catastrophic failures will occur in the second half of the year 2002?
(2) Determine the same probability given that 2 catastrophic failures have occurred in the first half of the year 2002.

3.2) By making use of the independence and homogeneity of its increments, show that the covariance function of a homogeneous Poisson process with intensity λ is given by $C(s, t) = \lambda \min(s, t)$.

3.3) The number of cars which pass a certain intersection daily between 12:00 and 14:00 follows a homogeneous Poisson process with intensity $\lambda = 40$ per hour. Among these there are 0.8% which disregard the STOP-sign.
What is the probability that at least one car disregards the STOP-sign between 12:00 and 13:00?

3.4) A Geiger counter is struck by radioactive particles according to a homogenous Poisson process with intensity $\lambda = 1$ per 12 seconds. On average, the Geiger counter only records 4 out of 5 particles.

(1) What is the probability that the Geiger counter records at least 2 particles per minute?
(2) What are the expected value and variance of the of the random time Y between the occurrence of two successive particles which are recorded?

3.5) An electronic system is subject to two types of shocks which arrive independently of each other according to homogeneous Poisson processes with intensities $\lambda_1 = 0.002\,[h]$ and $\lambda_2 = 0.01\,[h]$, respectively. A shock of type 1 always causes a system failure, whereas a shock of type 2 causes a system failure with probability 0.4.
What is the probability that the system fails within a day due to a shock?

3.6) Let $\{N(t), t \geq 0\}$ be a homogeneous Poisson process with intensity λ.
Prove that for any positive τ the stochastic process $\{X(t), t \geq 0\}$ defined by

$$X(t) = N(t + \tau) - N(t)$$

is wide-sense stationary.

3.7)* Prove *Campbell's theorem* by using the independence and the homogeneity of the increments of a homogeneous Poisson process.

3.8) Let $\{N(t), t \geq 0\}$ be a homogeneous Poisson process with intensity λ and T_i the arrival time of the i-th Poisson event.

For $t \to \infty$, determine and sketch the covariance function $C(\tau)$ of the shot noise process $\{X(t), t \geq 0\}$ given by

$$X(t) = \sum_{i=1}^{N(t)} h(t - T_i) \quad \text{with} \quad h(t) = \begin{cases} \sin t & \text{for } 0 \leq t \leq 2\pi \\ 0, & \text{elsewhere} \end{cases}.$$

3.9) There are two independent homogeneous Poisson processes **I** and **II** with intensities λ_1 and λ_2, respectively.
Determine the expected value of the random number of events of process **II**, which occur between any two successive events of process **I**.

3.10) Let $\{N(t), t \ge 0\}$ be an inhomogeneous Poisson process with intensity function $\lambda(t) = 0.8 + 2t$.
(1) What is the trend function of this process?
(2) Determine the probability that at least 500 Poisson events occur in the interval [20, 30].
Hint Approximate the Poisson distribution by the normal distribution.

3.11) (1) Determine the optimal renewal interval and the corresponding cost rate for policy 2 given that the system lifetime has a Weibull distribution with form parameter β and scale parameter θ.
(2) Check whether the cost rates are sensitive to deviations from the optimal renewal interval.

3.12)* Let $\{N(t), t \ge 0\}$ be an inhomogeneous Poisson process with intensity function $\lambda(t)$ and arrival time T_i of the ith Poisson event.
Show that, given $N(t) = n$, the random vector $(T_1, T_2, ..., T_n)$ has the same probability distribution as n ordered, independent, and identically distributed random variables with distribution function

$$F(x) = \begin{cases} \dfrac{\lambda(x)}{\lambda(t)} & \text{for } 0 \le x < t \\ 1, & \text{elsewhere} \end{cases}$$

Hint Compare with theorem 3.4.

4 Renewal Processes

As already introduced in the previous chapter, after a *renewal* a system is "as good as new"; i.e. after its renewal a system has the same lifetime distribution as at the moment of its installation. The formal subject of *renewal theory* is a simple maintenance policy: A system is renewed on every failure in negligible time and after that it immediately starts continuing its work. In what follows, this policy is referred to as *policy* 0. It reflects practical situations when failed (sub-) systems are completely replaced by equivalent (sub-) systems. In such cases renewal theory provides the mathematical tools for organising a stable production process. At the same time it yields the mathematical foundations for analyzing the behaviour of complicated systems in the running time of which renewal points are "imbedded" in a sense that will be defined more precisely later.

4.1 Foundations

From the mathematical point of view, the main subject of renewal theory consists in investigating properties of the stochastic process which counts the number of renewals within a given time interval when applying the above-mentioned maintenance policy 0. This point process not only plays an important role in engineering, but also in the natural and social sciences as well as in risk theory (particle counting, population development, arrival time points of claims at an insurance company). As mentioned in section 3.1 a point process can equivalently be characterized by the number of events (renewals) occurring in $[0, t]$, by the time points when the events occur, or by the interarrival times between two neighbouring events. In renewal theory, the latter characterization is preferred.

Definition 4.1 (*renewal process*) A *renewal process* is a sequence of nonnegative, independent random variables $\{Y_i;\, i = 1, 2, ...\}$ with $Y_2, Y_3, ...$ being identically distributed. ●

If policy 0 is the practical background o f a renewal process, then Y_i denotes the time between the $(i-1)$-th and the i-th renewal. If at time $t = 0$ policy 0 has already been in effect for a while, then Y_1 is a residual lifetime in the sense of section 1.2.3. However, the age of the system working at time $t = 0$ need not to be known. But if at time $t = 0$ a new system started working, then all the random variables $Y_1, Y_2, ...$ are identically distributed.

Let the random variables $Y_2, Y_3, ...$ be identically distributed as Y with distribution function $F(t) = P(Y \le t)$, whereas Y_1 has distribution function $F_1(t) = P(Y_1 \le t)$.

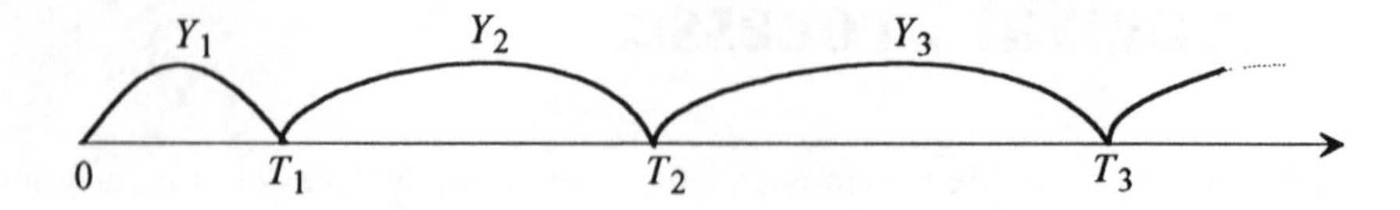

Figure 4.1 Illustration of a renewal process

Definition 4.2 A renewal process is called *delayed* if $F_1(t) \not\equiv F(t)$ and *ordinary* if $F_1(t) \equiv F(t)$. ●

Since, by assumption, the renewals occur in negligible time, T_n defined by

$$T_n = \sum_{i=1}^{n} Y_i; \quad n = 1, 2, ...;$$

is the time point at which the nth failure (renewal) takes place. Hence, T_n is called a *renewal time*. The time intervals between two neighbouring renewals are called *renewal cycles*.

Let the *renewal counting process* $\{N(t), t \geq 0\}$ be defined by

$$N(t) = \begin{cases} \max(n; \ T_n \leq t) \\ 0 \quad \text{for} \quad t < T_1 \end{cases}$$

$N(t)$ is the random number of renewals occuring in $(0, t]$. Since $N(t) \geq n$ if and only if $T_n \leq t$,

$$F_{T_n}(t) = P(T_n \leq t) = P(N(t) \geq n), \tag{4.1}$$

where, because of the independence of the Y_i, $F_{T_n}(t)$ is the convolution of F_1 with the $(n-1)$-th convolution power of F (see section 1.4):

$$F_{T_n}(t) = F_1 * F^{*(n-1)}(t), \quad F^{*(0)}(t) \equiv 1, \quad t \geq 0; \quad n = 1, 2, ... \tag{4.2}$$

If the densities $f_1(t) = F_1'(t)$ and $f(t) = F'(t)$ exist, then the density of T_n is

$$f_{T_n}(t) = f_1 * f^{*(n-1)}(t), \quad f^{*(0)}(t) \equiv 1, \quad t \geq 0; \ n = 1, 2, ... \tag{4.3}$$

Using

$$P(N(t) \geq n) = P(N(t) = n) + P(N(t) \geq n+1)$$

and (4.1), the probability distribution of $N(t)$ is seen to be

$$P(N(t) = n) = F_{T_n}(t) - F_{T_{n+1}}(t), \quad F_{T_0}(t) \equiv 1; \quad n = 0, 1, ... \tag{4.4}$$

Formulas (4.1) and (4.2) can be used to solve the so-called *spare part problem*: How many spare systems (spare parts) are necessary for making sure that the renewal process can be maintained over the intervall $[0, t]$ with probability $1-\alpha$? This requires the determination of the smallest n satisfying

$$1 - F_{T_n}(t) \geq 1 - \alpha$$

Example 4 Consider an ordinary renewal process with exponentially distributed cycle lengths: $F(t) = 1 - e^{-\lambda t}$, $t \geq 0$. According to theorem 3.2, this renewal process is nothing else than the homogeneous Poisson process with intensity λ. The renewal times T_n are, therefore, Erlang-distributed with parameters n and λ. Because of of (3.5),

$$P(N(t) = n) = F_{T_n}(t) - F_{T_{n+1}}(t) = \frac{(\lambda t)^n}{n!} e^{-\lambda t}; \quad n = 0, 1, \ldots$$

It can easily be verified that for $\lambda = 0.05$; $t = 200$ and $1 - \alpha = 0.99$ the smallest $n = n_{\min}$ satisfying

$$1 - F_{T_n}(200) \geq 0,99$$

is $n_{\min} = 18$. Thus, at least 18 spare systems have to be in stock to ensure that with probability 0.99 every failed system can be replaced by a new one over the interval $(0, 200]$. □

The exact computation of the probability distribution of $N(t)$ using (4.4) is only possible with distribution functions F for which the convolution powers $F^{*(n)}$ can be given explicitly. But only to a few types of probability distributions have this property. Besides the exponential distribution, this type includes the normal and the Erlang distributions.

1) ***Normal distribution*** Let $F(t) = \Phi\left(\frac{t-\mu}{\sigma}\right)$, where $\Phi(x)$ is given by (1.51). Since the sum of independent normally distributed random variables is again normally distributed, where the parameters of the sum are obtained by summing up the parameters of the summands (example 1.7),

$$F^{*(n)}(t) = P(N(t) \geq n) = \Phi\left(\frac{t - n\mu}{\sigma\sqrt{n}}\right) \tag{4.5}$$

Thus, for $t > 0$

$$P(N(t) = 0) = 1 - \Phi\left(\frac{t-\mu}{\sigma}\right)$$

$$P(N(t) = n) = \Phi\left(\frac{t - n\mu}{\sigma\sqrt{n}}\right) - \Phi\left(\frac{t - (n+1)\mu}{\sigma\sqrt{n+1}}\right); \quad n = 1, 2, \ldots$$

2) ***Erlang distribution*** Let $F(t)$ be the distribution function of an Erlang-distributed random variable with parameters m and λ. Then $F^{*(n)}$ is the distribution function of a sum of mn independent, identically distributed exponential random variables with parameter λ (example 1.6). Therefore, from (3.5),

$$F^{*(n)}(t) = e^{-\lambda t} \sum_{i=mn}^{\infty} \frac{(\lambda t)^i}{i!}$$

and, hence,

$$P(N(t) = n) = e^{-\lambda t} \sum_{i=mn}^{m(n+1)-1} \frac{(\lambda t)^i}{i!}$$

4.2 Renewal Function

4.2.1 Renewal Equations

The expected number of renewals which occur in a given time interval is of great practical and theoretical importance.

Definition 4.3 (***renewal function***) The expected value of the random number of renewals occurring in $(0, t]$ as a function of t is called the *renewal function.* ●

In what follows, the renewal functions belonging to delayed and ordinary renewal processes are denoted by $H_1(t)$ and $H(t)$, respectively.

The renewal function is most easily obtained by using formula (1.11):

$$H_1(t) = E(N(t)) = \sum_{n=1}^{\infty} P(N(t) \geq n)$$

In view of (4.1) and (4.2),

$$H_1(t) = \sum_{n=1}^{\infty} F_1 * F^{*(n-1)}(t) \tag{4.6}$$

or, by definition of the convolution power in (1.47),

$$\begin{aligned} H_1(t) &= \sum_{n=0}^{\infty} F_1 * F^{*(n)}(t) \\ &= F_1(t) + \sum_{n=1}^{\infty} \int_0^t F_1 * F^{*(n-1)}(t-x)\, dF(x) \\ &= F_1(t) + \int_0^t \sum_{n=1}^{\infty} \left(F_1 * F^{*(n-1)}(t-x)\right) dF(x) \end{aligned}$$

By (4.6), the integrand is equal to $H_1(t-x)$. Hence, the renewal function satisfies the integral equation

$$H_1(t) = F_1(t) + \int_0^t H_1(t-x)\,dF(x) \tag{4.7}$$

In particular, the renewal function of an ordinary renewal process satisfies the integral equation

$$H(t) = F(t) + \int_0^t H(t-x)\,dF(x) \tag{4.8}$$

The integral equations (4.7) and (4.8) are called *renewal equations.* They have unique solutions (*Feller* (1971)).

The sum representation of $H_1(t)$ given in (4.6) can be used in the equivalent form

$$H_1(t) = F_1(t) + \sum_{n=1}^{\infty} \int_0^t F^{*(n)}(t-x)\,dF_1(x)$$

$$= F_1(t) + \int_0^t \left(\sum_{n=1}^{\infty} F^{*(n)}(t-x) \right) dF_1(x)$$

This yields the integral equation

$$H_1(t) = F_1(t) + \int_0^t H(t-x)\,dF_1(x) \tag{4.9}$$

If the densities $f_1(t)$ and $f(t)$ exist, then, by differentiation of (4.6) with respect to t, one obtains a sum representation of the *renewal density* $h_1(t) = dH_1(t)/dt$:

$$h_1(t) = \sum_{n=1}^{\infty} f_1 * f^{*(n-1)}(t) \tag{4.10}$$

Remark From this representation a useful probabilistic interpretation of the renewal density emerges: For Δt sufficiently small, $h_1(t)\Delta t$ is approximately equal to the probability of the occurrence of a renewal in $[t, t+\Delta t]$.

By differentiating the renewal equations with respect to t one obtains the following integral equations for $h_1(t)$ and for the renewal density $h(t) = dH(t)/dt$ of an ordinary renewal process, respectively:

$$h_1(t) = f_1(t) + \int_0^t h_1(t-x) f(x)\,dx \tag{4.11}$$

$$h(t) = f(t) + \int_0^t h(t-x) f(x)\,dx \tag{4.12}$$

Further, differentiating (4.9) with respect to t yields the integral equation

$$h_1(t) = f_1(t) + \int_0^t h(t-x) f_1(x)\,dx \tag{4.13}$$

Generally, solutions of the integral equations (4.7) and (4.8) as well as (4.11) and (4.12) can only approximately be obtained. However, since the Laplace-transform of the convolution of two functions is equal to the product of the Laplace-transforms of these functions (section 1.5.2), it is easily possible to solve these integral equations in the image-space of the Laplace transformation.

Let $\hat{h}_1(s)$, $\hat{h}(s)$, $\hat{f}_1(s)$ and $\hat{f}(s)$ be the Laplace-transforms of $h_1(t)$, $h(t)$, $f_1(t)$ and $f(t)$, respectively. Then the application of the Laplace-transform to the integral equations (4.11) and (4.12) yields the following algebraic equations for $\hat{h}_1(s)$ and $h(s)$:

$$\hat{h}_1(s) = \hat{f}_1(s) + \hat{h}_1(s) \cdot \hat{f}(s)$$

$$\hat{h}(s) = \hat{f}(s) + \hat{h}(s) \cdot \hat{f}(s)$$

Their solutions are

$$\hat{h}_1(s) = \frac{\hat{f}_1(s)}{1 - \hat{f}(s)}, \qquad \hat{h}(s) = \frac{\hat{f}(s)}{1 - \hat{f}(s)} \tag{4.14}$$

Thus, for ordinary renewal processes there is a one-to-one correspondence between the renewal function and the probability distribution of the cycle length. The Laplace-transforms of the corresponding renewal functions are (section 1.5.2)

$$\hat{H}_1(s) = \frac{\hat{f}_1(s)}{s(1 - \hat{f}(s))}, \qquad \hat{H}(s) = \frac{\hat{f}(s)}{s(1 - \hat{f}(s))} \tag{4.15}$$

Example 4.2 Let $F_1(t) = F(t) = (1 - e^{-\lambda t})^2$, $t \geq 0$. Thus, $F(t)$ is the survival function of a parallel system consisting of two subsystems, the lifetimes of which are independent, identically distributed exponential random variables with parameter λ. The corresponding probability density and its Laplace-transform are

$$f(t) = 2\lambda(e^{-\lambda t} - e^{-2\lambda t}) \quad \text{and} \quad \hat{f}(s) = \frac{2\lambda^2}{(s+\lambda)(s+2\lambda)}$$

From (4.14), the Laplace-transform of the corresponding renewal density is

$$\hat{h}(s) = \frac{2\lambda^2}{s(s+3\lambda)}$$

The pre-image can easily be determined by decomposing the fraction into partial fractions to obtain

$$h(t) = \frac{2}{3}\lambda(1 - e^{-3\lambda t})$$

Integration yields the renewal function

$$H(t) = \frac{2}{3}\lambda\left[t + \frac{1}{3\lambda}\left(e^{-3\lambda t} - 1\right)\right] \tag{4.16}$$

□

In what follows, such cycle length distributions of renewal processes are considered which allow explicit statements on the corresponding renewal functions.

1) ***Exponential distribution*** Let be $f(t) = \lambda e^{-\lambda t},\ t \geq 0$. The Laplace-transform of $f(t)$ is (see example 1.10):

$$\hat{f}(s) = \frac{\lambda}{s+\lambda}$$

Hence,

$$\hat{H}(s) = \frac{\lambda}{s+\lambda} \Big/ \left(s - \frac{\lambda s}{s+\lambda}\right) = \frac{\lambda}{s^2}$$

The corresponding pre-image is

$$H(t) = \lambda t$$

Corollary An ordinary renewal process has exponentially distributed cycle lengths with parameter λ if and only if its renewal function is given by $H(t) = \lambda t$.

2) ***Erlang-distribution*** Let the cycle length be Erlang-distributed with parameters m and λ. Then the corresponding renewal function can again be obtained by applying the Laplace-transform (see *Beichelt/Franken* (1984)). The result is

$$H(t) = e^{-\lambda t} \sum_{n=1}^{\infty} \sum_{i=mn}^{\infty} \frac{(\lambda t)^i}{i!}$$

In particular,

$$m = 1: \quad H(t) = \lambda t \qquad \textit{(homogeneous Poisson process)}$$

$$m = 2: \quad H(t) = \frac{1}{2}\left[\lambda t - \frac{1}{2} + \frac{1}{2} e^{-2\lambda t}\right]$$

$$m = 3: \quad H(t) = \frac{1}{3}\left[\lambda t - 1 + \frac{2}{\sqrt{3}} e^{-1{,}5\lambda t} \sin\left(\frac{\sqrt{3}}{2}\lambda t + \frac{\pi}{3}\right)\right]$$

$$m = 4: \quad H(t) = \frac{1}{4}\left[\lambda t - \frac{3}{2} + \frac{1}{2} e^{-2\lambda t} + \sqrt{2}\, e^{-\lambda t} \sin\left(\lambda t + \frac{\pi}{4}\right)\right]$$

3) ***Normal distribution*** Let the cycle lengths be normally distributed with expected value μ and variance σ^2, $\mu > 3\sigma$. From (4.5) and (4.6),

$$H(t) = \sum_{n=1}^{\infty} \Phi\left(\frac{t - n\mu}{\sigma\sqrt{n}}\right)$$

This sum representation is very convenient for numerical computations since already the sum of the first few terms approximates the renewal function with sufficient accuracy.

For the practically important case of Weibull-distributed cycle lengths no explicit formulas of the renewal function can be given. However, tables are available (see *Beichelt/ Franken* (1984)).

As pointed out in the above corollary, an ordinary renewal process has renewal function $H(t) = \lambda t = t/\mu$ if and only if $f(t) = \lambda e^{-\lambda t},\ t \geq 0$. Hence an interesting question is whether, for a given $F(t)$, a delayed renewal process exists which also has the renewal function $H_1(t) = t/\mu$.

Theorem 4.1 Let $\mu = E(Y) = \int_0^\infty \overline{F}(t)\,dt < \infty$. Then the delayed renewal process has renewal function

$$H_1(t) = t/\mu \tag{4.17}$$

if and only if $f_1(t) \equiv f_S(t)$, where

$$f_S(t) = \frac{1}{\mu}(1 - F(t)) \tag{4.18}$$

Equivalently, the delayed renewal process has renewal function (4.17) if and only if $F_1(t) \equiv F_S(t)$, where

$$F_S(t) = \frac{1}{\mu}\int_0^t (1 - F(x))\,dx,\ t \geq 0 \tag{4.19}$$

Proof Let $\hat{f}(s)$, $\hat{f}_S(s)$, and $\hat{f}_1(s)$ be the Laplace-transforms of $f(t)$, $f_S(t)$, and $f_1(t)$, respectively. Then, from section 1.5.2,

$$\hat{f}_S(s) = \frac{1}{\mu s}(1 - \hat{f}(s))$$

If in the first equation of (4.15) $\hat{f}_1(s)$ is replaced by $\hat{f}_S(s)$, then the Laplace-transform of the corresponding renewal function $H_1(t) = H_S(t)$ is seen to be

$$\hat{H}_S(s) = 1/(\mu s^2)$$

But this equation is equivalent to $H_S(t) = t/\mu$. ■

The random variable Y_S with density (4.18) (or distribution function (4.19)) plays an important role in characterizing stationary renewal processes (section 4.5). The first two moments of Y_S are

$$E(Y_S) = \frac{\mu^2 + \sigma^2}{2\mu} \quad \text{and} \quad E(Y_S^2) = \frac{\mu_3}{3\mu}, \tag{4.20}$$

where $\sigma^2 = Var(Y)$ and $\mu_3 = E(Y^3)$.

The higher moments of $N(t)$ also have some importance, for instance when investigating the behaviour of the renewal function (first moment) as $t \to \infty$. In case of ordinary renewal processes, the moments of higher order can principally be derived from the binomial moments. The *binomial moment of the order n* of $N(t)$ is defined by

$$E\binom{N(t)}{n} = \frac{1}{n!}E\{[N(t)][N(t)-1]\cdots[N(t)-(n-1)]\} \tag{4.21}$$

Note that, apart from the factor $1/n!$, the binomial moment of order n of $N(t)$ is equal to the nth *factorial moment of* $N(t)$. According to (*Franken* (1963), for ordinary renewal processes, the binomial moment of order n of $N(t)$ is equal to the nth convolution power of the renewal function:

$$E\binom{N(t)}{n} = H^{*(n)}(t)$$

In particular, for $n = 2$,

$$E\binom{N(t)}{2} = \frac{1}{2}E\{[N(t)][N(t)-1]\} = \frac{1}{2}\left\{E[N(t)]^2 - H(t)\right\} = H^{*(2)}(t)$$

so that, from (1.10), the variance of $N(t)$ is equal to

$$Var(N(t)) = 2\int_0^t H(t-x)\,dH(x) + H(t) - [H(t)]^2$$

Since $H(t-x) \le H(t)$ for $0 \le x \le t$, this equation implies an upper bound for the variance of $N(t)$:

$$Var(N(t)) \le [H(t)]^2 + H(t)$$

4.2.2 Bounds for the Renewal Function

This section presents some bounds for the renewal function of ordinary renewal processes. Since an explicit representation of the renewal function exists only for a small number of lifetime distributions, bounds for $H(t)$ which only require information on one or more distribution parameters, have a great practical importance.

1) Elementary bounds The obvious inequality

$$\max_{1\le i\le n} Y_i \le \sum_{i=1}^{n} Y_i = T_n$$

implies for any t with $F(t) < 1$

$$F^{*(n)}(t) = P(T_n \le t) \le P\left(\max_{1\le i\le n} Y_i \le t\right) = [F(t)]^n$$

Summing from $n = 1$ to ∞ on both sides of this inequality, using (4.6) and the geometric series, yields bounds which are only useful for small t:

$$F(t) \le H(t) \le \frac{F(t)}{1-F(t)}$$

2) Linear bounds of *Marshall* (1973) Let $\mathbf{F} = \{t;\ t \ge 0, F(t) < 1\}$, $\mu = E(Y)$, $\overline{F}(t) = 1 - F(t)$, and

$$a_0 = \inf_{t \in \mathbf{F}} \frac{F(t) - F_S(t)}{\overline{F}(t)}, \qquad a_1 = \sup_{t \in \mathbf{F}} \frac{F(t) - F_S(t)}{\overline{F}(t)},$$

where $F_S(t)$ is given by (4.19). Then,

$$\frac{t}{\mu} + a_0 \le H(t) \le \frac{t}{\mu} + a_1 \tag{4.22}$$

The derivation of these bounds is quite straightforward: According to the definition of the a_i,

$$a_0\,\overline{F}(t) \le F(t) - F_S(t) \le a_1\,\overline{F}(t)$$

Convolution with $F^{*(n)}(t)$ leads to

$$a_0\left[F^{*(n)}(t) - F^{*(n+1)}(t)\right] \le F^{*(n+1)}(t) - F_S * F^{*(n)}(t) \le a_1\left[F^{*(n)}(t) - F^{*(n+1)}(t)\right]$$

Summing from $n = 0$ to $n = \infty$ and taking into account (4.6) and theorem 4.1 yields (4.22). Since

$$\frac{F(t) - F_S(t)}{\overline{F}(t)} \ge -F_S(t) \ge -1$$

for all $t \ge 0$, formula (4.22) implies a simpler lower bound on $H(t)$:

$$H(t) \ge \frac{t}{\mu} - F_S(t) \ge \frac{t}{\mu} - 1$$

These bounds were found by *Butterworth* and *Marshall* (1964).

Let $\lambda_S(t) = f_S(t)/\overline{F}_S(t)$ be the failure rate belonging to $F_S(t)$:

$$\lambda_S(t) = \frac{\overline{F}(t)}{\int_t^\infty \overline{F}(x)\,dx}$$

Then a_0 and a_1 can be written in the form

$$a_0 = \frac{1}{\mu}\inf_{t \in \mathbf{F}} \frac{1}{\lambda_S(t)} - 1 \quad \text{and} \quad a_1 = \frac{1}{\mu}\sup_{t \in \mathbf{F}} \frac{1}{\lambda_S(t)} - 1$$

Thus, (4.22) becomes

$$\frac{t}{\mu} + \frac{1}{\mu}\inf_{t \in \mathbf{F}} \frac{1}{\lambda_S(t)} - 1 \le H(t) \le \frac{t}{\mu} + \frac{1}{\mu}\sup_{t \in \mathbf{F}} \frac{1}{\lambda_S(t)} - 1 \tag{4.23}$$

Since

$$\inf_{t \in \mathbf{F}} \lambda(t) \le \inf_{t \in \mathbf{F}} \lambda_S(t) \quad \text{and} \quad \sup_{t \in \mathbf{F}} \lambda(t) \ge \sup_{t \in \mathbf{F}} \lambda_S(t),$$

the bound (4.23) can also be simplified:

$$\frac{t}{\mu}+\frac{1}{\mu}\inf_{t\in\mathbf{F}}\frac{1}{\lambda(t)}-1\le H(t)\le\frac{t}{\mu}+\frac{1}{\mu}\sup_{t\in\mathbf{F}}\frac{1}{\lambda(t)}-1 \qquad (4.24)$$

Example 4.3 As in example 4.2, let $F(t)=(1-e^{-t})^2,\ t\ge 0$. In this case,

$$\mu=\int_0^\infty \overline{F}(t)\,dt=3/2$$

and

$$\overline{F}_S(t)=1-\frac{1}{\mu}\int_0^t \overline{F}(x)\,dx$$

$$=\frac{2}{3}\left(2-\frac{1}{2}e^{-t}\right)e^{-t},\quad t\ge 0$$

The failure rates belonging to $F(t)$ and $F_S(t)$ are, therefore,

$$\lambda(t)\ =\frac{2(1-e^{-t})}{2-e^{-t}}$$

$$\lambda_S(t)=2\frac{2-e^{-t}}{4-e^{-t}},\quad t\ge 0$$

Both failure rates are strictly increasing in t (Figure 4.2) and have, moreover, the properties

$$\lambda(0)\ =0,\quad \lambda(\infty)\ =1$$

$$\lambda_S(0)=2/3,\quad \lambda_S(\infty)=1$$

Hence the bounds (4.23) and (4.24) are

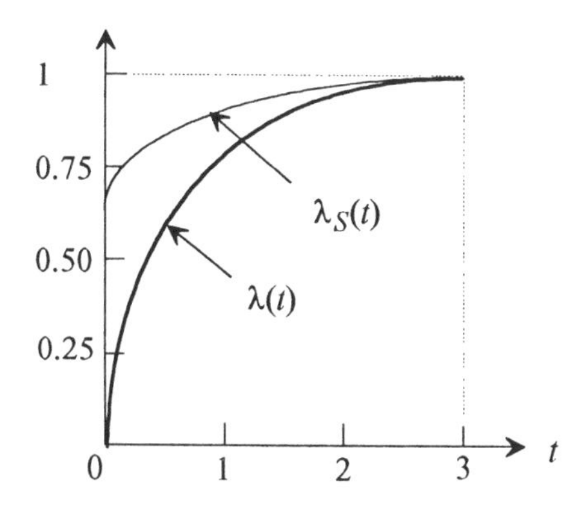

Figure 4.2 Failure rates

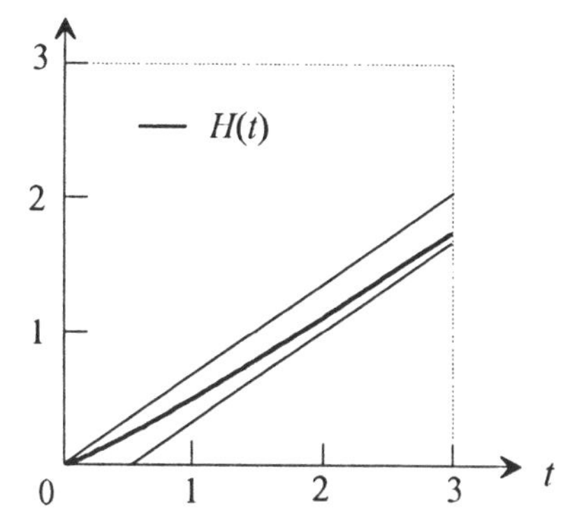

Figure 4.3 Bounds for the renewal function

$$\frac{2}{3}t - \frac{1}{3} \le H(t) \le \frac{2}{3}t \tag{4.25}$$

and

$$t - \frac{1}{3} \le H(t) \le \infty,$$

respectively. In this case, the bounds (4.24) give no information on the renewal function at all. Figure 4.3 compares the bounds (4.25) with the exact graph of the renewal function as given by (4.16) for $\lambda = 1$. The deviation of the lower bound from $H(t)$ is negligibly small for $t > 3$. □

3) Upper bound of *Lorden* If μ_2 is the second moment of the cycle length, then

$$H(t) \le \frac{t}{\mu} + \frac{\mu_2}{\mu^2} - 1$$

4) Upper bound of *Brown* If the failure rate belonging to $F(t)$ is nondecreasing, then the upper bound of *Lorden* can be improved:

$$H(t) \le \frac{t}{\mu} + \frac{\mu_2}{2\mu^2} - 1$$

5) Bounds of *Barlow* and *Proschan* If the failure rate belonging to $F(\mathrm{t})$ is nondecreasing, then

$$\frac{t}{\int_0^t \bar{F}(x)\,dx} - 1 \le H(t) \le \frac{t F(t)}{\int_0^t \bar{F}(x)\,dx}$$

Figure 4.4 compares the exact graph of the renewal function (4.16) for $\lambda = 1$ with the corresponding *Barlow/Proschan* bounds.

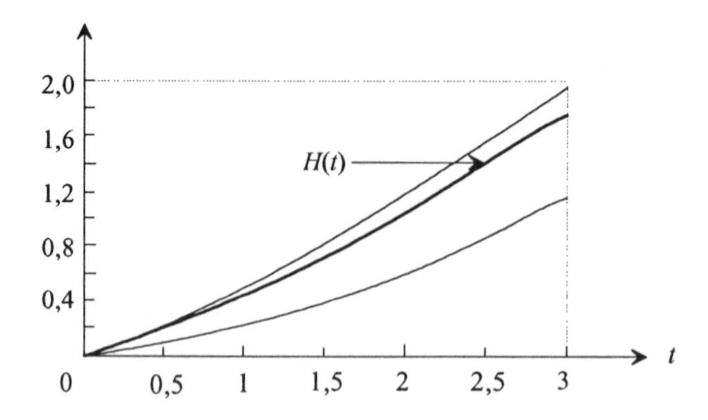

Figure 4.4 Barlow/Proschan bounds for the renewal function

4.3 Recurrence Times

Besides the process of renewal times $\{T_n;\ n = 1, 2, ...\}$ and the renewal counting process $\{N(t),\ t \geq 0\}$ there are still two other important stochastic processes which are associated with the renewal process $\{Y_n;\ n = 1, 2, ...\}$. These are the process of backward recurrence times $\{R(t),\ t \geq 0\}$ and the process of forward recurrence time $\{V(t),\ t \geq 0\}$, where

$$R(t) = t - T_{N(t)} \tag{4.26}$$

is called *backward recurrence time* and

$$V(t) = T_{N(t)+1} - t \tag{4.27}$$

is called *forward recurrence time.* $R(t)$ is the age and $V(t)$ the residual lifetime of the system which is working at time t (Figure 4.5).

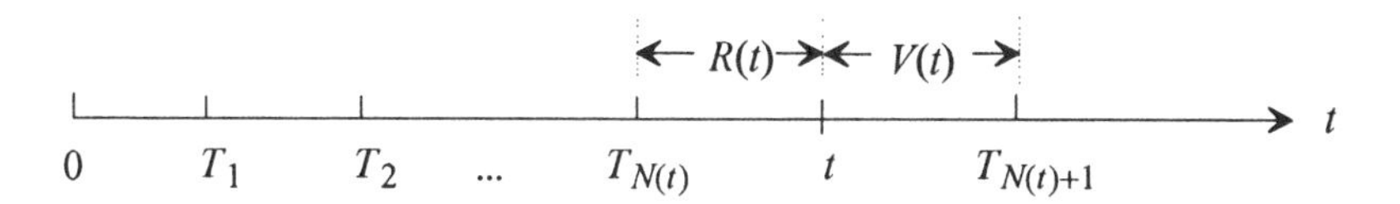

Figure 4.5 Illustration of the recurrence times

The processes $\{Y_n;\ n = 1, 2, ...\}$ $\{T_n;\ n = 1, 2, ...\}$, $\{N(t),\ t \geq 0\}$, $\{R(t),\ t \geq 0\}$, and $\{V(t),\ t \geq 0\}$ are statistically equivalent in the following sense: There is a one-to-one correspondence between the sample paths of these processes (Figure 4.6).

Let

$$F_{R(t)}(x) = P(R(t) \leq x) \quad \text{and} \quad F_{V(t)}(x) = P(V(t) \leq x)$$

be the distribution function of the backward and the forward recurrence time, re-

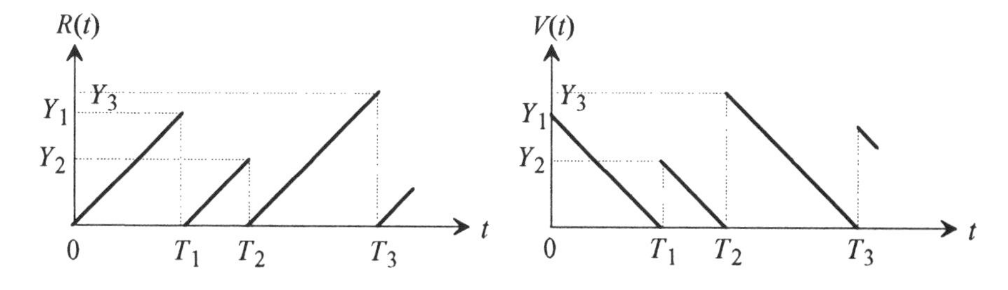

Figure 4.6 Sample paths of the backward and forward recurrence times

spectively. In view of (1.7) and (4.6), for $0 < x < t$,

$$\begin{aligned}
F_{R(t)}(x) &= P(t - x \le T_{N(t)}) \\
&= \sum_{n=1}^{\infty} P(t - x \le T_n,\ N(t) = n) \\
&= \sum_{n=1}^{\infty} P(t - x \le T_n \le t < T_{n+1}) \\
&= \sum_{n=1}^{\infty} \int_{t-x}^{t} \overline{F}(t-u)\, dF_{T_n}(u) \\
&= \int_{t-x}^{t} \overline{F}(t-u) \sum_{n=1}^{\infty} d\left(F_1 * F^{*(n-1)}(u)\right) \\
&= \int_{t-x}^{t} \overline{F}(t-u)\, dH_1(u)
\end{aligned}$$

Hence,

$$F_{R(t)}(x) = \begin{cases} \int_{t-x}^{t} \overline{F}(t-u)\, dH_1(u) & \text{for } \ 0 \le x \le t \\ 1 & \text{for } \ t > x \end{cases} \tag{4.28}$$

The corresponding probability density $f_{R(t)}(x) = dF_{R(t)}(x)/dx$ is

$$f_{R(t)}(x) = \begin{cases} \overline{F}(x)\, h_1(t-x) & \text{for } \ 0 \le x \le t \\ 0, & \text{elsewhere} \end{cases} \tag{4.29}$$

The distribution function of the forward recurrence time is obtained analogously:

$$\begin{aligned}
F_{V(t)}(x) &= P(T_{N(t)+1} \le t + x) = \sum_{n=0}^{\infty} P(T_{n+1} \le t + x,\ N(t) = n) \\
&= F_1(t+x) - F_1(t) + \sum_{n=1}^{\infty} \int_0^t [F(t+x-u) - F(t-u)]\, dF_{T_n}(u) \\
&= F_1(t+x) - F_1(t) + \int_0^t [F(t+x-u) - F(t-u)]\, dH_1(u)
\end{aligned}$$

If using the equation (derived from (4.7))

$$F_1(t) = H_1(t) - \int_0^t F(t-u)\, dH_1(u),$$

then $F_{V(t)}(x)$ can be written in the form

$$F_{V(t)}(x) = F_1(t+x) - \int_0^t \overline{F}(t+x-u)\,dH_1(u) \tag{4.30}$$

The corresponding probability density

$$f_{V(t)}(x) = dF_{V(t)}(x)/dx$$

is given by

$$f_{V(t)}(x) = f_1(t+x) + \int_0^t f(t+x-u)\,h_1(u)\,du \tag{4.31}$$

Remark $\overline{F}_{V(t)}(x) = 1 - F_{V(t)}(x)$ is the probability that the system which is working at time t does not fail in $(t, t+x]$. Therefore, $\overline{F}_{V(t)}(x)$ is also called *interval reliability*.

The backward and forward recurrence times of an ordinary renewal process with cycle lengths being exponentially distributed with parameter λ are also exponentially distributed with parameter λ:

$$f_{R(t)}(x) = f_{V(t)}(x) = \lambda\, e^{-\lambda x},\ x \geq 0$$

Note that the probability distributions of $R(t)$ and $V(t)$ do not depend on t. In view of the *memoryless property* of the exponential distribution (example 1.2), this fact is not surprising.

In order to compute the expected value of the forward recurrence time by means of Wald's equation (theorem 1.3) itis necessary to show that $N(t)+1$ is a stopping time for the renewal process $\{Y_1, Y_2, \ldots\}$. This can be done quite easily: If "$\approx$" denotes the equivalence of two random events, then

$$\text{"}N(t)+1 = n\text{"} \approx \text{"}N(t) = n-1\text{"}$$

$$\approx \text{"}Y_1 + Y_2 + \ldots + Y_{n-1} \leq t < Y_1 + Y_2 + \ldots + Y_n\text{"}$$

Thus, the event "$N(t)+1 = n$" is independent of all $Y_{n+1}, Y_{n+2}, \ldots$ so that $N(t)+1$ is indeed a stopping time for $\{Y_1, Y_2, \ldots\}$. (Analogously it can be shown that $N(t)$ is not a stopping time for $\{Y_1, Y_2, \ldots\}$.) Consequently, Wald's equation is not applicable for determining the expected value of $R(t)$.)

Applying Wald's equation to (4.27) yields

$$E(V(t)) = E(Y_1) + \mu H_1(t) - t$$

In particular, for ordinary renewal processes,

$$E(V(t)) = \mu[H(t)+1] - t$$

Based on the distribution function of the forward recurrence time, statements can be made about the corresponding renewal counting process $\{N(t),\ t \geq 0\}$, e.g.

$$P(N(t+x) - N(t) \leq n) = V(t) * F^{*(n-1)}(x); \quad n = 1, 2, \ldots$$

4.4 Asymptotic Behaviour

This section investigates the behaviour of the renewal counting process $\{N(t), t \geq 0\}$ and, in particular, that of the renewal function $H_1(t)$ as $t \to \infty$. The results obtained allow the construction of estimates of $H_1(t)$ and of the probability distribution of $N(t)$ as t tends to infinity if $E(Y_1) < \infty$. The proof of theorem 4.2 is analogously to the one given in *Tijms* (1994).

Theorem 4.2 (*elementary renewal theorem*) For any $F_1(t)$,

$$\lim_{t\to\infty} \frac{H_1(t)}{t} = \frac{1}{\mu}$$

Proof Let $\mu < \infty$. Taking the expected value on both sides of the inequality $T_{N(t)+1} \geq t$ yields $E(Y_1) + \mu H_1(t) \geq t$. Consequently,

$$\lim_{t\to\infty} \inf\left(\frac{E(Y_1)}{\mu t} + \frac{H_1(t)}{t}\right) = \lim_{t\to\infty} \inf \frac{H_1(t)}{t} \geq \frac{1}{\mu} \tag{4.32}$$

For a given finite number A, a "truncated renewal process" $\{\tilde{Y}_1, \tilde{Y}_2, ...\}$ is now defined as follows:

$$\tilde{Y}_i = \begin{cases} Y_i & \text{for } Y_i \leq A \\ A & \text{for } Y_i > A \end{cases}; \quad i = 1, 2, ...$$

Let $\tilde{\mu}$, $\tilde{T}_n$, $\tilde{N}(t)$, and $\tilde{H}_1(t)$ be the characteristica belonging to the truncated renewal process. Since the cycle lengths of this process are less than or equal to A, it follows that $\tilde{T}_{\tilde{N}(t)+1} \leq t + A$. Taking the expected value on both sides of this inequality yields $E(\tilde{Y}_1) + \tilde{\mu}\tilde{H}_1(t) \leq t + A$. Hence,

$$\lim_{t\to\infty} \sup\left(\frac{E(\tilde{Y}_1)}{\tilde{\mu} t} + \frac{\tilde{H}_1(t)}{t}\right) = \lim_{t\to\infty} \sup\left(\frac{\tilde{H}_1(t)}{t}\right) \leq \frac{1}{\tilde{\mu}}$$

The inequality $\tilde{T}_n \leq T_n$ implies $\tilde{N}(t) \geq N(t)$ and, therefore, $\tilde{H}_1(t) \geq H_1(t)$. Consequently,

$$\lim_{t\to\infty} \sup\left(\frac{H_1(t)}{t}\right) \leq \frac{1}{\tilde{\mu}} \tag{4.33}$$

Letting $A \to \infty$ yields

$$\lim_{t\to\infty} \sup\left(\frac{H_1(t)}{t}\right) \leq \frac{1}{\mu} \tag{4.34}$$

Combining (4.32) and (4.34) proves the theorem for $\mu < \infty$. If $\mu = \infty$, then $\tilde{\mu} = \tilde{\mu}(A) \to \infty$ for $A \to \infty$ so that the theorem follows from (4.33). ■

Definition 4.4 A random variable ξ or its distribution function is called *arithmetic* if there exists a positive constant a such that

$$\sum_{i=0}^{\infty} P(\xi = a\,i) = 1$$ •

Thus, the realizations of an arithmetic random variable are multiples of a positive constant. The following theorem is given without proof (see, for example, *Feller* (1971)).

Theorem 4.3 (*key renewal theorem*) If $F(t)$ is nonarithmetic and if $g(t)$ is integrable on $[0,\infty)$, then, for all $F_1(t)$,

$$\lim_{t\to\infty} \int_0^t g(t-x)\,dH_1(x) = \frac{1}{\mu}\int_0^{\infty} g(x)\,dx$$ ■

The key renewal theorem, also called *fundamental renewal theorem* or *theorem of Smith*, has proved a useful tool for solving many problems of applied probability.

If $g(x)$ is assumed to be given by

$$g(x) = \begin{cases} 1 & \text{for } 0 \le x \le h \\ 0, & \text{elsewhere} \end{cases},$$

then the key renewal theorem becomes the *renewal theorem of Blackwell.*

Theorem 4.4 (*Blackwell's renewal theorem*) If $F(t)$ is nonarithmetic, then, for any $F_1(t)$ and $h > 0$,

$$\lim_{t\to\infty} [H_1(t+h) - H_1(t)] = \frac{h}{\mu}$$ ■

Whereas the elementary renewal theorem refers to "a global transition" in the stationary regime, Blackwell's renewal theorem expresses the corresponding "local behaviour".

Theorem 4.5 Suppose $F(t)$ is not arithmetic and $\sigma^2 = Var(Y) < \infty$. Then,

$$\lim_{t\to\infty} \left(H_1(t) - \frac{t}{\mu}\right) = \frac{\sigma^2}{2\mu^2} - \frac{E(Y_1)}{\mu} + \frac{1}{2}$$

Proof The renewal equation (4.7) is equivalent to

$$H_1(t) = F_1(t) + \int_0^t F_1(t-x)\,dH(x)$$

According to theorem 4.1, in the special case $F_1(t) \equiv F_S(t)$, this equation can be written in the form

$$\frac{t}{\mu} = F_S(t) + \int_0^t F_S(t-x)\,dH(x)$$

Subtraction of the second integral equation from the first gives

$$H_1(t) - \frac{t}{\mu} = \overline{F}_S(t) - \overline{F}_1(t) + \int_0^t \overline{F}_S(t)\, dH(x) - \int_0^t \overline{F}_1(t)\, dH(x)$$

Applying the key renewal theorem yields

$$\lim_{t\to\infty} \left(H_1(t) - \frac{t}{\mu}\right) = \frac{1}{\mu}\int_0^\infty \overline{F}_S(t)\, dt - \frac{1}{\mu}\int_0^\infty \overline{F}_1(t)\, dt$$

The theorem now follows from (4.20):

$$\lim_{t\to\infty} \left(H_1(t) - \frac{t}{\mu}\right) = \frac{1}{\mu}\,\frac{\mu^2+\sigma^2}{2\mu} - \frac{E(Y_1)}{\mu}$$

■

For ordinary renewal processes this result becomes

$$\lim_{t\to\infty} \left(H(t) - \frac{t}{\mu}\right) = \frac{1}{2}\left(\frac{\sigma^2}{\mu^2} - 1\right)$$

Corollary If the assumptions of theorem 4.5 are satisfied, then the key renewal theorem also implies the elementary renewal theorem.

Another direct consequence of the key renewal theorem is the fact that $F_S(x)$ is the limiting distribution of both the backward- and the forward recurrence times as t tends to infinity:

$$\lim_{t\to\infty} F_{R(t)}(x) = F_S(x) \tag{4.35}$$

$$\lim_{t\to\infty} F_{V(t)}(x) = F_S(x) \tag{4.36}$$

In view of the definition of the forward recurrence time one expects the equation

$$\lim_{t\to\infty} E(V(t)) = \frac{\mu}{2}$$

to be valid. However, according to (4.20),

$$\lim_{t\to\infty} E(V(t)) = E(X_S) = \frac{\mu+\sigma^2}{2\mu} > \frac{\mu}{2}$$

This "contradiction" is known as the *paradox of the renewal theory.* The intuitive explanation of this phenomenon is that on average the reference time point t is contained more frequently in longer renewal cycles than in shorter ones.

The following property of the renewal density corresponds to theorems 4.2 and 4.5:

$$h_1(t) = \frac{1}{\mu}$$

The next theorem is given without proof.

Theorem 4.6 Let $E(Y_1) < \infty$ and $\mu < \infty$. Then,

$$\lim_{t\to\infty} P\left(\frac{N(t) - t/\mu}{\sigma\sqrt{t\mu^{-3}}} \le x\right) = \Phi(x)$$ ■

An important consequence of this theorem is that, for t sufficiently large, $N(t)$ is approximately normally distributed with expected value t/μ and variance $\sigma^2 t\mu^{-3}$. Thus, this theorem can be used to construct approximate probability limits for $N(t)$: For t sufficiently large, $N(t)$ satisfies the inequality

$$\frac{t}{\mu} - u_{\alpha/2}\,\sigma\sqrt{t\mu^{-3}} \le N(t) \le \frac{t}{\mu} + u_{\alpha/2}\,\sigma\sqrt{t\mu^{-3}}$$

with probability $1-\alpha$, where $u_{\alpha/2}$ is the $(1-\alpha/2)$th percentile of the standard normal distribution:

$$\Phi(u_{\alpha/2}) = 1 - \alpha/2$$

Example 4.4 Let

$$t = 1000,\ \mu = 10,\ \sigma = 2 \ \text{ and } \alpha = 0.05$$

Since $u_{0.025} \approx 2$, $N(1000)$ is contained in the interval $[96, 104]$ with an approximate probability of 0.95. □

Knowledge of the asymptotic distribution of $N(t)$ makes it possible, without knowing the exact distribution of Y, to answer approximately a question that arose in section 4.1: How many spare systems (spare parts) are necessary for guaranteeing that the (ordinary) renewal process can be maintained over an interval $[0, t]$ with a given probability of $1-\alpha$? Since with probability $1-\alpha$

$$\frac{N(t) - t/\mu}{\sigma\sqrt{t\mu^{-3}}} \le u_\alpha,$$

the required number $n_{\min}$ is for large t approximately equal to

$$\frac{t}{\mu} + u_\alpha \sigma\sqrt{t\mu^{-3}}$$

Example 4.5 Given the numerical values $t = 2000$, $\mu = 20$, $\sigma = 0.05$ and $\alpha = 0.01$. Since $u_{0.01} = 2.32$,

$$n_{\min} \approx \frac{2000}{20} + 11.6\sqrt{\frac{2000}{20^3}} = 105.8 \approx 106$$

Thus, about 106 spare systems are at least required in order to be able to maintain the renewal process over the interval $(0, 2000]$ with probability 0.99.

If $t = 200$, $\mu = 1/\lambda = 20$, $\sigma = 20$ and $\alpha = 0.01$, then, using the normal approximation,

$$n_{\min} = 17$$

In example 4.1, the same numerical values, assuming that Y is exponentially distributed, gave the result $n_{\min} = 18$. The difference of one spare system can be explained by the fact that $t = 200$ is not large enough compared with $\mu = 20$ so that the deviation of the exact value from the approximate one cannot be expected to be negligibly small. □

4.5 Stationary Renewal Processes

As already pointed out, the renewal process $\{Y_1, Y_2, ...\}$ is statistically equivalent to the process of forward recurrence times $\{V(t), t \geq 0\}$ and to the process of backward recurrence times $\{R(t), t \geq 0\}$. Thus, the stationarity of the renewal process $\{Y_1, Y_2, ...\}$ can be introduced via the stationarity of, for example, the process of forward recurrence times, using definition 2.2. In the sequel, stationarity refers to strict stationarity.

Definition 4.5 A renewal process $\{Y_1, Y_2, ...\}$ is said to be *stationary* if the corresponding process of the forward recurrence times $\{V(t), t \geq 0\}$ is stationary. ●

It is stated here without proof that the process $\{V(t), t \geq 0\}$ has the Markov property (definition 2.7). Hence it is stationary if its one-dimensional distribution functions $F_{V(t)}(x)$ do not depend on t (theorem 2.1):

$$F_{V(t)}(x) = F_V(x) \quad \text{for all} \quad t \geq 0$$

A stationarity criterion and the corresponding distribution function of $V(t)$ yields the following theorem.

Theorem 4.7 Let $F(t)$ be nonarithmetic and $\mu = \int_0^\infty \overline{F}(t)\, dt < \infty$. Then the renewal process given by $F_1(t)$ and $F(t)$ is stationary if and only if

$$H_1(t) = \frac{t}{\mu} \tag{4.37}$$

A consequence of theorems 4.1 and 4.7 is the

Corollary A renewal process is stationary if and only if

$$F_1(t) = F_S(t) = \frac{1}{\mu} \int_0^t \overline{F}(x)\, dx\,, \quad t \geq 0 \tag{4.38}$$

Proof If (4.37) and (4.38) hold, then, in view of theorem 4.1, formula (4.30) implies that

$$F_{V(t)}(x) = \frac{1}{\mu}\int_0^{t+x} \overline{F}(u)\,du - \frac{1}{\mu}\int_0^t \overline{F}(t+x-u)\,du$$

$$= \frac{1}{\mu}\int_0^{t+x} \overline{F}(u)\,du - \frac{1}{\mu}\int_x^{t+x} \overline{F}(u)\,du$$

$$= \frac{1}{\mu}\int_0^x \overline{F}(u)\,du,$$

so that $F_{V(t)}(x)$ does not depend on t.

Conversely, if $F_{V(t)}(x)$ does not depend on t, then (4.36) implies

$$F_{V(t)}(x) = \lim_{t\to\infty} F_{V(t)}(x)$$

$$= F_S(x)$$

In view of the corollary, this result proves the theorem . ■

Theorem 4.7 allows the following interpretation of the elementary renewal theorem: After a sufficiently large time span (*transient response time*) every renewal process with nonarithmetic distribution function $F(t)$ and finite mean cycle lengths behaves like a stationary renewal process.

If $\{N_S(t),\ t \ge 0\}$ denotes the renewal counting process which belongs to a stationary renewal process, then it can easily be shown that $\{N_S(t),\ t \ge 0\}$ is stationary in the sense of definition 3.2. Determining its binomial moments according to (4.21) yields (*Beichelt, Franken* (1984))

$$E\binom{N_S(t)}{n} = \frac{1}{\mu}\int_0^t H^{*(n-1)}(x)\,dx; \quad n = 1, 2, \ldots; \quad H^{*(0)} \equiv 1$$

From this one obtains the variance of $N_S(t)$ to be

$$Var(N_S(t)) = \frac{2}{\mu}\int_0^t H(x)\,dx + \frac{t}{\mu}\left(1 - \frac{t}{\mu}\right)$$

Moreover, the following bounds hold:

$$\frac{\mu_2 t}{\mu^3} - \frac{t}{\mu} - \frac{4\mu_3}{3\mu^3} \le Var(N_S(t)) \le \frac{\mu_2 t}{\mu^3} - \frac{t}{\mu} + \frac{\mu_2^2}{4\mu^4},$$

where $\mu_i = E(Y^i)$, $i = 2, 3$.

4.6 Alternating Renewal Processes

Up till now it has been assumed that renewals take negligible time. In order to be able to model practical situations where this assumption does not hold, the concept of a renewal process has to be generalized: the renewal time of the system after the i th failure is assumed to be a positive random variable Z_i; $i = 1, 2, ...$ Immediately after a renewal the system starts working again. In this way a sequence of two-dimensional random vectors $\{(Y_i, Z_i); i = 1, 2, ...\}$ is generated where as before Y_i denotes the lifetime of the system after the i th renewal. Thus,

$$S_1 = Y_1; \quad S_n = \sum_{i=1}^{n-1} (Y_i + Z_i) + Y_n; \;\; n = 2, 3, ...$$

are the time points, when a failures occur and

$$T_n = \sum_{i=1}^{n} (Y_i + Z_i); \;\; n = 1, 2, ...$$

are the time points when a renewed system starts working.

Definition 4.6 If $\{Y_1, Y_2, ...\}$ and $\{Z_1, Z_2, ...\}$ are two independent sequences of independent random variables, then the random sequence $\{(Y_1, Z_1), (Y_2, Z_2), ...$ is said to be an *alternating renewal process.* ●

Sometimes the random sequence $\{(S_1, T_1), (S_2, T_2), ...\}$ is referred to as an alternating renewal process. Both definitions are obviously equivalent.

If a working system is assigned a 1 and a failed system a 0, then a binary indicator variable of the system state is given by

$$X(t) = \begin{cases} 0, & \text{if } \; t \in [S_n, T_n), \;\; n = 1, 2, ... \\ 1, & \text{elsewhere} \end{cases} \tag{4.39}$$

The stochastic process $\{X(t), t \geq 0\}$ can also serve as a basis for defining alternating renewal processes since there is a one-to-one correspondence between the sample paths of the process $\{X(t), t \geq 0\}$ and the realizations of the underlying alternating renewal process (Figure 4.7).

In what follows, all Y_i and Z_i are assumed to be distributed as Y and Z, respectively, with $F(y) = P(Y \leq y)$ and $G(z) = P(Z \leq z)$. By definition of an alternating renewal process: $P(X(+0) = 1) = 1$.

Remark Analogously to the concept of a delayed renewal process, the alternating renewal process can be generalized by assigning the random lifetime Y_1 a probability distribution different from that of Y. However, this generalization and some other possibilities will not be discussed here, although no principal difficulties would arise.

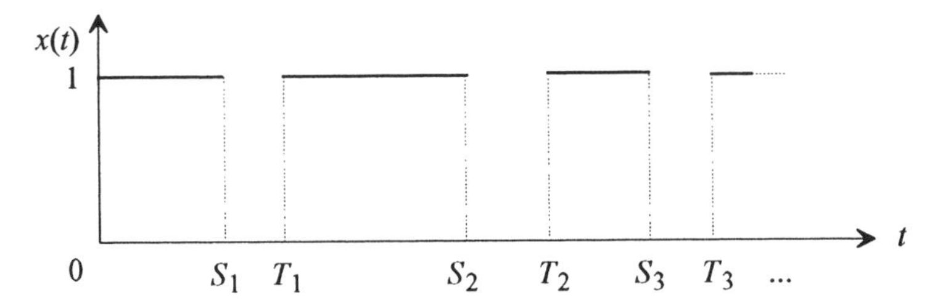

Figure 4.7 Sample path of an alternating renewal process

Let $N_a(t)$ and $N_e(t)$ be the random numbers of failures and renewals in $(0, t]$, respectively. Since S_n and T_n are sums of independent random variables (compare with (4.2),

$$\begin{aligned} F_{S_n}(t) &= P(S_n \le t) \\ &= P(N_a(t) \ge n) \\ &= F * (G * F)^{*(n-1)}(t) \end{aligned}$$

$$\begin{aligned} F_{T_n}(t) &= P(T_n \le t) \\ &= P(N_e(t) \ge n) \\ &= (F * G)^{*(n)}(t) \end{aligned}$$

Hence, analogously to (4.6), the expected values

$$H_a(t) = E(N_a(t)) \text{ and } H_e(t) = E(N_e(t))$$

are given by

$$H_a(t) = \sum_{n=1}^{\infty} F * (G * F)^{*(n-1)}(t)$$

and

$$H_e(t) = \sum_{n=1}^{\infty} (F * G)^{*(n)}(t),$$

where $(F * G)^{0*}(t) \equiv 1$. $H_a(t)$ and $H_e(t)$ are referred to as the *renewal functions* of the alternating renewal process. They satisfy equations which can immediately be derived from (4.7) to (4.9).

Remark Note that $H_a(t)$ may be interpreted as the renewal function of a delayed renewal process whose first system lifetime is distributed according to $F(x)$, whereas the following "system lifetimes" are identically distributed as $Y + Z$; that is, they have distribution function $(F * G)(t)$. Analogously, $H_e(t)$ may be interpreted as the renewal function of an ordinary renewal process whose cycle lengths have distribution function $(F * G)(t)$.

Let $V_a(t)$ be the residual lifetime of the system that is working at time t. Then $P(X(t)=1, V_a(t)>x)$ is the probability that the system is working at time t and does not fail in the interval $(t, t+x]$. As in section 4.3, this probability is called *interval reliability*. It can be determined by the total probability rule:

$$P(X(t)=1, V_a(t)>x) = \sum_{n=0}^{\infty} P(T_n \le t,\ T_n + Y_{n+1} > t+x)$$

$$= \overline{F}(t+x) + \sum_{n=1}^{\infty} \int_0^t P(t+x < T_n + Y_{n+1} \,|\, T_n = u)\, dF_{T_n}(u)$$

$$= \overline{F}(t+x) + \int_0^t P(t+x-u < Y) \sum_{n=1}^{\infty} d(F * G)^{*(n)}(u)$$

Hence,

$$P(X(t)=1,\ V_a(t)>x) = \overline{F}(t+x) + \int_0^t \overline{F}(t+x-u)\, dH_e(u) \tag{4.40}$$

The probability $A(t) = P(X(t)=1)$ that the system is working at time t is called *point availability* of system. This important characteristic of an alternating renewal process is obtained from (4.40) by letting $x=0$:

$$A(t) = \overline{F}(t) + \int_0^t \overline{F}(t-u)\, dH_e(u) \tag{4.41}$$

By equation (1.9), $A(t)$ is equal to the expected value of the indicator variable of the system state:

$$A(t) = E(X(t))$$

The *average availability* of the system in the interval $[0, t]$ is

$$\overline{A}(t) = \frac{1}{t}\int_0^t A(x)\, dx$$

Let $U(t)$ denote the random *total working time* of the system in the interval $[0, t]$. Then,

$$U(t) = \int_0^t X(x)\, dx \tag{4.42}$$

By changing the order of integration,

$$E(U(t)) = E\left(\int_0^t X(x)\, dx\right) = \int_0^t E(X(x))\, dx$$

Thus,

$$E(U(t)) = \int_0^t A(x)\, dx = t\,\overline{A}(t)$$

The following theorem provides information on the limiting behaviour of the interval reliability and the point availability as t tends to infinity. A proof need not be given, since the assertions are an immediate consequence of the key renewal theorem. (Mind also the *remark* at page 130.)

Theorem 4.8 If the distribution function $(F * G)(t)$ of $Y+Z$ is nonarithmetic and $E(Y)+E(Z)<\infty$, then

$$A_x = \lim_{t\to\infty} P(X(t)=1,\ V_a(t)>x) = \frac{1}{E(Y)+E(Z)}\int_x^\infty \overline{F}(u)\,du$$

$$A = \lim_{t\to\infty} A(t) = \lim_{t\to\infty} \overline{A}(t) = \frac{E(Y)}{E(Y)+E(Z)} \tag{4.43}$$

■

A_x said to be the *long-run* or *stationary interval reliability* and A the *long-run* or *stationary availability*. Clearly, $A = A_0$. If, analogously to renewal processes, the time between two neighbouring renewal times is called a *renewal cycle*, then the long-run availability is equal to the expected share of the working time in the expected renewal cycle length. It should be mentioned that equation (4.43) is also valid if within the renewal cycles the Y_i and Z_i are dependent of each other. Note that, in general,

$$E\left(\frac{Y}{Y+Z}\right) \neq \frac{E(Y)}{E(Y)+E(Z)} \tag{4.44}$$

Example 4.6 Let lifetimes and renewal times be exponentially distributed with densities

$$f(y) = \lambda_1 e^{-\lambda_1 y},\ y \ge 0, \quad \text{and} \quad g(z) = \lambda_0 e^{-\lambda_0 z},\ z \ge 0,$$

respectively. Application of the Laplace-transform to (4.41) yields

$$\hat{A}(s) = \overline{F}(s) + \overline{F}(s)\cdot \hat{h}_e(s)$$
$$= \frac{1}{s+\lambda_1} + \frac{1}{s+\lambda_1}\hat{h}_e(s)$$

The Laplace-transform of the convolution of f and g is

$$L\{f*g\} = \frac{\lambda_0\lambda_1}{(s+\lambda_0)(s+\lambda_1)}$$

Hence, according to (4.14),

$$\hat{h}_e(s) = \frac{\dfrac{\lambda_0\lambda_1}{(s+\lambda_0)(s+\lambda_1)}}{1-\dfrac{\lambda_0\lambda_1}{(s+\lambda_0)(s+\lambda_1)}} = \frac{1}{s}\,\frac{\lambda_0\lambda_1}{s+\lambda_0+\lambda_1}$$

By inserting $\hat{h}_e(s)$ in $\hat{A}(s)$ and expanding $\hat{A}(s)$ into partial fractions,

$$\hat{A}(s) = \frac{1}{s+\lambda_1} + \frac{\lambda_1}{s} \cdot \frac{1}{s+\lambda_1} - \frac{\lambda_1}{s} \cdot \frac{1}{s+\lambda_0+\lambda_1}$$

Transforming back yields the point availability:

$$A(t) = \frac{\lambda_0}{\lambda_0+\lambda_1} + \frac{\lambda_1}{\lambda_0+\lambda_1} e^{-(\lambda_0+\lambda_1)t}, \quad t \geq 0$$

Since

$$E(Y) = 1/\lambda_1 \text{ and } E(Z) = 1/\lambda_0$$

this result verifies equation (4.43). On the other hand, if $\lambda_0 \neq \lambda_1$, then (*Beichelt* (1995), example 6.7)

$$E\left(\frac{Y}{Y+Z}\right) = \frac{\lambda_0}{\lambda_0 - \lambda_1}\left(1 + \frac{\lambda_1}{\lambda_0 - \lambda_1} \ln \frac{\lambda_1}{\lambda_0}\right),$$

so that the statement (4.44) is verified. For instance, if

$$E(Z) = 0.25\, E(Y),$$

then

$$A = \frac{E(Y)}{E(Y)+E(Z)} = 0.8$$

and

$$E\left(\frac{Y}{Y+Z}\right) = 0.7172$$

□

If Y is exponentially distributed with parameter λ and if Z is uniformly distributed over the interval $[0, d]$, then there also exists an explicit formula for the point availability:

$$A(t) = \sum_{n=0}^{[t/d]} \frac{(\lambda(t-nd))^n}{n!} e^{-\lambda(t-nd)},$$

where $[t/d]$ is the greatest integer less than or equal to t/d. The proof is left to the reader.

Numerical methods have generally to be applied to determine interval reliability and point availability when applying formulas (4.40) and (4.41). This is again due to the fact that there are either no explicit or rather complicated representations of the renewal function for most of the common lifetime distributions. However, formulas (4.40) and (4.41) may be applied to obtaining approximate values of interval reliability and point availability if they are used in connection with the bounds and approximations for the renewal function given in sections 4.2 and 4.4.

4.7 Cumulative Stochastic Processes

Cumulative stochastic processes arise by additive superposition of random variables at random time points. Consider the following illustrative example:

Example 4.7 Mechanical wear of an item is caused by shocks. (For instance, for the brake disks of a car every application of the brakes is a shock.) After the i th shock the degree of wear of the item increases by C_i units, where the $C_1, C_2, \ldots$ are independent random variables, identically normal distributed as C with $E(C) =$ 9.2 and $\sqrt{Var(C)} = 2.8 \left[\text{in } 10^{-4} mm\right]$. Assuming the initial degree of wear of the item to be zero, the item is replaced by an equivalent new one if the total degree of wear exceeds $0.1mm$. What is the probability of replacing the system although no more than 100 shocks have occurred?

If

$$X_{100} = \sum_{i=1}^{100} C_i$$

denotes the degree of wear after 100 shocks, then the distribution function of X_{100} is

$$P(X_{100} \le x) = \Phi\left(\frac{x - 9.2 \cdot 100}{\sqrt{2.8^2 \cdot 100}}\right) = \Phi\left(\frac{x - 920}{28}\right)$$

Hence the desired probability is

$$P(X_{100} > 1000) = 1 - \Phi\left(\frac{1000 - 920}{28}\right) = 1 - \Phi(2.86)$$

$$= 0.021$$

In other words, the item survives the first 100 shocks with probability 0.979. □

Besides the individual contributions of shocks to the degree of wear, the times between successive occurances of shocks have to be taken into account in order to characterize the speed of the wear. This leads to the following general definition.

Definition 4.7 ***(cumulative stochastic process)*** Let $\{N(t), t \ge 0\}$ be a point process with sequence of jump times $T_1, T_2, \ldots$ Each jump time T_i is assigned a random variable C_i. Then the stochastic process $\{X(t), t \ge 0\}$ defined by

$$X(t) = \sum_{i=1}^{N(t)} C_i \tag{4.45}$$

is called a *cumulative stochastic process*. ●

Thus, C_i is the height of the jump occuring at time T_i (Figure 4.8).

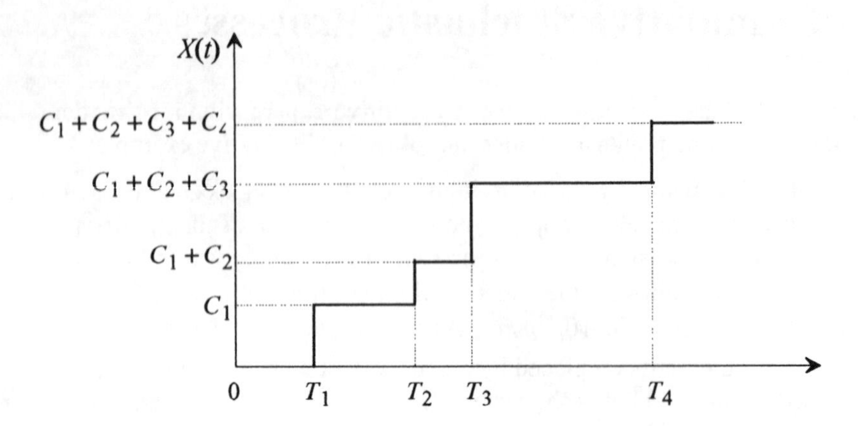

Figure 4.8 Sample path of a cumulative process with nonnegative increments

Besides its interpretation as a degree of wear, $X(t)$ can, depending on the meaning of C_i, be a a profit, loss or any other criterion. For instance, if $\{N(t), t \geq 0\}$ is a renewal counting process and C_i denotes the cost of the renewal at time T_i, then $X(t)$ is the total (cumulative) renewal cost arising in $(0, t]$. If C_i denotes the profit the system makes in $(T_{i-1}, T_i]$, then $X(t)$ is the total profit the system makes in the interval $(0, t]$. (However, a possible profit made in the interval $(T_{N(t)}, t]$ is not taken into account.)

Cumulative stochastic processes are considered in this section under the following assumptions:

1) $\{N(t), t \geq 0\}$ is a renewal counting process belonging to an ordinary renewal process $\{Y_1, Y_2, ...\}$.

2) The $C_1, C_2, ...$ are independent random variables. Y_i and C_j are independent of each other for all $i \neq j$.

3) The random vectors (Y_i, C_i) are identically distributed as (Y, C) for all $i = 1,2,...$ Y and C have finite, positive expected values and variances.

Under these assumptions Wald's equation (theorem 1.3) immediately yields the trend function $m(t) = E(X(t))$ of a cumulative stochastic process:

$$m(t) = E(C)\,H(t),$$

where $H(t) = E(N(t))$ is the renewal function of the underlying renewal process $\{Y_1, Y_2, ...\}$. From this and the elementary renewal theorem an important asymptotic property of cumulative stochastic processes follows:

$$\lim_{t\to\infty} \frac{E(X(t))}{t} = \frac{\nu}{\mu} \tag{4.46}$$

with

$$\mu = E(Y) \quad \text{and} \quad \nu = E(C \tag{4.47}$$

Equation (4.46) means that the expected long-run (stationary) loss or profit per unit time is equal to the expected loss or profit per unit time within a renewal cycle.

Generally, explicit formulas do not exist for the distribution function or density of $X(t)$. Hence the following explicit formula for the asymptotic probability distribution of $X(t)$ as $t \to \infty$ is practically important. It is a consequence of the central limit theorem (theorem 1.2); for details see *Gut* (1990).

Theorem 4.9 If

$$\gamma^2 = Var(\mu C - \nu Y) > 0, \tag{4.48}$$

then

$$\lim_{t\to\infty} P\left(\frac{X(t) - \frac{\nu}{\mu}t}{\mu^{-3/2}\gamma\sqrt{t}} \le u\right) = \Phi(u),$$

where $\Phi(u)$ is given by (1.51). ■

This theorem implies that for large t the random variable $X(t)$ has approximately a normal distribution with expected value $(\nu/\mu)\,t$ and variance $\mu^{-3}\gamma^2 t$, i.e.

$$X(t) \approx N\left(\frac{\nu}{\mu}\,t,\ \mu^{-3}\gamma^2 t\right) \tag{4.49}$$

If C and Y are independent, then the parameter γ^2 can be written in the following form:

$$\gamma^2 = \mu^2 Var(C) + \nu^2 Var(Y) \tag{4.50}$$

In this case, in view of assumption 3, the condition (4.48) is always fulfilled. This condition actually only excludes the case $\gamma^2 = 0$, i.e. linear dependence between Y and C. The following example illustrates the practical application of theorem 4.9.

Example 4.8 Given an alternating renewal process $\{(Y_i, Z_i);\ i = 1, 2, ...\}$ as defined in section 4.6, the total renewal time in (0, t] is given by (a possible renewal time running at time t is not taken into account)

$$X(t) = \sum_{i=1}^{N(t)} Z_i,$$

where $N(t)$ is the number of renewal times T_n in (0, t] (notation and assumptions as in section 4.6). Hence, the development of the total renewal time is governed by a cumulative stochastic process. In order to investigate the asymptotic behaviour of $X(t)$ as $t \to \infty$, using theorem 4.9, C has to be replaced by Z and Y by $Y+Z$. Consequently, if t is sufficiently large, then $X(t)$ has approximately a normal distribution with parameters

$$E(X(t)) = \frac{E(Z)}{E(Y)+E(Z)}\,t \quad \text{and} \quad Var(X(t)) = \frac{\gamma^2}{[E(Y)+E(Z)]^3}\,t$$

Because of the independence of Y and Z,

$$\begin{aligned}\gamma^2 &= Var[Z\,E(Y+Z) - (Y+Z)E(Z)] \\ &= Var[Z\,E(Y) - Y\,E(Z)] \\ &= [E(Y)]^2\,Var(Z) + [E(Z)]^2\,Var(Y) > 0,\end{aligned}$$

so that the assumption (4.48) is satisfied. In particular, let (all parameters in *hours*)

$$E(Y) = 120, \ \sqrt{Var(Y)} = 40 \quad \text{and} \quad E(Z) = 4, \ \sqrt{Var(Z)} = 2$$

Then,

$$\gamma^2 = 120^2 \cdot 4 + 16 \cdot 1600 = 83200$$

and

$$\gamma = 288.4$$

Consider, for example, the total renewal time in the interval $\left[0,\ 10^4\,hours\right]$. The probability that $X(10^4)$ does not exceed the nominal value of $350\,hours$ is

$$P(X(10^4) \le 350) = \Phi\left(\frac{350 - \frac{4}{124}\,10^4}{124^{-3/2}\cdot 288.4 \cdot \sqrt{10^4}}\right)$$

$$= \Phi(1.313)$$

Hence,

$$P(X(10^4) \le 350) = 0.905$$

□

First passage time The examples considered so far motivate an investigation of the random time $L(x)$ when the stochastic process $\{X(t),\ t \ge 0\}$ exceeds a given nominal value x for the first time:

$$L(x) = \inf_t \{t,\ X(t) > x\}$$

If, for instance, x is the critical wear limit of an item, then crossing the level x is commonly referred to as the occurrence of a *drift failure*. In this case it is justified to denote L as the lifetime of the item. An obvious relationship exists between the distribution functions of the *first passage time* $L(x)$ and of $X(t)$ if $\{X(t),\ t > 0\}$ has nondecreasing sample paths, i.e. if the C_i are nonnegative random variables:

$$P(L(x) \le t) = P(X(t) > x)$$

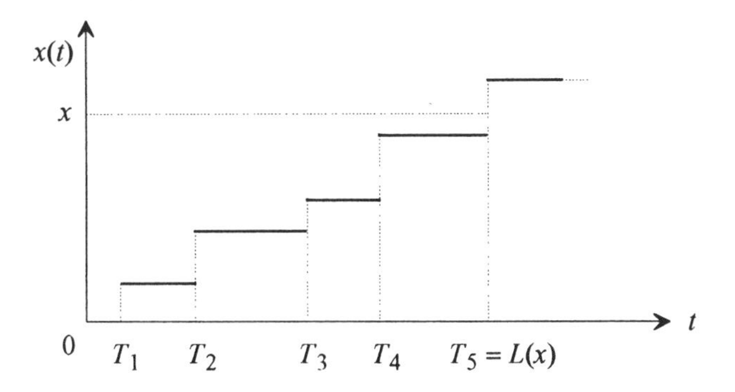

Figure 4.9 Level crossing of a cumulative stochastic process

Figure 4.9 illustrates this fact on a given sample path of a cumulative stochastic process.

Example 4.9 A system is subject to shocks according to a homogeneous Poisson process $\{N(t), t \geq 0\}$ with intensity λ. C is assumed to be a nonnegative random variable with distribution function $G(t)$. $X(t)$ is interpreted as the degree of wear of the system at time t. The system has failed if and only if $X(t) > x$. Given n shocks in $(0, t]$, the probability that the system does not fail in $(0, t]$ is $G^{*(n)}(x)$. Here $G^{*(n)}$ is the n-th convolution power of G as defined in (1.49). By the total probability rule, the probability that the system does not fail in $(0, t]$ is

$$\bar{F}(t) = e^{-\lambda t} \sum_{n=0}^{\infty} \frac{(\lambda t)^n}{n!} G^{*(n)}(x); \quad x \geq 0, \ t \geq 0$$

Given the notation in (4.47), letting $\lambda = 1/\mu$ and using the fact that the renewal function of a homogeneous Poisson process is $H(t) = t/\mu$, the trend function of the wear process $\{X(t), t \geq 0\}$ is seen to be

$$m(t) = \frac{\nu}{\mu} t$$

If, in particular, C is normally distributed with expected value ν and standard deviation σ ($\nu > 3\sigma$), then the probability that the system does not fail in the interval $(0, t]$ is

$$\bar{F}(t) = e^{-\lambda t} \sum_{n=0}^{\infty} \frac{(\lambda t)^n}{n!} \Phi\left(\frac{x - n\nu}{\sigma\sqrt{n}}\right); \quad x \geq 0, \ t \geq 0 \qquad \square$$

The probability distribution of $L(x)$ is generally not explicitly available. Hence the following theorem is important for practical applications since it provides information on the asymptotic behaviour of the distribution of $L(x)$ as x tends to infinity. The analogy of this theorem to theorem 4.9 is obvious.

Theorem 4.10 If

$$\gamma^2 = Var(\mu C - \nu Y) > 0,$$

then

$$\lim_{x\to\infty} P\left(\frac{L - \frac{\mu}{\nu}x}{\nu^{-3/2}\,\gamma\,\sqrt{x}} \le u\right) = \Phi(u),$$

where $\Phi(u)$ is given by (1.51). ■

For a recent proof see *Gut* (1990). (Moreover, *Gut* proved that this theorem is also valid for arbitrary Y_i and C_i if in addition to (4.48) the condition $\nu > 0$ holds.) As in theorem 4.9, the random variables Y_i and C_i need not be independent.

A consequence of theorem 4.10 is that for large x the first passage time $L = L(x)$ has approximately a normal distribution with expected value $(\mu/\nu)x$ and variance $\nu^{-3}\gamma^2 x$:

$$L(x) \approx N\left(\frac{\mu}{\nu}x,\ \nu^{-3}\gamma^2 x\right) \tag{4.51}$$

The probability distribution given in (4.51) is called *Birnbaum-Saunders-distribution.*

Continuation of example 4.7 In addition to the parameters

$$\nu = E(C) = 9.2 \quad \text{and} \quad \sqrt{Var(C)} = 2.8 \quad [\text{in } 10^{-4}\,mm]$$

it will be assumed that expected value and standard deviation of the cycle length are given by

$$\mu = E(Y) = 6 \quad \text{and} \quad \sqrt{Var(Y)} = 2 \quad [\text{in } hours]$$

Provided C and Y are independent, the parameter γ is

$$\gamma = 0.\,0024916$$

What is the probability that a nominal value of $0.1\,mm$ is only exceeded after 600 *hours*? Theorem 4.10 can be applied to solve this problem, since $0.1\,mm$ is sufficiently large in comparision to the shock parameter ν. Hence,

$$P(L(0.1) > 600) = 1 - \Phi\left(\frac{600 - \frac{6}{9.2}10^3}{(9.2)^{-3/2}\cdot 24.916\cdot\sqrt{0.1}}\right)$$

$$= 1 - \Phi(-1.848)$$

Thus, the desired probability is $P(L(0.1) > 600) = 0.967$. □

4.8 Regenerative Stochastic Processes

At the beginning of this chapter it was been pointed out that, apart from its own significance, renewal theory provides mathematical foundations for analyzing the behaviour of complicated systems which have renewal points imbedded in their running times. This is always the case if the running time of a system is partitioned by so-called *regeneration points* into *regeneration cycles* with the following characteristic properties:

1) After every regeneration point the future operation of the system is independent of its past operation.
2) Within every regeneration cycle the operation of the system is governed by the same stochastic rules.

Thus, regeneration points are nothing else but renewal points of a system and, hence, generate a renewal process. However, now it is not only the distances between regeneration points that are interesting, but also the behaviour of the system within a regeneration cycle. For a mathematical definition of a regenerative stochastic process, an ordinary renewal process $\{L_1, L_2, ...\}$ is introduced, where the L_i are identically distributed as L. L_i denotes the random length of the i-th regeneration cycle. Thus, if $T_n = \Sigma_{i=1}^{n} L_i$, then $\{T_1, T_2, ...\}$ is the random sequence of regeneration points. Let $\{N(t), t \geq 0\}$ be the corresponding renewal counting process. A *regeneration cycle* is given by $\{[L, W(x)], 0 \leq x < L\}$, where $W(x)$ denotes the state of the system at time x (with respect to the preceeding regeneration point). The renewal process with cycle lengths $L_1, L_2, ...$ determines the sequence of regeneration cycles

$$\{[L_i, W_i(x)], 0 \leq x < L_i\}, ; \; i = 1, 2, ...$$

The characteristic properties of regeneration points and regeneration cycles given verbally above become mathematically precise by assuming that the regeneration cycles are independent of each other and are identically distributed as the *typical regenerating cycle* $\{[L, W(x)], 0 \leq x < L\}$.

The probability distribution of the typical regenerating cycle is called the *cycle distribution.*

Definition 4.8 Using the notation and assumptions introduced earlier, the stochastic process $\{X(t), t \geq 0$, defined by

$$X(t) = W_{N(t)}(t - T_{N(t)}), \tag{4.52}$$

is said to be a *regenerative stochastic process.* The T_n; $n = 1, 2, ...$; are its *regeneration points.* ●

Intuitively speaking, definition 4.8 means that $T_{N(t)}$, the regeneration point before t, is is declared to be the new origin. After it the process $\{W(x), x \geq 0\}$ with $x = t - T_{N(t)}$ develops from $x = 0$ to the following regeneration point $T_{N(t)+1}$, which is reached at time $x = L_{N(t)+1} = T_{N(t)+1} - T_{N(t)}$. Thus, a regenerative process restarts at every regeneration point.

Example 4.10 The alternating renewal process $\{(Y_i, Z_i); i = 1, 2, ...\}$ is a simple example of a regenerative process. In this special case the cycle lengths L_i are given by $L_i = Y_i + Z_i$, where the random vectors (Y_i, Z_i) are independent of each other and identically distributed as (Y, Z). The stochastic process $\{W(x), x \geq 0\}$ characterizes the working and renewal phases within a cycle:

$$W(x) = \begin{cases} 1 & \text{for} \quad 0 \leq x < Y \\ 0 & \text{for} \quad Y \leq x < Y + Z \end{cases}$$

In this way the typical regeneration cycle $\{[L, W(x)], 0 \leq x < L$, where $L = Y + Z$, is completely determined and (4.52) defines the corresponding regenerative stochastic process. Thus, not only the lengths L_i of the regeneration cycles are of interest, but also the relationship between the working and renewal phases. □

In order to determine the one-dimensional distribution of regenerative processes the renewal function $H(t)$ of the ordinary renewal process $\{L_1, L_2, ...\}$ and the probabilities

$$Q(x, B) = P(W(x) \in B, L > x)$$

have to be introduced, where B is a subset of the state space of $\{W(x), x \geq 0$. Analogously to the derivation of (4.40) it can be shown that

$$P(X(t) \in B) = Q(t, B) + \int_0^t Q(t - x, B)\, dH(x) \tag{4.53}$$

The following theorem considers the behaviour of these probabilities as $t \to \infty$.

Theorem 4.11 (Theorem of *Smith*) If L is nonarithmetic and $E(L) > 0$, then

$$\lim_{t \to \infty} P(X(t) \in B) = \frac{1}{E(L)} \int_0^\infty Q(x, B)\, dx$$

For any L with $E(L) > 0$, a somewhat weaker statement holds:

$$\lim_{t \to \infty} \frac{1}{t} \int_0^t P(X(x) \in B)\, dx = \frac{1}{E(L)} \int_0^\infty Q(x, B)\, dx$$ ■

(For the definition of (non-) arithmetic random variables see definition 4.4.) The practical application of these *stationary state probabilities* of a regenerative stochastic process is illustrated by the following maintenance problem.

Example 4.11 ***(age renewal policy)*** The system is renewed upon failure by an *emergency renewal* or at age τ by a *preventive renewal*, whichever occurs first.

Note that this maintenance policy is a special case of policy 4 in section 3.2 (see page 102) if letting there $p = 1$.

A renewed system is assumed to have the same lifetime distribution as the original one, i.e. it is "as good as new". Emergency and preventive renewals are assumed to require constant times d_e and d_p, respectively. Furthermore, let $F(t) = P(T \leq t)$ be the distribution function of the system lifetime T, $\overline{F}(t) = 1 - F(t)$ the survival probability and $\lambda(t)$ the failure rate of the system.

To specify an underlying regenerative stochastic process, the time points at which the renewed system starts resuming its work are declared to be the regeneration points. Therefore, the random length L of the typical renewal cycle has structure

$$L = \min(T, \tau) + Z,$$

where the random renewal time Z is given by

$$Z = \begin{cases} d_e & \text{for} \quad T < \tau \\ d_p & \text{for} \quad T \geq \tau \end{cases}$$

Hence the expected length of a regeneration cycle is

$$E(L) = \int_0^\tau \overline{F}(x)\,dx + d_e F(\tau) + d_p \overline{F}(\tau)$$

Let

$$W(x) = \begin{cases} 1 & \text{if the system is working} \\ 0 & \text{otherwise} \end{cases}$$

Then, for $B = \{1\}$,

$$Q(x, B) = P(W(x) = 1,\ L > x) = \begin{cases} 0 & \text{for} \quad \tau < x \leq L \\ P(T > x) & \text{for} \quad 0 \leq x \leq \tau \end{cases}$$

Thus,

$$\int_0^\infty Q(x, B)\,dx = \int_0^\tau P(T > x)\,dx = \int_0^\tau \overline{F}(x)\,dx$$

Theorem 4.11 now immediately yields the stationary availability of the system:

$$A(\tau) = \lim_{t \to \infty} P(X(t) = 1) = \frac{\int_0^\tau \overline{F}(x)\,dx}{\int_0^\tau \overline{F}(x)\,dx + d_e F(\tau) + d_p \overline{F}(\tau)}$$

Policy 4 can also be described by an alternating renewal process. Hence, applying theorem 4.8 (formula (4.43)) would yield the same result.

Let $\tau*$ denote a renewal interval τ maximizing $A(\tau)$. The necessary condition $dA(\tau)/d\tau = 0$ yields the equation

$$\lambda(\tau)\int_0^\tau \overline{F}(x)\,dx - F(\tau) = \frac{d}{1-d}, \tag{4.54}$$

where $d = d_p/d_e$. A unique solution $\tau*$ exists if $\lambda(t)$ is strictly increasing to infinity and $d < 1$. The corresponding maximum availability is

$$A(\tau*) = \frac{1}{1 + (d_e - d_p)\lambda(\tau*)}$$

If, in particular, T has a Rayleigh-distribution with distribution function

$$F(t) = 1 - e^{-(t/\theta)^2}, \quad t \geq 0,$$

then equation (4.54) becomes

$$\frac{\sqrt{\pi}}{\theta}\tau\left[2\Phi\left(\frac{\sqrt{2}}{\theta}\tau\right) - 1\right] + e^{-(\tau/\theta)^2} = \frac{1}{1-d},$$

where $\Phi(x)$ is defined in (1.51). For instance, let $d_p = 2$ and $d_e = 10$. Then,

$$\tau* = 0.511\,\theta \quad \text{and} \quad A(\tau*) = \theta/(\theta + 8.18)$$

By comparision, the stationary availability of the system without preventive renewals, that is, only emergency renewals are performed, is

$$A(\infty) = \theta/(\theta + 11.28)$$

□

Exercises

Note: Exercises 4.1 to 4.12 refer to ordinary renewal processes and $f(t)$, $F(t)$, μ, and μ_2 denote in this order the density, distribution function, expected value and second moment of the cycle length. $N(t)$ and $H(t)$ are the renewal counting function and renewal function, respectively.

4.1) A system starts working at time $t = 0$. Its lifetime has a normal distribution with expected value $\mu = 120$ and standard deviation $\sigma = 24$ [in *hours*]. After its failure it is replaced by an equivalent new one in negligible time and immediately resumes its work. How many spare systems must be available in order to maintain the replacement process over the interval [0, 10 000 h] (1) with probability 0.90, and (2) with probability 0.99, respectively?

4.2) (1) Use the Laplace-transformation to find the renewal function $H(t)$ of an ordinary renewal process whose cycle lengths have an Erlang distribution with parameters $n = 2$ and λ.
(2) Compare the exact graph of the renewal function with the linear bounds of *Marshall* and the *Barlow/Proschan*-bounds (if the latter are applicable).

4.3) The probability density function of the cycle lengths of an ordinary renewal process is given by

$$f(t) = p\lambda_1 e^{-\lambda_1 t} + (1-p)\lambda_2 e^{-\lambda_2 t}, \quad 0 \le p \le 1, \; t \ge 0.$$

(1) Determine the expected value μ and the second moment μ_2 of the cycle length.
(2) Verify that the corresponding renewal function is given by

$$H(t) = \frac{t}{\mu} + \left(\frac{\mu_2}{2\mu^2} - 1\right)\left(1 - e^{-(p\lambda_1 + (1-p)\lambda_2)t}\right).$$

4.4)* (1) Verify that the probability $p(t) = P(N(t) \text{ is odd})$ satisfies the integral equation

$$p(t) = F(t) - \int_0^t p(t-x)\,dF(t).$$

(2) Determine this probability if the cycle lengths are exponential with parameter λ.

4.5) Let the renewal function of an ordinary renewal process be given by $H(t) = t/10$. Determine the probability $P(N(10) \ge 2)$.

4.6)* By applying the Laplace-transform verify that $H_2(t) = E(N^2(t))$ satisfies the integral equation

$$H_2(t) = 2H(t) - F(t) + \int_0^t H_2(t-x) f(x)\,dx.$$

4.7) Prove equations (4.20).

4.8) Sketch the *Barlow/Proschan*-bounds on the renewal function in the interval $[0, 4]$ if $F(t) = 1 - e^{-t^2}$, $t \ge 0$ (Rayleigh-distribution).

4.9) (1) What is the statement of theorem 4.3 and (2) what is the statement of theorem 4.5 if $F(t) = 1 - e^{-t^2}$, $t \ge 0$?

4.10) The times between the arrivals of successive particles at a counter generate an ordinary renewal process. After recording 10 particles the counter is blocked for τ (= const.) time units. Particles arriving during a blocked period are not registered.
What is the distribution function of the time from the end of a blocked period to the arrival of the first particle in the steady state?

4.11) Let $R(t)$ be the backward- and $V(t)$ be the forward recurrence times of an ordinary renewal process.
For $x > y/2$, determine functional relationships between $F(t)$ and the conditional probabilities (1) $P(V(t) > y \mid R(t) = x)$ and (2) $P(V(t) > y \mid R(t + y/2) = x)$.

4.12)* Show that the second moment of the forward recurrence time is given by

$$E(V^2(t)) = t^2 + \mu_2[1 + H(t)] - 2\mu\left(t + \int_0^t H(x)\,dx\right).$$

4.13) Let (Y, Z) be the typical cycle of an alternating renewal process, where Y and Z have an Erlang distribution with parameters $n = 2$ and $n = 1$, respectively, and λ. Determine the probability that the system is in state 1 over the interval $[t, t+x]$, given that it is in state 1 at time t, as t tends to infinity.

4.14) The interarrival times of claims at an insurance company can be modeled by an ordinary renewal process $\{X_1, X_2, ...\}$. The corresponding claim sizes C_i generate the ordinary renewal process $\{C_1, C_2, ...\}$ which is assumed to be independent of $\{X_1, X_2, ...\}$. Let X be the typical interarrival time and C be the typical claim size. From statistical evaluations, it is known that

$$\mu = E(X) = 2\,[h], \quad Var(X) = 3, \quad \nu = E(C) = \$\,900, \quad Var(C) = 360\,000$$

Find approximate solutions to

(1) What total minimum premium per unit time $\kappa_{\min,\alpha}$ has the insurance company to take in so that it will achieve a profit of at least $\$1000\,000$ within $10\,000$ hours with probability $\alpha = 0.95$?
(2) What is the probability that the total claim reaches level $\$4\,000\,000$ not before $10\,000$ hours?
Hint Note that the random profit of the company at time t is $G(t) = t\kappa_{\min,\alpha} - C(t)$, where $C(t)$ is the cumulative claim size at time t.

4.15) The time intervals between successive repairs of a system generate an ordinary renewal process with typical cycle length X. The cost of repairs are mutually independent and independent of the repair time points. Let C be the typical repair cost and

$$\mu = E(X) = 180, \quad \sigma = \sqrt{Var(X)} = 30 \ \text{[in } days\text{]}, \quad \nu = 200, \ \sqrt{Var(C)} = 40 \ \text{[in \$]}.$$

Determine approximately the probabilities that
(1) the cumulative repair cost arising in [0, 3600 days] does not exceed $\$4500$,
(2) a cumulative repair cost of $\$3000$ is exceeded not before 2200 days.

4.16) A system is subject to an age renewal policy with renewal interval τ as described in example 4.11.
Determine the stationary availability of the system by modeling its operation by an alternating renewal process.

4.17) A system is subject to an age renewal policy with renewal interval τ. However, unlike example 4.11, it is now assumed that the renewals occur in negligible time. On the other hand, it is assumed that preventive and emergency renewals incur constant costs c_p and c_e, respectively. Let $F(t)$ be the distribution function of the lifetime X of the system and $\lambda(t)$ the corresponding failure rate.
(1) Determine the expected total maintenance cost $K(\tau)$ per unit time for an unbounded running time of the system.
(2) Give a necessary condition for an optimal renewal interval $\tau = \tau^*$ with respect to $K(\tau)$.
(3) Determine τ^* if T is uniformly distributed over the interval $[0, T]$ and $0 < c_p < c_h$.

4.18) A system is preventively renewed at fixed time points $\tau, 2\tau, ...$. Failures between these time points are removed by emergency renewals.
(1) With the notation and assumptions of problem 4.17, determine the cost rate $K(\tau)$.
(2) Given $F(t) = (1 - e^{-\lambda t})^2, \ t \geq 0$, provide a necessary condition for a renewal interval $\tau = \tau^*$ which is optimum with respect to $K(\tau)$.

Hint Make use of the renewal function obtained in example 4.2.

5 Discrete-Time Markov Chains*

5.1 Foundations and Examples

This chapter also deals with sequences of random variables $\{X_0, X_1, ...\}$. However, the X_n are no longer assumed to be completely independent of each other. Instead, the assumption is made that, conditionally on $X_n = x_n$, the random variable X_{n+1} does not depend on $X_0, X_1, ..., X_{n-1}$; $n = 1, 2, ...$ Moreover, only sequences of discrete random variables X_n are considered.

Definition 5.1 Let $\{X_0, X_1, ...\}$ be a sequence of random variables and $\mathbf{Z} = \{0, \pm 1, \pm 2, ...\}$ be the union of the sets of their realizations. Then $\{X_0, X_1, ...\}$ is called a *discrete-time Markov chain* with state space $\mathbf{Z}$ if

$$P(X_{n+1} = i_{n+1} | X_n = i_n, ..., X_1 = i_1, X_0 = i_0) = P(X_{n+1} = i_{n+1} | X_n = i_n) \qquad (5.1)$$

holds for any $n = 1, 2, ...$ and any $i_0, i_1, ..., i_{n+1}$ with $i_k \in \mathbf{Z}$ ●

According to section 2.3, a discrete-time Markov chain is a Markov process with discrete state space $\mathbf{Z}$ and discrete parameter space $\mathbf{T}$. (Definition 5.1 assumes without loss of generality that $\mathbf{Z} = \{0, \pm 1, \pm 2, ...\}$ and $\mathbf{T} = \{0, 1, 2, ..\}$). If the time point $t = n$ is interpreted as the present, then $t = n + 1$ is a future time point and the time points $t = n - 1, ..., 1, 0$ are in the past. Thus, condition (5.1) allows the following interpretation:

The future development of a discrete Markov chain depends only on its present state, and not on the sequence of previous states.

Comment Without the condition $X_n = i_n$, X_{n+1} and X_{n-1} will, in general, be dependent random variables.

It is usually rather complicated to check whether a particular stochastic process has the Markovian property (5.1). Statistical tests, which may be found in statistical software packages, require an enormous amount of data. Hence one should first try to confirm or to reject this hypothesis using reasoning based on properties of the underlying technical, physical, economical or other practical situation. For instance, the final profit of a gambler usually depends on his present profit, but not on the way he has obtained it. If it is known that up to the end of the n-th month a supplier has sold a total of $X_n = i_n$ personal computers, then for predicting the total number of computers X_{n+1}, sold at the end of the next month, knowledge about the number of computers sold within the first n - 1 months will provide negligible information. A car driver regularly checks the tread depth of his tyres after every

5000 *km*. For predicting the tread depth after a further 5000 *km,* he will only need the present tread depth, not how the tread depth has evolved to its present value. On the other hand, for predicting the future concentration of noxious substances in the air, it has been proved necessary to take into account not only the present value of the concentration, but also the development leading to this value.

The conditional probabilities

$$p_{ij}(n) = P(X_{n+1} = j | X_n = i); \quad n = 0, 1, ...$$

are referred to as the *one-step transition probabilities* of the Markov chain.

A Markov chain is called *homogeneous* if it has homogeneous (stationary) increments (definition 2.4). Thus, the one-step transition probabilities of a homogeneous Markov chain do not depend on n:

$$p_{ij}(n) = p_{ij} \quad \text{for all} \quad n = 0, 1, ...$$

Note: This chapter only deals with homogeneous Markov chains. For the sake of brevity, the attribute *homogeneous* is generally omitted.

The one-step transition probabilities are combined in the *matrix of the one-step transition probabilities* (shortly: *transition matrix*) **P**:

$$\mathbf{P} = \begin{pmatrix} p_{00} & p_{01} & p_{02} & \cdots \\ p_{10} & p_{11} & p_{12} & \cdots \\ \cdot & \cdot & \cdot & \cdots \\ \cdot & \cdot & \cdot & \cdots \\ p_{i0} & p_{i1} & p_{i2} & \cdots \\ \cdot & \cdot & \cdot & \cdots \\ \cdot & \cdot & \cdot & \cdots \end{pmatrix}$$

p_{ij} is the probability of transition from state i to state j in one step (or, equivalently, *in one time unit, in one jump*). However, with probability p_{ii} the Markov chain remains in state i for another time unit. The one-step transition probabilities have some obvious properties:

$$\begin{aligned} & p_{ij} \geq 0; \quad i, j \in \mathbf{Z} \\ & \sum_{j \in \mathbf{Z}} p_{ij} = 1 \end{aligned} \tag{5.2}$$

The *m-step transition probabilities* of a Markov chain are defined by

$$p_{ij}^{(m)} = P(X_{n+m} = j | X_n = i); \quad m = 1, 2, ...$$

Thus, $p_{ij}^{(m)}$ is the probability of transition from state i to state j in m steps (in m time units). Note that $p_{ij} = p_{ij}^{(1)}$. Furthermore, it is convenient to define

$$p_{ij}^{(0)} = \begin{cases} 1 & \text{if } i = j \\ 0 & \text{if } i \neq j \end{cases}$$

The following relationships between transition probabilities of a Markov chain are called ***Chapman-Kolmogorov equations***:

$$p_{ij}^{(m)} = \sum_{k \in \mathbf{Z}} p_{ik}^{(r)} p_{kj}^{(m-r)}; \quad r = 0, 1, ..., m \tag{5.3}$$

The proof of (5.3) is easy: Since after r time units the Markov chain must be in some state $k \in \mathbf{Z}$, applying the total probability rule and making use of the Markovian property yields

$$\begin{aligned} p_{ij}^{(m)} &= P(X_m = j | X_0 = i) = \sum_{k \in \mathbf{Z}} P(X_m = j, X_r = k | X_0 = i) \\ &= \sum_{k \in \mathbf{Z}} P(X_m = j | X_r = k, X_0 = i) P(X_r = k | X_0 = i) \\ &= \sum_{k \in \mathbf{Z}} P(X_m = j | X_r = k) P(X_r = k | X_0 = i) \\ &= \sum_{k \in \mathbf{Z}} p_{ik}^{(r)} p_{kj}^{(m-r)} \end{aligned}$$

so that (5.3) is proved. Let

$$\mathbf{P}^{(m)} = \left(\left(p_{ij}^{(m)}\right)\right); \quad m = 0, 1, ...$$

be the *matrix of the m-step transition probabilities*. Then the Chapman-Kolmogorov equations can be written in the equivalent form

$$\mathbf{P}^{(m)} = \mathbf{P}^{(r)} \mathbf{P}^{(m-r)}; \quad r = 0, 1, ..., m$$

By induction, this relationship implies that

$$\mathbf{P}^{(m)} = \mathbf{P}^m$$

Thus, the matrix of the m-step transition probabilities is equal to the m-fold product of the matrix of the one-step transition probabilities.

A probability distribution $\mathbf{p}^{(0)}$ of X_0 is said to be an *initial distribution* of the Markov chain:

$$\mathbf{p}^{(0)} = \left\{ p_i^{(0)} = P(X_0 = i),\ i \in \mathbf{Z},\ \sum_{i \in \mathbf{Z}} p_i^{(0)} = 1 \right\} \tag{5.4}$$

A Markov chain is completely characterized by its transition matrix $\mathbf{P}$ and an initial distribution $\mathbf{p}^{(0)}$. To prove this one has to show that given $\mathbf{P}$ and $\mathbf{p}^{(0)}$ all finite-dimensional probabilities can be determined:

By the Markov property, for any finite set of states $i_0, i_1, ..., i_n$

$$P(X_0 = i_0, X_1 = i_1, ..., X_n = i_n)$$

$$= P(X_n = i_n | X_0 = i_0, X_1 = i_1, ..., X_{n-1} = i_{n-1}) \cdot P(X_0 = i_0, X_1 = i_1, ..., X_{n-1} = i_{n-1})$$

$$= P(X_n = i_n | X_{n-1} = i_{n-1}) \cdot P(X_0 = i_0, X_1 = i_1, ..., X_{n-1} = i_{n-1})$$

$$= p_{i_{n-1} i_n} \cdot P(X_0 = i_0, X_1 = i_1, ..., X_{n-1} = i_{n-1})$$

The second factor in the last row is now treated in the same way and so on so that the final result is

$$P(X_0 = i_0, X_1 = i_1, ..., X_n = i_n) = p_{i_0}^{(0)} \cdot p_{i_0 i_1} \cdot p_{i_1 i_2} \cdots p_{i_{n-1} i_n} \tag{5.5}$$

This proves the assertion.

Given an initial distribution $\mathbf{p}^{(0)} = \left\{ p_i^{(0)}, \; i \in \mathbf{Z} \right\}$, the *absolute* or *one-dimensional state probabilities* of the Markov chain after m steps

$$p_j^{(m)} = P(X_m = j), \quad j \in \mathbf{Z}$$

are obtained by applying the total probability rule:

$$p_j^{(m)} = \sum_{i \in \mathbf{Z}} p_i^{(0)} \, p_{ij}^{(m)}, \quad m = 1, 2, ... \tag{5.6}$$

Definition 5.2 An initial distribution $\{\pi_i = P(X_0 = i); \; i \in \mathbf{Z}\}$ is called *stationary* if it satisfies the system of linear equations

$$\pi_j = \sum_{i \in \mathbf{Z}} \pi_i \, p_{ij}; \quad j \in \mathbf{Z} \tag{5.7}$$

●

When comparing (5.6) and (5.7) the intuitive meaning of this definition becomes clear: The absolute state probabilities of the Markov chain after one time unit are the same as at $t = 0$. Furthermore, it can easily be shown by induction that in this case the absolute state probabilities after m steps are also the same as in the beginning:

$$p_j^{(m)} = \sum_{i \in \mathbf{Z}} \pi_i \, p_{ij}^{(m)} = \pi_j, \quad m = 1, 2, ... \tag{5.8}$$

Thus, state probabilities π_i satisfying (5.7) are time-independent absolute state probabilities, which together with **P** fully characterize a stationary probability distribution of the Markov chain. Hence they are also called *equilibrium state probabilities* of the Markov chain. Moreover, the structure of the n-dimensional state probabilities given by (5.5) verifies theorem 2.1 in this particular case:

> *A Markov chain is strictly stationary if and only if its absolute state probabilities do not depend on time.*

Example 5.1 (***random walk***) A particle moves along the real axis in one step from any integer-valued coordinate i either to $i+1$ or to i-1 with equal probabilities. The single steps are independent of each other. If X_0 denotes the starting position of the particle and X_n the position of the particle after n steps, then $\{X_0, X_1, ...\}$ is a discrete Markov chain with state space $\mathbf{Z} = \{0, \pm 1, \pm 2, ...\}$ and one-step transition probabilities

$$p_{ij} = \begin{cases} 1/2 & \text{for } j = i+1 \text{ or } j = i-1 \\ 0 & \text{otherwise} \end{cases}$$ □

Example 5.2 (***random walk with absorbing barriers***) The random walk considered in the previous example is modified in the following way: There are absorbing barriers at $x = 0$ and $x = 6$, so that if the particle arrives at one of them, it cannot leave it again. Provided that $0 < X_0 < 5$, the state space of the corresponding Markov chain $\{X_0, X_1, ...\}$ is $\mathbf{Z} = \{0, 1, ..., 6\}$ and the transition probabilities are

$$p_{ij} = \begin{cases} 1/2 & \text{for } j = i+1 \text{ or } j = i-1 \text{ and } 1 \le i \le 5 \\ 1 & \text{for } i = j = 0 \text{ or } i = j = 6 \\ 0 & \text{otherwise} \end{cases}$$

The matrices of the one and two-step transition probabilities are

$$\mathbf{P} = \begin{pmatrix} 1 & 0 & 0 & 0 & 0 & 0 & 0 \\ 1/2 & 0 & 1/2 & 0 & 0 & 0 & 0 \\ 0 & 1/2 & 0 & 1/2 & 0 & 0 & 0 \\ 0 & 0 & 1/2 & 0 & 1/2 & 0 & 0 \\ 0 & 0 & 0 & 1/2 & 0 & 1/2 & 0 \\ 0 & 0 & 0 & 0 & 1/2 & 0 & 1/2 \\ 0 & 0 & 0 & 0 & 0 & 0 & 1 \end{pmatrix}, \quad \mathbf{P}^{(2)} = \begin{pmatrix} 1 & 0 & 0 & 0 & 0 & 0 & 0 \\ 1/2 & 1/4 & 0 & 1/4 & 0 & 0 & 0 \\ 1/4 & 0 & 1/2 & 0 & 1/4 & 0 & 0 \\ 0 & 1/4 & 0 & 1/2 & 0 & 1/4 & 0 \\ 0 & 0 & 1/4 & 0 & 1/2 & 0 & 1/4 \\ 0 & 0 & 0 & 1/4 & 0 & 1/4 & 1/2 \\ 0 & 0 & 0 & 0 & 0 & 0 & 1 \end{pmatrix}$$

If the starting position of the particle X_0 is uniformly distributed over $\{1, 2, ..., 5\}$, that is if the initial distribution $p_i^{(0)} = P(X_0 = i) = 1/5$; $i = 1, 2, ..., 5$; is given, then, by (5.6), the absolute distribution of the position of the particle after 2 steps is

$$\mathbf{p}^{(2)} = \left\{ \frac{3}{20}, \frac{2}{20}, \frac{3}{20}, \frac{3}{20}, \frac{3}{20}, \frac{3}{20}, \frac{3}{20} \right\}$$ □

Example 5.3 (***random walk with reflecting barriers***) Given the finite state space $\mathbf{Z} = \{0, 1, \ldots, 2s\}$, a particle moves from position i to position j after one time unit with probability

$$p_{ij} = \begin{cases} \frac{2s-i}{2s} & \text{for } j = i+1 \\ \frac{i}{2s} & \text{for } j = i-1 \\ 0 & \text{otherwise} \end{cases} \tag{5.9}$$

Thus, the greater the distance of the particle from the central point s of $\mathbf{Z}$, the greater the probability that the particle moves in the direction of the central point after the next time unit. Once the particle has arrived at one of the end points $x=0$ or $x = 2s$, it will return after the next time unit with probability 1 to position $x = 1$ or $x = 2s$, respectively. (Hence the terminology *reflecting barriers.*) If the particle is at $x = s$, then the probabilities of moving to the left or to the right after the next time unit are equal, namely 1/2. In this sense, the particle is at $x = s$ in an *equilibrium state.* The situation may be illustrated by assuming that there is a force concentrated at the central point. Its attraction to the particle increases with the distance of the particle from the central point.
In view of this physical interpretation it is not surprising that *P.* and *T. Ehrenfest* found this random walk as early as 1907 when investigating the following diffusion model: In a closed container there are exactly $2s$ molecules of a particular type. The container is separated into two equal parts by a membrane, which is permeable to these molecules. Let X_n be the random number of the molecules in one part of the container after n transitions of any molecule from one part of the container to the other. If X_0 denotes the initial number of molecules in the specified part of the container, then *P.* and *T. Ehrenfest* observed that the random sequence $\{X_0, X_1, \ldots\}$ behaves approximately like a Markov chain with transition probabilities given by (5.9). Thus, the more molecules of the specified type there are in one part of the container, the more they want to move into the other part. In other words, the system tends to the equilibrium state, i.e. an equal number of particles in each part of the container.

The system of linear equations (5.7) for the stationary state probabilities is

$$\begin{aligned} \pi_0 &= \pi_1 p_{10} \\ \pi_j &= \pi_{j-1} p_{j-1\,j} + \pi_{j+1} p_{j+1\,j}; \quad j = 1, 2, \ldots, 2s-1 \\ \pi_{2s} &= \pi_{2s-1} p_{2s-1\,2s} \end{aligned}$$

The solution is

$$\pi_j = \binom{2s}{j} 2^{-2s}; \quad j = 0, 1, \ldots, 2s.$$

As expected, the state s has the greatest stationary probability. □

Example 5.4 Depending on its energy, an electron circles around the atomic nucleus in a trajectory from a countably infinite set of trajectories $\{1, 2, ...\}$. In one step the transition from trajectory i to trajectory j occurs with probability

$$p_{ij} = a_i e^{-b|i-j|}, \quad b > 0$$

Hence the two-step transition probabilities are

$$p_{ij}^{(2)} = a_i \sum_{k=1}^{\infty} a_k e^{-b(|i-k|+|k-j|)}$$

The a_i cannot be chosen arbitrarily. In view of (5.2) they must satisfy the conditions

$$a_i \left(e^{-b(i-1)} + e^{-b(i-2)} + \dots + e^{-b} \right) + a_i \sum_{k=0}^{\infty} e^{-bk} = 1$$

or, equivalently,

$$a_i \left(e^{-b} \frac{1 - e^{-b(i-1)}}{1 - e^{-b}} + \frac{1}{1 - e^{-b}} \right) = 1$$

Therefore,

$$a_i = \frac{e^b - 1}{1 + e^b - e^{-b(i-1)}}\,; \quad i = 1, 2, \dots$$

The structure of the p_{ij} implies that $a_i = p_{ii}$ for all $i = 1, 2, \dots$ □

Example 5.5 Customers arrive at random time points at a service station with a single server. Let Y_i denote the time between the arrival of the (i-1) th and the ith customer; $i = 1, 2, \dots$ The $Y_1, Y_2, \dots$ are assumed to be independent and identically distributed with density $g(t)$. (According to definition 4.2, the sequence $Y_1, Y_2, \dots$ is formally an ordinary renewal process.) Thus, the n th customer arrives at time

$$T_n = \sum_{i=1}^{n} Y_i\,; \quad n = 1, 2, \dots; \quad T_0 = 0$$

The service times are assumed to be independent exponential random variables with parameter μ and independent of all Y_i. A customer leaves the station immediately after completion of his service. If an arriving customer finds the server busy, he joins the queue. If no other customer is waiting, then he is first in the queue.

Let X_n denote number of customers in the station immediately before arrival of the $(n+1)$th customer, that is, the number of customers the $(n+1)$th customer finds in the station. Thus, $0 \le X_n \le n$; $n = 0, 1, , \dots$ Then $\{X_0, X_1, \dots\}$ is a discrete Markov chain with parameter space $\mathbf{T} = \{0, 1, \dots\}$ and state space $\mathbf{Z} = \{0, 1, \dots\}$ which starts in state 0 with probability 1: $P(X_0 = 0) = 1$. Furthermore, let A_n be the number of customers leaving the station in the interval $[T_n, T_{n+1})$ of length Y_{n+1}.

Then, $0 \le A_n \le X_n$ and

$$X_n = X_{n-1} - A_n + 1; \quad n = 1, 2, ...$$

According to theorem 3.2, given that $Y_{n+1} = t$ the random variable A_n is Poisson distributed with parameter μt as long as the server is busy. Thus, for $i \ge 0$ and $1 \le j \le i+1$,

$$P(X_n = j | X_{n-1} = i, \, Y_{n+1} = t) = \frac{(\mu t)^{i+1-j}}{(i+1-j)!} e^{-\mu t}; \quad n = 1, 2, ...$$

Therefore, the one-step transition probability $p_{ij} = P(X_n = j | X_{n-1} = i)$ of the Markov chain $\{X_0, X_1, ...\}$ is

$$p_{ij} = \int_0^\infty \frac{(\mu t)^{i+1-j}}{(i+1-j)!} e^{-\mu t} g(t)\, dt, \quad 1 \le j \le i+1$$

The normalizing condition $\sum_{j=0}^{i+1} p_{ij} = 1$ yields the transition probability p_{i0}:

$$p_{i0} = 1 - \sum_{j=1}^{i+1} p_{ij}$$

If $X(t)$ denotes the number of customers in the station at time t, then the stochastic process with continuous time $\{X(t), \, t \ge 0\}$ is generally not a Markov process, since the future behaviour of this process depends not only on its present state but also on the time a customer has already been served. (The only exception is the case when the service times are also exponentially distributed.) The discrete Markov chain $\{X_0, X_1, ...\}$ is *embedded* in the stochastic process $\{X(t), \, t \ge 0\}$ in the sense that $X(T_{n+1} - 0) = X_n$; $n = 0, 1, ...$ □

Example 5.6 Let X_n denote the number of traffic accidents over a period of n weeks in a particular area and let Y_i be the corresponding number in the i-th week. Then,

$$X_n = \sum_{i=1}^{n} Y_i$$

The Y_i are assumed to be independent and identically distributed as a random variable Y with

$$q_k = P(Y = k); \quad k = 0, 1, ...$$

Then $\{X_1, X_2, ...\}$ is a Markov chain with state space $\mathbf{Z} = \{0, 1, ... \}$ and transition probabilities

$$p_{ij} = \begin{cases} q_k & \text{if } j = i + k; \; k = 0, 1, ... \\ 0 & \text{otherwise} \end{cases}$$ □

Example 5.7 ***(sequence of moving averages)*** Let $\{Y_i;\ i = 0, 1, ...\}$ be a sequence of independent, identically distributed binary random variables with

$$P(Y_i = 1) = P(Y_i = -1) = 1/2$$

Let *moving averages* X_n be defined as follows (see example 2.12)

$$X_n = \tfrac{1}{2}(Y_n + Y_{n-1}); \quad n = 1, 2, ...$$

Then X_n has range {-1, 0, +1} and probability distribution

$$\left\{P(X_n = -1) = \tfrac{1}{4},\ P(X_n = 0) = \tfrac{1}{2},\ P(X_n = +1) = \tfrac{1}{4}\right\}$$

Since X_n and X_{n+m} are independent for $m > 1$, the matrix of the m-step transition probabilities $p_{ij}^{(m)} = P(X_{n+m} = j | X_n = i)$ is

$$\mathbf{P}^{(m)} = \begin{array}{c} \\ -1 \\ 0 \\ +1 \end{array} \begin{array}{c} \begin{array}{ccc} -1 & 0 & +1 \end{array} \\ \begin{pmatrix} 1/4 & 1/2 & 1/4 \\ 1/4 & 1/2 & 1/4 \\ 1/4 & 1/2 & 1/4 \end{pmatrix} \end{array}$$

The matrix of the one-step transition probabilities $p_{ij} = P(X_{n+1} = j | X_n = i)$ is

$$\mathbf{P}^{(1)} = \mathbf{P} = \begin{pmatrix} 1/2 & 1/2 & 0 \\ 1/4 & 1/2 & 1/4 \\ 0 & 1/2 & 1/2 \end{pmatrix}$$

Since obviously $\mathbf{P}^{(1)} \cdot \mathbf{P}^{(1)} \neq \mathbf{P}^{(2)}$, the Chapman-Kolmogorov equations do not hold. Therefore, $\{X_1, X_2, ...\}$ cannot be a Markov chain. □

5.2 Classification of States

5.2.1 Closed Sets of States

A subset $\mathbf{C}$ of the state space $\mathbf{Z}$ of a Markov chain is said to be *closed* if

$$\sum_{j \in \mathbf{C}} p_{ij} = 1 \quad \text{for all } i \in \mathbf{C} \tag{5.10}$$

If a Markov chain is in a closed set of states, then it cannot leave this set since (5.10) is equivalent to

$$p_{ij} = 0 \quad \text{for all } i \in \mathbf{C},\ j \notin \mathbf{C}$$

Furthermore, (5.10) even implies that

$$p_{ij}^{(m)} = 0 \quad \text{for all} \quad i \in \mathbf{C},\, j \notin \mathbf{C} \text{ and } m \geq 1 \tag{5.11}$$

If $m = 2$, then formula (5.11) can be proved as follows: From (5.3),

$$p_{ij}^{(2)} = \sum_{k \in \mathbf{C}} p_{ik} p_{kj} + \sum_{k \notin \mathbf{C}} p_{ik} p_{kj} = 0,$$

since $j \notin \mathbf{C}$ implies $p_{kj} = 0$ in the first sum and $p_{ik} = 0$ in the second sum. Formula (5.11) follows now inductively from the Chapman-Kolmogorov equations.

A closed set of states is called *minimal* if it does not contain a proper closed subset. In particular, a Markov chain is said to be *irreducible* if its state space $\mathbf{Z}$ is minimal. Otherwise the Markov chain is said to be *reducible.*

A state i is said to be absorbing if $p_{ii} = 1$. Thus, if a Markov chain has arrived in an absorbing state, it cannot leave this state again. Hence, a set of states consisting of one absorbing state is a minimal closed set of states. For instance, the states 0 and 6 of example 5.2 are absorbing.

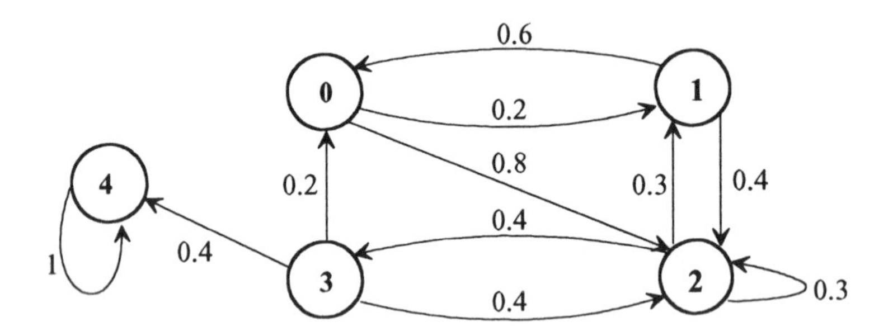

Figure 5.1 Transition graph of the Markov chain of example 5.8

Example 5.8 Let $\mathbf{Z} = \{0, 1, 2, 3, 4\}$ and

$$\mathbf{P} = \begin{pmatrix} 0 & 0.2 & 0.8 & 0 & 0 \\ 0.6 & 0 & 0.4 & 0 & 0 \\ 0 & 0.3 & 0.3 & 0.4 & 0 \\ 0.2 & 0 & 0.4 & 0 & 0.4 \\ 0 & 0 & 0 & 0 & 1 \end{pmatrix}$$

It is helpful to illustrate the possible transitions between the states of a Markov chain by *transition graphs.* The nodes of these graphs represent the states of the Markov chain. A directed edge from node i to node j exists if and only if $p_{ij} > 0$, that is, if a one-step transition from state i to state j is possible. The corresponding one-step transition probabilities are attached to the edges.

Figure 5.1 shows that {0, 1, 2, 3} is not a closed set of states since the condition (5.10) is not fulfilled for $i = 3$. State 4 is absorbing so that {4} is a minimal closed set of states. This Markov chain is, therefore, reducible. □

5.2.2 Equivalence Classes

State j is said to be *accessible* from state i (symbolically: $i \Rightarrow j$) if there exists an $m \geq 1$ such that $p_{ij}^{(m)} > 0$.

The relation " $\Rightarrow$ " is transitive: if $i \Rightarrow k$ and $k \Rightarrow j$, then there exist $m > 0$ and $n > 0$ with $p_{ik}^{(m)} > 0$ and $p_{kj}^{(n)} > 0$. Therefore,

$$p_{ij}^{(m+n)} = \sum_{r \in \mathbf{Z}} p_{ir}^{(m)} p_{rj}^{(n)} \geq p_{ik}^{(m)} p_{kj}^{(n)} > 0$$

Consequently, $i \Rightarrow k$ and $k \Rightarrow j$ imply $i \Rightarrow j$, that is, the transitivity.

The set $\mathbf{M}(i) = \{k,\ i \Rightarrow k\}$ consisting of all those states which are accessible from i is closed. In order to prove this assertion it is to show that $k \in \mathbf{M}(i)$ and $j \notin \mathbf{M}(i)$ imply that $k \not\Rightarrow j$. The proof is carried out indirectly: If under the assumptions stated $k \Rightarrow j$, then $i \Rightarrow k$ and the transitivity would imply $i \Rightarrow j$. But this contradicts the definition of $\mathbf{M}(i)$.

If both $i \Rightarrow j$ and $j \Rightarrow i$ hold, then i and j are said to *communicate* (symbolically: $i \Leftrightarrow j$).

Communication " $\Leftrightarrow$ " is an *equivalence relation* since it satisfies the three characteristic properties:

(1) $i \Leftrightarrow i$ *reflexivity*

(2) If $i \Leftrightarrow j$, then $j \Leftrightarrow i$ *commutativity*

(3) If $i \Leftrightarrow j$ and $j \Leftrightarrow k$, then $i \Leftrightarrow k$ *associativity*

Properties (1) and (2) are an immediate consequence of the definition of " $\Leftrightarrow$ ". To verify property (3) note that $i \Leftrightarrow j$ and $j \Leftrightarrow k$ imply the existence of m and n so that $p_{ij}^{(m)} > 0$ and $p_{jk}^{(n)} > 0$, respectively. Hence, by (5.3),

$$p_{ik}^{(m+n)} = \sum_{r \in \mathbf{Z}} p_{ir}^{(m)} p_{rk}^{(n)} \geq p_{ij}^{(m)} p_{jk}^{(n)} > 0$$

Likewise there exist M and N with

$$p_{ki}^{(M+N)} \geq p_{kj}^{(M)} p_{ji}^{(N)} > 0$$

so that the associativity is proved.

The state space $\mathbf{Z}$ is partitioned by the equivalence relation "$\Leftrightarrow$" into disjoint, but not necessarily closed, classes in the following way:

Two states i and j belong to the same class if and only if they communicate.

In what follows, the class containing state i will be denoted as $\mathbf{C}(i)$. Clearly, any state in a class can be used to characterize this class. All properties of states still to be introduced will be *class properties*, i.e. if state i has one of these properties, all states in $\mathbf{C}(i)$ have same property, too.

A state i is called *essential* if any state j which is accessible from i has the property that i is also accessible from j. In this case, $\mathbf{C}(i)$ is called an *essential class.*

A state i is called *inessential* if it is not essential. In this case, $\mathbf{C}(i)$ is called an *inessential class.* If i is inessential, then there exists a state j for which $i \Rightarrow j$ but not $j \Rightarrow i$.

It is easily verified that *essential* and *inessential* are indeed class properties. In example 5.8, the states 0, 1, 2 and 3 are inessential since state 4 is accessible from each of these states but none of the states 0, 1, 2 or 3 is accessible from state 4.

Theorem 5.1 (1) Essential classes are minimal closed classes.
(2) Inessential classes are not closed.

Proof (1) The assertion is a direct consequence of the definition of essential classes.
(2) If i is inessential, then there exists a state j with $i \Rightarrow j$ and $j \nRightarrow i$. Hence, $j \notin \mathbf{C}(i)$. Assuming $\mathbf{C}(i)$ is closed implies that $p_{kj}^{(m)} = 0$ for all $m \geq 1$, $k \in \mathbf{C}(i)$ and $j \notin \mathbf{C}(i)$. Therefore, $\mathbf{C}(i)$ cannot be closed. (According to the definition of the relation $i \Rightarrow j$ there exists an m with $p_{ij}^{(m)} > 0$.) ■

Let

$$p_i^{(m)}(\mathbf{C}) = \sum_{j \in \mathbf{C}} p_{ij}^{(m)}$$

$p_i^{(m)}(\mathbf{C})$ is the probability that the Markov chain starting from state i is in state set $\mathbf{C}$ after m time units. Furthermore, let $\mathbf{C}_w$ and $\mathbf{C}_u$ be the sets of all essential and inessential states of a Markov chain. The following theorem asserts that a Markov chain with finite state space which starts from an inessential state will leave the set of inessential states with probability 1 and never return (for a proof see *Chung*

(1960)). This theorem justifies the notation *essential* and *inessential states.* However, depending on the transition probabilites, the Markov chain may in the initial phase return more or less frequently to the set of inessential states if it has started there.

Theorem 5.2 Let the state space set **Z** be finite. Then,

$$\lim_{m\to\infty} p_i^{(m)}(\mathbf{C}_u) = 0 \qquad \blacksquare$$

Example 5.9 If the number of states in a Markov chain is small, the essential and inessential states can immediately be specified from the transition matrix. However, it may be useful to create a more suitable form of this matrix by rearranging its rows and colums, or, equivalently, by changing the notation of the states, For instance, consider a Markov chain with state space $\mathbf{Z} = \{0, 1, 2, 3\}$ and transition matrix

$$\mathbf{P} = \begin{pmatrix} 3/5 & 0 & 2/5 & 0 \\ 0 & 3/4 & 0 & 1/4 \\ 1/3 & 0 & 2/3 & 0 \\ 0 & 1/2 & 0 & 1/2 \end{pmatrix}$$

Then, by changing the order of rows and columns, the following equivalent representation of $\mathbf{P}$ is obtained:

$$\mathbf{P} = \begin{pmatrix} 3/5 & 2/5 & 0 & 0 \\ 1/3 & 2/3 & 0 & 0 \\ 0 & 0 & 3/4 & 1/4 \\ 0 & 0 & 1/2 & 1/2 \end{pmatrix} = \begin{pmatrix} \mathbf{Q}_{11} & \mathbf{0} \\ \mathbf{0} & \mathbf{Q}_{22} \end{pmatrix},$$

where $\mathbf{Q}_{11}$ and $\mathbf{Q}_{22}$ are square matrices of order 2 and $\mathbf{0}$ is a square matrix whose elements are all zero. Hence this Markov chain is reducible. Its state space (in new notation) consists of two essential classes $\mathbf{C}(0) = \{0, 1\}$ and $\mathbf{C}(2) = \{2, 3\}$ with transition matrices $\mathbf{Q}_{11}$ and $\mathbf{Q}_{22}$, respectively. □

Example 5.10 Let the transition matrix of a Markov chain with state space $\mathbf{Z} = \{0, 1, ..., 5\}$ be given by

$$\mathbf{P} = \begin{pmatrix} 1/3 & 2/3 & 0 & 0 & 0 & 0 \\ 1/2 & 1/2 & 0 & 0 & 0 & 0 \\ 0 & 0 & 1/3 & 2/3 & 0 & 0 \\ 0 & 0 & 2/3 & 1/3 & 0 & 0 \\ 0.4 & 0 & 0.2 & 0.1 & 0.1 & 0.2 \\ 0.1 & 0.2 & 0.1 & 0.2 & 0.3 & 0.1 \end{pmatrix} = \begin{pmatrix} \mathbf{Q}_{11} & \mathbf{0} & \mathbf{0} \\ \mathbf{0} & \mathbf{Q}_{22} & \mathbf{0} \\ \mathbf{Q}_{31} & \mathbf{Q}_{32} & \mathbf{Q}_{33} \end{pmatrix}$$

where the symbolic representation of the transition matrix uses the notation introduced in the previous example. This Markov chain has the two essential classes $\mathbf{C}(0) = \{0, 1\}$ and $\mathbf{C}(2) = \{2, 3\}$ and the inessential class $\mathbf{C}(4) = \{4, 5\}$. It is evident that from the class of inessential states transitions both to essential and inessential states are possible. However, according to theorem 5.2, the Markov chain will sooner or later leave the inessential class for one of the essential classes. □

5.2.3 Periodicity

Let d_i be the greatest common divisor of the indices $m \geq 1$ for which $p_{ii}^{(m)} > 0$. Then d_i is said to be the *period* of state *i*. If $p_{ii}^{(m)} = 0$ for all $m > 0$, then the period of *i* is defined to be infinite.

State i is said to be *aperiodic* if $d_i = 1$.

If *i* has period d_i, then $p_{ii}^{(m)} > 0$ holds if and only if *m* is of the form $m = n \cdot d_i$; $n = 1, 2, ...$ Hence, returning to state *i* is only possible after such numbers of steps which are multiple of d_i. The following theorem shows that the period is a class property.

Theorem 5.3 All states of a class have the same period.

Proof Let $i \Leftrightarrow j$. Then there exist *m* and *n* with $p_{ij}^{(m)} > 0$ and $p_{ji}^{(n)} > 0$. If the inequality $p_{ii}^{(r)} > 0$ holds for a positive integer *r*, then, in view of the Chapman-Kolmogorov equation,

$$p_{jj}^{(n+r+m)} \geq p_{ji}^{(n)} p_{ii}^{(r)} p_{ij}^{(m)} > 0$$

Since $p_{ii}^{(2r)} \geq p_{ii}^{(r)} \cdot p_{ii}^{(r)} > 0$, this inequality also holds if *r* is replaced by 2*r*:

$$p_{jj}^{(n+2r+m)} > 0$$

Thus, d_j divides the difference

$$(n + 2r + m) - (n + r + m) = r$$

Since this holds for all *r* for which $p_{ii}^{(r)} > 0$, d_j must divide d_i. Changing the roles of *i* and *j* shows that d_i also divides d_j. Thus, $d_i = d_j$. ■

Example 5.11 Let a Markov chain have state space $\mathbf{Z} = \{0, 1, ..., 6\}$ and transition matrix

$$\mathbf{P} = \begin{pmatrix} 1/3 & 2/3 & 0 & 0 & 0 & 0 & 0 \\ 1/3 & 1/3 & 1/3 & 0 & 0 & 0 & 0 \\ 1 & 0 & 0 & 0 & 0 & 0 & 0 \\ 0 & 1/3 & 0 & 1/3 & 1/3 & 0 & 0 \\ 0 & 0 & 0 & 0 & 1 & 0 & 0 \\ 0 & 0 & 0 & 0 & 0 & 1/2 & 1/2 \\ 0 & 0 & 0 & 0 & 1/2 & 0 & 1/2 \end{pmatrix}$$

Clearly, $\{0, 1, 2\}$ is a closed set of essential states. State 4 is absorbing, so $\{4\}$ is another closed set. Having once arrived in a closed set of states the Markov chain cannot leave it again. $\{3, 5, 6\}$ is a set of inessential states. When starting in its set of inessential states the Markov chain will at some stage leave this set and never return since it will be caught by a closed set of states. All states in $\{0, 1, 2\}$ have period 1. □

Theorem 5.4 (*Chung* (1960) The state space $\mathbf{Z}$ of an irreducible Markov chain with period $d > 1$ can be partitioned into disjoint subsets $\mathbf{Z}_1, \mathbf{Z}_1, ..., \mathbf{Z}_d$ with

$$\mathbf{Z} = \bigcup_{k=1}^{d} \mathbf{Z}_k$$

such that from any state $i \in \mathbf{Z}_k$ a transition can only be made to a state $j \in \mathbf{Z}_{k+1}$ ($j \in \mathbf{Z}_1$ if $i \in \mathbf{Z}_d$). ■

This theorem implies a characteristic structure of the transition matrix of a periodic Markov chain. For instance, if $d = 3$, then the transition matrix $\mathbf{P}$ looks like

$$\mathbf{P} = \begin{matrix} \mathbf{Z}_1 \\ \mathbf{Z}_2 \\ \mathbf{Z}_3 \end{matrix} \overset{\begin{matrix} \mathbf{Z}_1 & \mathbf{Z}_2 & \mathbf{Z}_3 \end{matrix}}{\begin{pmatrix} \mathbf{0} & \mathbf{Q}_1 & \mathbf{0} \\ \mathbf{0} & \mathbf{0} & \mathbf{Q}_2 \\ \mathbf{Q}_3 & \mathbf{0} & \mathbf{0} \end{pmatrix}}$$

where $\mathbf{P}$ may be rotated by 90^0. ($\mathbf{Q}_i$ and $\mathbf{0}$ refer to the notation introduced in example 5.9.) According to the definition of a period, if a Markov chain with period d starts in $\mathbf{Z}_i$, it will again be in $\mathbf{Z}_i$ after d transitions. Hence the corresponding d-step transition matrix is given by

$$\mathbf{P}^{(d)} = \begin{matrix} \mathbf{Z}_1 \\ \mathbf{Z}_2 \\ \mathbf{Z}_3 \end{matrix} \overset{\begin{matrix} \mathbf{Z}_1 & \mathbf{Z}_2 & \mathbf{Z}_3 \end{matrix}}{\begin{pmatrix} \mathbf{R}_1 & \mathbf{0} & \mathbf{0} \\ \mathbf{0} & \mathbf{R}_2 & \mathbf{0} \\ \mathbf{0} & \mathbf{0} & \mathbf{R}_3 \end{pmatrix}}$$

This structure of the transition matrix allows the following interpretation: A Markov chain $\{X_0, X_1, ...\}$ with period d becomes a Markov chain with period 1 and closed equivalence classes $\mathbf{Z}_1, \mathbf{Z}_2, ..., \mathbf{Z}_d$ if, with respect to transitions within the Markov chain $\{X_0, X_1, ...\}$, only the states after every d steps are registered.

Example 5.12 Let a Markov chain have state space $\mathbf{Z} = \{0, 1, ..., 5\}$ and transition matrix

$$\mathbf{P} = \begin{pmatrix} 0 & 0 & 2/5 & 3/5 & 0 & 0 \\ 0 & 0 & 1 & 0 & 0 & 0 \\ 0 & 0 & 0 & 0 & 1/2 & 1/2 \\ 0 & 0 & 0 & 0 & 2/3 & 1/3 \\ 1/2 & 1/2 & 0 & 0 & 0 & 0 \\ 1/4 & 3/4 & 0 & 0 & 0 & 0 \end{pmatrix}$$

This Markov chain has period 3. One-step transitions are possible in the following order:

$$\mathbf{Z}_1 = \{0, 1\} \to \mathbf{Z}_2 = \{2, 3\} \to \mathbf{Z}_1 = \{4, 5\} \to \mathbf{Z}_1$$

The 3-step transition matrix is

$$\mathbf{P}^{(3)} = \begin{pmatrix} 2/5 & 3/5 & 0 & 0 & 0 & 0 \\ 3/8 & 5/8 & 0 & 0 & 0 & 0 \\ 0 & 0 & 31/40 & 9/40 & 0 & 0 \\ 0 & 0 & 3/4 & 1/4 & 0 & 0 \\ 0 & 0 & 0 & 0 & 11/20 & 9/20 \\ 0 & 0 & 0 & 0 & 21/40 & 19/40 \end{pmatrix}$$

5.2.4 Recurrence and Transience

This section deals with the return of a Markov chain to an initial state. This requires the definition of the *first-passage time probabilities*

$$f_{ij}^{(m)} = P(X_m = j;\, X_k \neq j;\, k = 1, 2, ..., m-1 \,|\, X_0 = i);\;\; i, j \in \mathbf{Z}$$

Thus, $f_{ij}^{(m)}$ is the probability that the Markov chain, starting from state i, makes its first transition into state j after m steps. Note that $p_{ij}^{(m)}$ is the probability that the Markov chain, starting from state i, is in state j after m steps, but it may have been in state j in between. Consequently,

$$f_{ij}^{(1)} = p_{ij}^{(1)} = p_{ij}$$

The total probability rule yields a relationship between the m-step transition probabilities and the first-passage time probabilities:

$$p_{ij}^{(m)} = \sum_{k=1}^{m} f_{ij}^{(k)} p_{jj}^{(m-k)}, \tag{5.12}$$

where, by convention, $p_{jj}^{(0)} = 1$ for all $j \in \mathbf{Z}$. Thus, the first- passage time probabilities can be determined recursivley from the following formula:

$$f_{ij}^{(m)} = p_{ij}^{(m)} - \sum_{k=1}^{m-1} f_{ij}^{(k)} p_{jj}^{(m-k)}; \quad m = 2, 3, ... \tag{5.13}$$

The random variable which has probability distribution $\left\{f_{ij}^{(m)}; m = 1, 2, ...\right\}$ is denotated as Y_{ij}. It is called a *first-passage time*. By definition,

$$\mu_{ij} = E(Y_{ij}) = \sum_{m=1}^{\infty} m f_{ij}^{(m)}$$

The probability of ever making a transition into state j when the process starts in state i is

$$f_{ij} = \sum_{m=1}^{\infty} f_{ij}^{(m)} \tag{5.14}$$

In particular, f_{ii} is the probability of ever returning to state i. This motivates the following important definition.

Definition 5.3 State i is *recurrent* if $f_{ii} = 1$ and *transient* if $f_{ii} < 1$.

Clearly, if state i is transient, then $\mu_{ii} = \infty$. But, if i is recurrent, then $\mu_{ii} = \infty$ is also possible. Therefore, recurrent states are classified as follows:

> A recurrent state i is said to be *positive recurrent* if $\mu_{ii} < \infty$ and *null-recurrent* if $\mu_{ii} = \infty$.
>
> An aperiodic and positive recurrent state is called *ergodic*.

The random time points $T_{i,n}$; $n = 1, 2, ...$; at which the n *th* return occurs in the starting state i, are *regeneration points* of the Markov chain (see definition 4.8). By convention, $T_{i,0} = 0$. The corresponding times between neighbouring regeneration points $T_{i,n} - T_{i,n-1}$; $n = 1, 2, ...$; are called *recurrence times*. (Note the difference from the concept of recurrence times defined in section 4.3.) They are independent and identically distributed as Y_{ii}. Hence the random sequence of recurrence times is an ordinary renewal process.

Let

$$N_i(t) = \max\left(n;\ T_{i,n} \le t\right),\quad H_i(t) = E(N_i(t))$$

and

$$N_i(\infty) = \lim_{t\to\infty} N_i(t),\quad H_i(\infty) = \lim_{t\to\infty} H_i(t)$$

Theorem 5.5 State i is recurrent if and only if at least one of the following conditions holds:

(1) $H_i(\infty) = \infty$

(2) $\sum_{m=1}^{\infty} p_{ii}^{(m)} = \infty$

Proof (1) If i is recurrent, then $P(T_{i,n} = \infty) = 0$ for $n = 1, 2, \ldots$. $N_i(\infty)$ is finite if and only if there exists an n with $T_{i,n} = \infty$. Therefore,

$$P(N_i(\infty) < \infty) \le \sum_{n=1}^{\infty} P(T_{i,n} = \infty) = 0$$

Thus, assumption $f_{ii} = 1$ implies that $N_i(\infty) = \infty$ and, therefore, $H_i(\infty) = \infty$ holds with probability 1. On the other hand, if $f_{ii} < 1$, then a return to state i will not occur with positive probability $1 - f_{ii}$. In this case, $N_i(\infty)$ is geometrically distributed with expected value $E(N_i(\infty)) = H_i(\infty) = f_{ii}/(1 - f_{ii})$. Both results together are equivalent to assertion (1).

(2) Let

$$I_{m,i} = \begin{cases} 1 & \text{for } X_m = i \\ 0 & \text{for } X_m \ne i \end{cases};\quad m = 1, 2, \ldots$$

be indicator variables for the random events "$X_m = i$", that is, for the events that the Markov chain is in state i at time $t = m$. Then,

$$N_i(\infty) = \sum_{m=1}^{\infty} I_{m,i}$$

Hence,

$$\begin{aligned} H_i(\infty) &= E\left(\ \sum_{m=1}^{\infty} I_{m,i}\right) \\ &= \sum_{m=1}^{\infty} E(I_{m,i}) = \sum_{m=1}^{\infty} P(I_{m,i} = 1) \\ &= \sum_{m=1}^{\infty} p_{ii}^{(m)} \end{aligned}$$

The assertion now follows from (1). ■

By adding up both sides of (5.13) from $m = 1$ to ∞ and changing the order of summation according to formula (1.55), theorem 5.5 yields the following corollary.

Corollary If state j is transient, then, for any $i \in \mathbf{Z}$,

$$\sum_{m=1}^{\infty} p_{ij}^{(m)} < \infty \quad \text{and, therefore,} \quad \lim_{m\to\infty} p_{ij}^{(m)} = 0$$

Theorem 5.6 Let i be a recurrent state and $i \Leftrightarrow j$. Then state j is also recurrent.

Proof By definition of the equivalence relation "$\Leftrightarrow$" there exist m and n with

$$p_{ij}^{(m)} > 0 \ \text{ and } \ p_{ji}^{(n)} > 0$$

By (5.3),

$$p_{jj}^{n+r+m} \geq p_{ji}^{(n)} p_{ii}^{(r)} p_{ij}^{(m)}$$

so that

$$\sum_{r=1}^{\infty} p_{jj}^{n+r+m} \geq p_{ij}^{(m)} p_{ji}^{(n)} \sum_{r=1}^{\infty} p_{ii}^{(r)} = \infty$$

The assertion is now a consequence of theorem 5.5. ■

Comment Because of theorem 5.6, recurrence and transience are class properties. Hence, an irreducible Markov chain is either *recurrent* or *transient*.

The following statement is elementary, but important.

An irreducible Markov chain with finite state space is recurrent.

It is easy to see that an inessential state is transient. Therefore, each recurrent state is essential. But not each essential state is recurrent. This assertion is proved by the following example.

Example 5.13 (*unbounded random walk*) Starting from 0 a particle jumps a unit distance along the x-axis to the right with probability p or to the left with probability $1-p$. The transitions occur independently of each other. Let X_n denote the location of the particle after the n th jump. Then the Markov chain $\{X_0, X_1, X_2, ...\}$ with $X_0 = 0$ has period $d = 2$. Thus,

$$p_{00}^{(2m+1)} = 0; \quad m = 0, 1, ...$$

In order to be again in state $x = 0$ after $2m$ steps, the particle must jump m times to the left and m times to the right. There are $\binom{2m}{m}$ sample paths which satisfy this condition. Hence,

$$p_{00}^{(2m)} = \binom{2m}{m} p^m (1-p)^m; \quad m = 1, 2, ...$$

Making use of the well-known series

$$\sum_{i=0}^{\infty} \binom{2i}{i} x^i = \frac{1}{\sqrt{1-4x}}, \quad 0 \le x < 1/4,$$

yields

$$\sum_{m=1}^{\infty} p_{00}^{(m)} = \sum_{m=1}^{\infty} \binom{2m}{m} [p(1-p)]^m$$

$$= \frac{1}{|1-2p|} - 1, \quad p \ne 1/2$$

This sum is bounded if $p \ne 1/2$. In this case state 0 is transient (theorem 5.5). Thus, since the Markov chain is irreducible, it is also transient (theorem 5.6).

If $p = 1/2$ (*symmetric random walk*, example 5.1) , then

$$\sum_{m=1}^{\infty} p_{00}^{(m)} = \lim_{p \to 1/2} \frac{1}{|1-2p|} - 1 = \infty,$$

so that in this case all states are recurrent. However, in both cases, $p \ne 1/2$ and $p = 1/2$, all states are essential since there is always a positive probability of making a transition to any state irrespective of the starting position. □

The symmetric random walk along a straight line can be easily generalized to n-dimensional Euclidian spaces: In the plane, the particle jumps one unit to the West, South, East, or North, respectively, each with probability 1/4. In the 3-dimensional Euclidian space, the particle jumps one unit to the West, South, East, North, upward, or downward, respectively, each with probability 1/6. When analyzing these random walks analogously to the one-dimensional case, an interesting phenomenon is obtained: the symmetric two-dimensional random walk (more exactly, the underlying Markov chain) is recurrent like the one-dimensional symmetric random walk, but all n- dimensional symmetric random walks with $n > 2$ are transient. Thus, somebody who chooses one of the six possibilities in a 3-dimensional labyrinth at random, will with positive probability never return to his starting position.

Example 5.14 A particle jumps from $x = i$; $i = 0, 1, 2, ..$; to $x = 0$ with probability p_i or to $i+1$ with probability $1-p_i$, $0 < p_i < 1$. The jumps are independent of each other.

Let X_n denote the position of the particle after the n-th jump. Then the transition matrix of the Markov chain $\{X_0, X_1, ...\}$ is given by

$$\mathbf{P} = \begin{pmatrix} p_0 & 1-p_0 & 0 & 0 & 0 & . & 0 & 0 & . & . \\ p_1 & 0 & 1-p_1 & 0 & 0 & . & 0 & 0 & . & . \\ p_2 & 0 & 0 & 1-p_2 & 0 & . & 0 & 0 & . & . \\ . & . & . & . & . & . & 0 & 0 & . & . \\ p_i & 0 & . & . & 0 & . & 1-p_i & 0 & . & . \\ . & . & . & . & . & . & . & . & . & . \\ . & . & . & . & . & . & . & . & . & . \end{pmatrix}.$$

The Markov chain $\{X_0, X_1, ...\}$ is, therefore, irreducible and aperiodic. Hence, to find the conditions under which this Markov chain is recurrent or transient, respectively, it is sufficient to consider state 0, say. It is not difficult to determine $f_{00}^{(m)}$:

$$f_{00}^{(1)} = p_0$$

$$f_{00}^{(m)} = \left(\prod_{i=0}^{m-2}(1-p_i)\right)p_{m-1}; \quad m = 2, 3, ...$$

If p_{m-1} is replaced by $(1-(1-p_{m-1}))$, then $f_{00}^{(m)}$ becomes

$$f_{00}^{(m)} = \left(\prod_{i=0}^{m-2}(1-p_i)\right) - \left(\prod_{i=0}^{m-1}(1-p_i)\right); \quad m = 2, 3, ...$$

Hence,

$$\sum_{n=1}^{m+1} f_{00}^{(n)} = 1 - \left(\prod_{i=0}^{m}(1-p_i)\right), \quad m = 1, 2, ...$$

Thus, state 0 is recurrent if and only if

$$\lim_{m\to\infty} \prod_{i=0}^{m}(1-p_i) = 0 \tag{5.15}$$

Assertion Condition (5.15) is true if and only if

$$\sum_{i=0}^{\infty} p_i = \infty \tag{5.16}$$

Proof Since $1 - p_i \le e^{-p_i}$; $i = 0, 1, ...$,

$$\prod_{i=0}^{m}(1-p_i) \le \exp\left(-\sum_{i=0}^{m} p_i\right)$$

Letting $m \to \infty$ in this inequality proves that (5.15) follows from condition (5.16).

The converse direction is proved indirectly: The assumption that (5.15) holds but (5.16) does not imply the existence of a positive integer k satisfying

$$0 < \sum_{i=k}^{m} p_i < 1$$

It can easily be shown by induction that

$$\prod_{i=k}^{m}(1-p_i) > 1 - p_k - p_{k+1} - \dots - p_m = 1 - \sum_{i=k}^{m} p_i$$

Therefore,

$$\lim_{m\to\infty} \prod_{i=k}^{m}(1-p_i) > \lim_{m\to\infty}\left(1 - \sum_{i=k}^{m} p_i\right) > 0$$

This contradicts the assumption that condition (5.15) holds so that the proof of the assertion is completed.

Thus, state 0 and with it the Markov chain are recurrent if and only if condition (5.16) holds. This is the case, for instance, if $p_i = p > 0;\ i = 0, 1, 2, \dots$ □

5.3 Limit Theorems and Stationary Distribution

Theorem 5.7 Let state i and j communicate, i.e. $i \Leftrightarrow j$. Then,

$$\lim_{n\to\infty} \frac{1}{n} \sum_{m=1}^{n} p_{ij}^{(m)} = \frac{1}{\mu_{jj}} \tag{5.17}$$

Proof Analogously to the proof of theorem 5.5 it can be shown that, given the Markov chain is in state i at time $t = 0$,

$$\sum_{m=1}^{n} p_{ij}^{(m)}$$

is equal to the expected number of transitions into state j in the time interval $(0, n]$. The theorem is, therefore, a direct consequence of the elementary renewal theorem (theorem 4.2). (If $i \neq j$, the corresponding renewal process is delayed.) ■

Theorem 5.7 holds even if the sequence $\{p_{ij}^{(m)};\ m = 1, 2, \dots\}$ has no limit. For instance, if $p_{ij}^{(1)} = 1,\ p_{ij}^{(2)} = 0,\ p_{ij}^{(3)} = 1, \dots$, then this sequence has no limit. However,

$$\lim_{n\to\infty} \frac{1}{n} \sum_{m=1}^{n} p_{ij}^{(m)} = \frac{1}{2}$$

But, if the limit $\lim_{m\to\infty} p_{ij}^{(m)}$ exists, then it coincides with the right hand side of (5.17). (This can easily be proved indirectly.) Since it can be shown that in case of an irreducible Markov chain the limits $\lim_{m\to\infty} p_{ij}^{(m)}$ exist for all $i,j \in \mathbf{Z}$, theorem 5.7 implies theorem 5.8.

Theorem 5.8 Let $p_{ij}^{(m)}$ be the m-step transition probabilities of an irreducible, aperiodic Markov chain. Then, for all $i,j \in \mathbf{Z}$,

$$\lim_{m\to\infty} p_{ij}^{(m)} = \frac{1}{\mu_{jj}}$$

In particular, if j is transient or null-recurrent, then

$$\lim_{m\to\infty} p_{ij}^{(m)} = 0 \qquad \blacksquare$$

Theorems 5.4 and 5.8 imply the following corollary.

Corollary For an irreducible Markov Chain with period d,

$$\lim_{m\to\infty} p_{ij}^{(md)} = \frac{d}{\mu_{jj}}$$

Theorem 5.9 Given an irreducible, aperiodic Markov chain, there exist two possibilities:

(1) If the Markov chain is transient or null-recurrent, then stationary distributions do not exist.

(2) If the Markov chain is positive recurrent, then there exists a unique stationary distribution $\{\pi_j, j \in \mathbf{Z}\}$, which, for any $i \in \mathbf{Z}$, is given by

$$\pi_j = \lim_{m\to\infty} p_{ij}^{(m)} = \frac{1}{\mu_{jj}}$$

Proof Without loss of generality, let $\mathbf{Z} = \{0, 1, ...\}$.

(1) By (5.8), a stationary distribution $\{p_j; j = 0, 1, ...\}$ satisfies for any $m = 1, 2, ...$ the system of linear algebraic equations

$$p_j = \sum_{i=0}^{\infty} p_i\, p_{ij}^{(m)}, \quad m = 1, 2, ... \tag{5.18}$$

If $\lim_{m\to\infty} p_{ij}^{(m)} = 0$, then there cannot exist a probability distribution $\{p_i; i = 0, 1, ...\}$ which is solution of this system of linear equations.

(2) Next the existence of a stationary distribution is shown. For $M < \infty$, any $i \in \mathbf{Z}$, and any $m = 1, 2, ...$,

$$\sum_{j=0}^{M} p_{ij}^{(m)} \le \sum_{j=0}^{\infty} p_{ij}^{(m)} = 1$$

Passing to the limit as $m \to \infty$ yields $\sum_{j=0}^{M} \pi_j \le 1$ for all M. Therefore,

$$\sum_{j=0}^{\infty} \pi_j \le 1 \tag{5.19}$$

Analogously, from

$$p_{ij}^{(m+1)} = \sum_{k=0}^{\infty} p_{ik}^{(m)} p_{kj} \ge \sum_{k=0}^{M} p_{ik}^{(m)} p_{kj}$$

it follows that

$$\pi_j \ge \sum_{k=0}^{\infty} \pi_k p_{kj} \tag{5.20}$$

Assuming there exists at least one state j for which (5.20) is a proper inequality, then summing inequalities (5.20) over all j yields

$$\sum_{j=0}^{\infty} \pi_j > \sum_{j=0}^{\infty} \sum_{k=0}^{\infty} \pi_k p_{kj} = \sum_{k=0}^{\infty} \pi_k \sum_{j=0}^{\infty} p_{kj}$$

$$= \sum_{k=0}^{\infty} \pi_k$$

But this is a contradiction to (5.19). Hence,

$$\pi_j = \sum_{k=0}^{\infty} \pi_k p_{kj}; \quad j = 0, 1, ...$$

Thus, at least one stationary distribution exists, namely $\{p_j ; j = 0, 1, ...\}$ where

$$p_j = \frac{\pi_j}{\sum_{i=0}^{\infty} \pi_i}, \quad j \in \mathbf{Z}$$

Letting $m \to \infty$ in (5.18) for any stationary distribution $\{p_j ; j = 0, 1, ...\}$, then, from theorem 5.8,

$$p_j = \sum_{i=0}^{\infty} p_i \pi_j = \pi_j \sum_{i=0}^{\infty} p_i = \pi_j, \quad j \in \mathbf{Z}$$

Thus, $\{\pi_j ; j = 0, 1, ...\}$ with $\pi_j = 1/\mu_{jj}$ is the only stationary distribution. ■

Example 5.15 A particle moves along the real axis. Starting from a position (state) $i = 1, 2, ...$, it jumps at the next time unit to state $i + 1$ with probability p and to state $i - 1$ with probability $q = 1 - p$. When the particle arrives at state 0, it remains there for a further time unit with probability q or jumps to state 1 with probability p. Let X_n denote the position of the particle after the n-th jump (time unit). Under which conditions has the Markov chain $\{X_0, X_1, ...\}$ a stationary distribution?

Since $p_{00} = q,\ p_{i\,i+1} = p$ and $p_{i\,i-1} = q;\ i = 1, 2, ...;\ q = 1 - p$, the system of linear equations (5.7) is

$$\pi_0 = \pi_0 q + \pi_1 q$$
$$\pi_i = \pi_{i-1} p + \pi_{i+1} q;\quad i = 1, 2, ...$$

Solving this system of equations recursively yields

$$\pi_i = \left(\frac{p}{q}\right)^i \pi_0;\quad i = 0, 1, ...$$

To guarantee that $\Sigma_{i=0}^{\infty} \pi_i = 1$, the condition $p < q$, or, equivalently, $p < 1/2$, must hold. In this case,

$$\pi_i = \frac{q-p}{q}\left(\frac{p}{q}\right)^i;\quad i = 0, 1, ...$$

The necessary condition $p < 1/2$ for the existence of a stationary distribution is intuitive, since otherwise the particle would tend to drift to infinity. But then no time-invariant behaviour of the Markov chain can be expected. □

Theorem 5.10 Let $\{X_0, X_1, ...\}$ be an irreducible, recurrent Markov chain with state space **Z** and stationary state probabilities $\pi_i,\ i \in \mathbf{Z}$. If $g(i)$ is any bounded function on **Z**, then

$$\lim_{n\to\infty} \frac{1}{n} \sum_{j=0}^{n} g(X_j) = \sum_{i\in\mathbf{Z}} \pi_i\, g(i)$$ ■

For example, if $c_i = g(i)$ denotes the profit which accrues from the Markov chain making a transition to state i, then $\Sigma_{i\in\mathbf{Z}} \pi_i c_i$ is the expected profit resulting from a state change of the Markov chain. Thus, theorem 5.10 is the analogue to formula (4.46) which refers to cumulative stochastic processes. A proof of theorem 5.10 under weaker assumptions can be found in *Tijms* (1994).

In particular, let

$$g(i) = \begin{cases} 1 & \text{for } i = k \\ 0 & \text{for } i \neq k \end{cases}$$

If, as generally assumed in this chapter, changes of state of the Markov chain occur after unit time intervals, then the limit

$$\lim_{n\to\infty} \frac{1}{n} \Sigma_{j=0}^{n} g(X_j)$$

is equal to the mean percentage of time the system is in state k. According to theorem 5.10, this percentage coincides with π_k. This property of the stationary state probabilities illustrates once more that they refer to an equilibrium state of the Markov chain.

Example 5.16 A system can be in one of the three states 1, 2, and 3: In state 1 it operates most efficiently. In state 2 it is still working but its efficiency is lower than in state 1. State 3 is the *down state*, the system is no longer operating and possibly has to be maintained. State changes can only occur after a fixed time unit of length, say, 1. Transitions into the same state are allowed. If $X(t)$ denotes the state of the system at time t, then $\{X(t), t \geq 0\}$ is assumed to be a Markov chain with transition matrix

$$\mathbf{P} = \begin{matrix} \\ 1 \\ 2 \\ 3 \end{matrix} \begin{matrix} \begin{matrix} 1 & \;\;2 & \;\;3 \end{matrix} \\ \begin{pmatrix} 0.8 & 0.1 & 0.1 \\ 0 & 0.6 & 0.4 \\ 0.8 & 0 & 0.2 \end{pmatrix} \end{matrix}$$

Note that from state 3 the system most probably makes a transition to state 1, but it may also stay in state 3 for one ore more time units (for example, if the maintenance measure has not been successful). The corresponding stationary state probabilities satisfy the system of linear equations

$$\pi_1 = 0.8\pi_1 \qquad\qquad +0.8\pi_3$$

$$\pi_2 = 0.1\pi_1 + 0.6\pi_2$$

$$\pi_3 = 0.1\pi_1 + 0.4\pi_2 + 0.2\pi_3$$

Only two of these equations are linearly independent. However, together with the normalizing constraint

$$\pi_1 + \pi_2 + \pi_3 = 1$$

the unique solution is

$$\pi_1 = \frac{4}{6}, \quad \pi_2 = \pi_3 = \frac{1}{6} \tag{5.21}$$

Let the system make profits $g(1) = \$1000$ and $g(2) = \$600$ per unit time in states 1 and 2, respectively, whereas it causes a loss of $g(3) = -\ \$100$ per unit time in state 3. According to theorem 5.10, after an infinite ($\approx$sufficiently long) running time of the system the expected profit per unit time is

$$\Sigma_{i=1}^{3}\, \pi_i\, g(i) = 1000 \cdot \frac{4}{6} + 600 \cdot \frac{1}{6} - 100 \cdot \frac{1}{6} = 250 \;\; [\$ \text{ per unit time}]$$

Now let Y be the random time during which the system is in the profitable states 1 and 2 (according to the structure of the transition matrix, such a time period always begins with state 1), and let Z be the random time during which the system is in the unprofitable state 3. The expected values $E(Y)$ and $E(Z)$ are to be determined. The random vector (Y, Z) can be interpreted as the typical cycle of an alternating renewal process. Therefore, by (4.43), the quotient $E(Y)/[E(Y) + E(Z)]$ is equal to the expected percentage of time the system is in states 1 or 2. As pointed out after theorem 5.10, this percentage must be equal to $\pi_1 + \pi_2$:

$$\frac{E(Y)}{E(Y)+E(Z)} = \pi_1 + \pi_2 \tag{5.22}$$

Since the expected time between transitions into state 3 is equal to $E(Y) + E(Z)$, the quotient $1 / [E(Y) + E(Z)]$ is equal to the rate of the occurence of transitions to state 3. On the other hand, this rate is also given by $\pi_1 p_{13} + \pi_2 p_{23}$. Hence,

$$\frac{1}{E(Y)+E(Z)} = \pi_1 p_{13} + \pi_2 p_{23} \tag{5.23}$$

From (5.22) and (5.23),

$$E(Y) = \frac{\pi_1 + \pi_2}{\pi_1 p_{13} + \pi_2 p_{23}}, \quad E(Z) = \frac{\pi_3}{\pi_1 p_{13} + \pi_2 p_{23}}$$

Substituting the numerical values (5.21) yields

$$E(Y) = 6.25 \quad \text{and} \quad E(Z) = 1.25$$ □

5.4 Birth- and Death Processes

In the examples considered so far frequently only direct transitions between "neighbouring " states were possible. More exactly, starting with state i, only transitions to states i - 1 or i +1 could be made in one step. In these cases the one-step transition probabilities have the principal structure

$$p_{i\,i+1} = p_i, \quad p_{i\,i-1} = q_i, \quad p_{i\,i} = r_i \quad \text{with} \quad p_i + q_i + r_i = 1 \tag{5.24}$$

Discrete Markov chains with such transition probabilities and state space $\mathbf{Z} = \{0, 1, ...\}$ are called *birth- and death processes.* (The state space implies that $q_0 = 0$.) The unbounded random walk considered in example 5.15 is a special birth- and death process with $p_i = p$ for $i = 0, 1, ...,$ $q_0 = 0$, $r_0 = q = 1 - p$, $q_i = q$ and $r_i = 0$ for $i = 1, 2, ...$ The random walk considered in example 5.13 also makes transitions only to neighbouring states, but its state space is $\mathbf{Z} = \{... -1, 0, +1, ...\}$.

Example 5.17 (***random walk with absorbing barriers***) A random walk with absorbing barriers 0 and s can be modeled by a birth- and death process. In addition to (5.24), its transition probabilities satisfy the conditions

$$r_0 = r_s = 1; \quad p_i > 0 \text{ and } q_i > 0 \text{ for } i = 1, 2, ..., s-1 \tag{5.25}$$

Let $p(k)$ be the probability that the random walk arrives at state 0 when starting from state k; $k = 1, 2, ..., s$-1. (Since s is absorbing the Markov chain cannot have been in this state before arriving at 0.) In view of the total probability rule,

$$p(k) = p_k\, p(k+1) + q_k\, p(k-1) + r_k\, p(k)\,,$$

or, replacing r_k by $r_k = 1 - p_k - q_k$,

$$p(k) - p(k+1) = \frac{q_k}{p_k}\,[p(k-1) - p(k)]\,; \quad k = 1, 2, ..., s-1$$

Repeated application of this difference equation yields

$$p(j) - p(j+1) = Q_j\,[p(0) - p(1)]\,; \;\; j = 0, 1, ..., s-1, \tag{5.26}$$

where $p(0) = 1$, $p(s) = 0$ and

$$Q_j = \frac{q_j\, q_{j-1} \cdots q_1}{p_j\, p_{j-1} \cdots p_1}\,; \quad j = 1, 2, ..., s-1\,; \quad Q_0 = 1$$

Summing the equations (5.26) from $j = k$ to $j = s-1$ yields

$$p(k) = \sum_{j=k}^{s-1} [p(j) - p(j+1)] = [p(0) - p(1)] \sum_{j=k}^{s-1} Q_j$$

In particular, for $k = 0$,

$$1 = [p(0) - p(1)] \sum_{j=0}^{s-1} Q_j$$

Combining the last two equations yields the desired probabilities to be

$$p(k) = \frac{\sum_{j=k}^{s-1} Q_j}{\sum_{j=0}^{s-1} Q_j}\,; \quad k = 0, 1, ..., s-1\,; \quad p(s) = 0 \tag{5.27}$$

Besides the interpretation of this birth- and death process as a random walk with absorbing barriers, the following application may be more interesting: Two gamblers begin a game with stakes of $\$k$ and $\$(s-k)$, respectively; k, s integers. After each move a gambler either wins or loses $1 or his stake remains constant. These possibilities are governed by transition probabilities satisfying (5.24) and (5.25). The game is finished if a gambler has won the entire stake of the other one or, equivalently, if one gambler has lost his entire stake. Hence this birth- and death process is also called *gambler's ruin problem.* □

To ensure that a birth- and death process is irreducible, assumptions (5.24) have to be supplemented by

$$p_i > 0 \text{ for } i = 0, 1, ... \quad \text{and} \quad q_i > 0 \text{ for } i = 1, 2, ... \tag{5.28}$$

Theorem 5.11 Under the additional assumptions (5.28) on the transition probabilities, a birth- and death process is recurrent if and only if

$$\sum_{j=1}^{\infty} \frac{q_j\, q_{j-1} \cdots q_1}{p_j\, p_{j-1} \cdots p_1} = \infty \tag{5.29}$$

Proof It is sufficient to show that state 0 is recurrent. This can be established by using the result (5.27) of example 5.17 since

$$\lim_{s \to \infty} p(k) = f_{k0}\,; \quad k = 1, 2, \ldots,$$

where the first-passage time probabilities f_{k0} are given by (5.14). If state 0 is recurrent, then it follows from the irreducibility of the Markov chain that $f_{00} = 1$ and $f_{k0} = 1$. However, $f_{k0} = 1$ holds if and only if (5.29) is valid.

Conversely, let (5.29) be true. Then, by the total probability rule,

$$\begin{aligned} f_{00} &= p_{00} + p_{01} f_{10} \\ &= r_0 + p_0 \cdot 1 \\ &= 1 \end{aligned}$$

This result proves the theorem. ∎

The notation *birth- and death process* results from the application of these processes to describing the development of biological populations. In this context, X_n is the number of individuals of a population after n time units assuming that the population does not increase or decrease, respectively, at more than one individual per unit time. Correspondingly, the p_i and q_i are respective called *birth-* or *death probabilities.* However, for practical purposes birth- and death processes are much more important in the case of continuous-time Markov chains (section 6.6).

Exercises

5.1) A Markov chain $\{X_0, X_1, \ldots\}$ has state space $\mathbf{Z} = \{0, 1, 2\}$ and transition matrix

$$\mathbf{P} = \begin{pmatrix} 0.5 & 0 & 0.5 \\ 0.4 & 0.2 & 0.4 \\ 0 & 0.4 & 0.6 \end{pmatrix}$$

(1) Determine $P(X_2 = 2 | X_1 = 0, X_0 = 1)$ and $P(X_2 = 2, X_1 = 0 | X_0 = 1)$.

(2) Determine $P(X_2 = 2, X_1 = 0 | X_0 = 0)$ and $P(X_{n+1} = 2, X_n = 0 | X_{n-1} = 0)$ for $n > 1$.

(3) Assuming the initial distribution $P(X_0 = 0) = 0.4;\ P(X_0 = 1) = P(X_0 = 2) = 0.3$, determine $P(X_1 = 2)$ and $P(X_1 = 1, X_2 = 2)$.

5.2) A Markov chain $\{X_0, X_1, ...\}$ has state space $\mathbf{Z} = \{0, 1, 2\}$ and transition matrix

$$\mathbf{P} = \begin{pmatrix} 0.2 & 0.3 & 0.5 \\ 0.8 & 0.2 & 0 \\ 0.6 & 0 & 0.4 \end{pmatrix}$$

(1) Determine the matrix of the 2-step transition probabilities $\mathbf{P}^{(2)}$.
(2) Assuming the initial distribution $P(X_0 = i) = 1/3$; $i = 0, 1, 2$; determine the probabilities $P(X_2 = 0)$ and $P(X_0 = 0, X_1 = 1, X_2 = 2)$

5.3) A Markov chain $\{X_0, X_1, ...\}$ has state space $\mathbf{Z} = \{0, 1, 2\}$ and transition matrix

$$\mathbf{P} = \begin{pmatrix} 0 & 0.4 & 0.6 \\ 0.8 & 0 & 0.2 \\ 0.5 & 0.5 & 0 \end{pmatrix}$$

(1) Given the initial distribution $P(X_0 = 0) = P(X_0 = 1) = 0.4$ and $P(X_0 = 2) = 0.2$, determine $P(X_3 = 2)$.
(2) Draw the corresponding transition graph.
(3) Compute the stationary distribution.

5.4) Let $\{Y_0, Y_1, ...\}$ be a sequence of independent, identically distributed binary random variables with $P(Y_i = 0) = P(Y_i = 1) = 1/2$; $i = 0, 2, ...$ Define a sequence of random variables $\{X_1, X_2, ...\}$ by

$$X_n = \frac{1}{2}(Y_n - Y_{n-1}); \quad n = 1, 2, ...$$

Check whether the random sequence $\{X_1, X_2, ...\}$ has the Markovian property.

5.5) A Markov chain $\{X_0, X_1, ...\}$ has state space $\mathbf{Z} = \{0, 1, 2, 3\}$ and transition matrix

$$\mathbf{P} = \begin{pmatrix} 0.1 & 0.2 & 0.4 & 0.3 \\ 0.2 & 0.3 & 0.1 & 0.4 \\ 0.4 & 0.1 & 0.3 & 0.2 \\ 0.3 & 0.4 & 0.2 & 0.1 \end{pmatrix}$$

Draw the corresponding transition graph and determine the stationary distribution.

5.6) Let an irreducible Markov chain with state space $\mathbf{Z} = \{1, 2, ..., n\}$, $n < \infty$, have a double stochastic transition matrix, that is, both

$$\sum_{j \in \mathbf{Z}} p_{ij} = 1 \text{ and } \sum_{i \in \mathbf{Z}} p_{ij} = 1$$

hold for all $i \in \mathbf{Z}$ and $j \in \mathbf{Z}$, respectively.
Verify that the stationary distribution of such a Markov chain is given by

$$\pi_j = 1/n, \; j \in \mathbf{Z}$$

5.7) A source emits symbols 0 and 1 for transmission to a receiver. Random noises $S_1, S_2, ...$ successively and independently affect the transmission process of a symbol in the following way: if a 0 (1) is to be transmitted, then S_i distorts it to a 1 (0) with probability p (q); $i = 1, 2, ...$

Let $X_0 = 0$ or $X_0 = 1$ denote whether 0 or a 1, respectively, is to be transmitted. Furthermore, let $X_i = 0$ or $X_i = 1$ denote whether, after the influence of noise S_i, a 0 or a 1 will be transmitted. The random sequence $\{X_0, X_1, ...\}$ is an irreducible Markov chain with state space $\mathbf{Z} = \{0, 1\}$ and transition matrix

$$\mathbf{P} = \begin{pmatrix} 1-p & p \\ q & 1-q \end{pmatrix}$$

(1) Verify: On condition that $0 < p+q \le 1$ the m-step transition matrix is given by

$$\mathbf{P}^{(m)} = \frac{1}{p+q}\begin{pmatrix} q & p \\ q & p \end{pmatrix} + \frac{(1-p-q)^m}{p+q}\begin{pmatrix} p & -p \\ -q & q \end{pmatrix}$$

(2) Let $p = q = 0.1$. The transmission of the symbols 0 and 1 is affected by the random noises $S_1, S_2, ..., S_5$. Determine the probability that a 0 emitted by the source is actually received.

5.8) A Markov chain has state space $\mathbf{Z} = \{0, 1, 2, 3, 4\}$ and transition matrix

$$\mathbf{P} = \begin{pmatrix} 0.5 & 0.1 & 0.4 & 0 & 0 \\ 0.8 & 0.2 & 0 & 0 & 0 \\ 0 & 1 & 0 & 0 & 0 \\ 0 & 0 & 0 & 0.9 & 0.1 \\ 0 & 0 & 0 & 1 & 0 \end{pmatrix}$$

(1) Determine the minimal closed sets.
(2) Check whether inessential states exist.

5.9) A Markov chain has state space $\mathbf{Z} = \{0, 1, 2, 3\}$ and transition matrix

$$\mathbf{P} = \begin{pmatrix} 0 & 0 & 1 & 0 \\ 1 & 0 & 0 & 0 \\ 0.4 & 0.6 & 0 & 0 \\ 0.1 & 0.4 & 0.2 & 0.3 \end{pmatrix}$$

Determine the classes of essential and inessential states.

5.10) A Markov chain has state space $\mathbf{Z} = \{0, 1, 2, 3, 4\}$ and transition matrix

$$\mathbf{P} = \begin{pmatrix} 0 & 0.2 & 0.8 & 0 & 0 \\ 0 & 0 & 0 & 0.9 & 0.1 \\ 0 & 0 & 0 & 0.1 & 0.9 \\ 1 & 0 & 0 & 0 & 0 \\ 1 & 0 & 0 & 0 & 0 \end{pmatrix}$$

(1) Draw the transition graph.
(2) Verify that the Markov chain is irreducible with period 3.
(3) Determine the stationary distribution.

5.11) A Markov chain has state space $\mathbf{Z} = \{0, 1, 2, 3, 4\}$ and transition matrix

$$\mathbf{P} = \begin{pmatrix} 0 & 1 & 0 & 0 & 0 \\ 1 & 0 & 0 & 0 & 0 \\ 0.2 & 0.2 & 0.2 & 0.4 & 0 \\ 0.2 & 0.8 & 0 & 0 & 0 \\ 0.4 & 0.1 & 0.1 & 0 & 0.4 \end{pmatrix}$$

(1) Find the essential and inessential states.
(2) Find the recurrent and transient states.

5.12) Determine the stationary distribution of the random walk considered in example 5.14 if $p_i = p,\ 0 < p < 1$.

5.13) Let the transition probabilities of a birth- and death process be given by

$$p_i = \frac{1}{1 + [i/(i+1)]^2} \quad \text{and} \quad q_i = 1 - p_i\,;\ i = 1, 2, \dots;\ p_0 = 1$$

Show that the process is transient.

5.14) Let i and j be two different states with $f_{ij} = f_{ji} = 1$.
Show that both i and j are recurrent.

5.15) The transition probabilities of two irreducible Markov chains with common state space $\mathbf{Z} = \{0, 1, \dots\}$ are given by

(1) $p_{i\,i+1} = 1/(i+2), \quad p_{i0} = (i+1)/(i+2); \quad i = 0, 1, \dots;$ and

(2) $p_{i\,i+1} = (i+1)/(i+2), \quad p_{i0} = 1/(i+2); \quad i = 0, 1, \dots;$ respectively.

Check in each case whether the process is recurrent, transient, null recurrent or positive recurrent.

5.16)* Assume that during its running time the behaviour of a technical system can be described by an irreducible Markov chain $\{X(t),\ t = 0, 1, \dots\}$ with finite state space $\mathbf{Z} = \{0, 1, \dots, n\}$, transition matrix $\mathbf{P}$ and stationary distribution $\{\pi_i,\ i \in \mathbf{Z}\}$. The system is operating if $X(t) = i$ with $i \in \mathbf{C} \subset \mathbf{Z}$ and not operating if $i \in \bar{\mathbf{C}} = \mathbf{Z} \setminus \mathbf{C}$. State changes of the Markov chain can only take place after every time unit. Let Y_t (Z_t) denote the t th working (downtime) period of the system. $(Y_t, Z_t);\ t = 1, 2, \dots$ is assumed to be an alternating renewal process with typical cycle (Y, Z).

Analogously to example 5.16, prove that in the long-run $E(Y)$ and $E(Z)$ are given by

$$E(Y) = \frac{\sum_{i \in \mathbf{C}} \pi_i}{\sum_{j \in \bar{\mathbf{C}}} \sum_{i \in \mathbf{C}} \pi_i p_{ij}}, \qquad E(Z) = \frac{\sum_{i \in \bar{\mathbf{C}}} \pi_i}{\sum_{j \in \bar{\mathbf{C}}} \sum_{i \in \mathbf{C}} \pi_i p_{ij}}$$

6 Continuous-Time Markov Chains

6.1 Foundations

This chapter deals with Markov processes having parameter set $\mathbf{T} = [0, \infty)$ and state space $\mathbf{Z} = \{0, \pm 1, \pm 2, ...\}$ or subsets of it.

Definition 6.1 A stochastic process $\{X(t), t \geq 0\}$ with parameter set $\mathbf{T}$ and state space $\mathbf{Z}$ is called a *continuous-time Markov chain* if, for any $n \geq 1$ and arbitrary sequences $\{t_0, t_1, ..., t_{n+1}\}$ with $t_0 < t_1 < ... < t_{n+1}$ and $\{i_0, i_1, ..., i_{n+1}\}$ with $i_k \in \mathbf{Z}$, the following relationship holds:

$$P(X(t_{n+1}) = i_{n+1} | X(t_n) = i_n, \ldots, X(t_1) = i_1, X(t_0) = i_0)$$

$$= P(X(t_{n+1}) = i_{n+1} | X(t_n) = i_n) \tag{6.1}$$

●

By definition 2.7, a continuous-time Markov chain is a continuous-time Markov process with discrete state space. The intuitive interpretation of the *Markov property* (6.1) is the same as for discrete-time Markov chains:

The future development of a continuous-time Markov chain depends only on the present and not on the past.

The conditional probabilities

$$p_{ij}(s, t) = P(X(t) = j | X(s) = i); \quad s < t; \; i, j \in \mathbf{Z};$$

are the *transition probabilities of the Markov chain.* A Markov chain is said to be *homogeneous* if for all $s, t \in \mathbf{T}$ and $i, j \in \mathbf{Z}$ the transition probabilities $p_{ij}(s, t)$ depend only on the difference $t - s$:

$$p_{ij}(s, t) = p_{ij}(0, t - s)$$

In this case the transition probabilities depend only on one variable. Hence they are written in the form

$$p_{ij}(t) = p_{ij}(0, t)$$

Note: This chapter considers only homogeneous Markov chains. Hence no confusion can arise if only *Markov chains* is referred to. Exceptions are made only when the property of homogeneity is of particular importance.

The transition probabilities are the elements of the *matrix of transition probabilities* (*transition matrix*)

$$\mathbf{P}(t) = ((p_{ij}(t))); \quad i, j \in \mathbf{Z}$$

Besides the trivial property $p_{ij}(t) \geq 0$, the transition probabilities are generally assumed to satisfy the conditions

$$\sum_{j \in \mathbf{Z}} p_{ij}(t) = 1; \quad t \geq 0,\ i \in \mathbf{Z} \tag{6.2}$$

Comment It is theoretically possible that for some $i \in \mathbf{Z}$ the inequality

$$\sum_{j \in \mathbf{Z}} p_{ij}(t) < 1; \quad t \geq 0,\ i \in \mathbf{Z}$$

holds. In this case, unboundedly many transitions between the states occur in the finite interval $[0, t)$ with positive probability $1 - \sum_{j \in \mathbf{Z}} p_{ij}(t)$. This situation approximately applies to nuclear chain reactions and population explosions of certain species of insects (e.g. locusts).

In the sequel it is assumed that

$$\lim_{t \to +0} p_{ii}(t) = 1 \tag{6.3}$$

By (6.2), this assumption is equivalent to

$$p_{ij}(0) = \lim_{t \to +0} p_{ij}(t) = \delta_{ij}; \quad i, j \in \mathbf{Z}; \tag{6.4}$$

where δ_{ij} denotes the *Kronecker-symbol*:

$$\delta_{ij} = \begin{cases} 1 & \text{for } i = j \\ 0 & \text{for } i \neq j \end{cases} \tag{6.5}$$

Analogously to (5.3), the ***Chapman-Kolmogorov equations*** are

$$p_{ij}(t + \tau) = \sum_{k \in \mathbf{Z}} p_{ik}(t) p_{kj}(\tau) \tag{6.6}$$

for any $t \geq 0$, $\tau \geq 0$, and $i, j \in \mathbf{Z}$.

By making use of the total probability rule, the homogeneity and the Markov property, (6.6) can be proved as follows:

$$p_{ij}(t + \tau) = P(X(t + \tau) = j \mid X(0) = i) = \frac{P(X(t + \tau) = j,\ X(0) = i)}{P(X(0) = i)}$$

$$= \sum_{k \in \mathbf{Z}} \frac{P(X(t + \tau) = j,\ X(t) = k,\ X(0) = i)}{P(X(0) = i)}$$

$$= \sum_{k \in \mathbf{Z}} \frac{P(X(t+\tau) = j | X(t) = k, X(0) = i)\, P(X(t) = k, X(0) = i)}{P(X(0) = i)}$$

$$= \sum_{k \in \mathbf{Z}} \frac{P(X(\tau+t) = j | X(t) = k)\, P(X(t) = k | X(0) = i)\, P(X(0) = i)}{P(X(0) = i)}$$

$$= \sum_{k \in \mathbf{Z}} P(X(\tau) = j | X(0) = k)\, P(X(t) = k | X(0) = i)$$

$$= \sum_{k \in \mathbf{Z}} p_{ik}(t) p_{kj}(\tau)$$

so that (6.6) is proved.

Absolute and stationary distributions Let $p_i(t) = P(X(t) = i)$ be the probability that the Markov chain is in state i at time t. $p_i(t)$ is called the *absolute state probability* at time t. Hence, $\{p_i(t),\ i \in \mathbf{Z}\}$ is said to be the *absolute* (*one-dimensional*) *probability distribution* of the Markov chain at time t. In particular, $\{p_i(0);\ i \in \mathbf{Z}\}$ is called an *initial distribution* of the Markov chain. By the total probability rule, given an initial distribution, the absolute probability distribution of the Markov chain at time t is given by

$$p_j(t) = \sum_{i \in \mathbf{Z}} p_i(0)\, p_{ij}(t), \quad j \in \mathbf{Z} \tag{6.7}$$

If the *n-dimensional distribution* of the Markov chain for arbitrary sequences $t_0, t_1, ..., t_n$ with $0 \le t_0 < t_1 < ... < t_n < \infty$ has to be computed, then its absolute distribution at time t_0 needs to be known, since, as it can be proved by repeated application of the definition of the conditional probability (1.6) and by making use of homogeneity,

$$P(X(t_0) = i_0, X(t_1) = i_1, ..., X(t_n) = i_n)$$

$$= p_{i_0}(t_0) p_{i_0 i_1}(t_1 - t_0) p_{i_1 i_2}(t_2 - t_1) \cdots p_{i_{n-1} i_n}(t_n - t_{n-1}) \tag{6.8}$$

Definition 6.2 An initial distribution $\{\pi_i = p_i(0),\ i \in \mathbf{Z}\}$ is said to be *stationary* if

$$\pi_i = p_i(t) \quad \text{for all } t \ge 0 \text{ and } i \in \mathbf{Z} \tag{6.9}$$

●

Thus, if at time $t = 0$ the initial state is determined by a stationary initial distribution, then the absolute state probabilities $p_j(t)$ do not depend on t and are equal to π_j. Consequently, the stationary initial probabilities π_j are the absolute probabilities of state j for all $j \in \mathbf{Z}$ and $t \ge 0$. Moreover, it follows from (6.8) that in this case all n-dimensional distributions of the Markov chain

$$\{P(X(t_1+h)=i_1, X(t_2+h)=i_2, \dots, X(t_n+h)=i_n),\ i_j \in \mathbf{Z}\} \tag{6.10}$$

do not depend on h, i.e. if the process starts with a stationary initial distribution, then the Markov chain is strictly stationary. (This result verifies the more general statement of theorem 2.1.) Moreover, it is justified to call $\{\pi_i, i \in \mathbf{Z}\}$ a *stationary distribution* of the Markov chain.

Example 6.1 An homogeneous Poisson process $\{N(t), t \geq 0\}$ with intensity λ is a homogeneous Markov chain with state space $\mathbf{Z} = \{0, 1, \dots\}$. Its transition probabilities are

$$p_{ij}(t) = \frac{(\lambda t)^{(j-i)}}{(j-i)!} e^{-\lambda t}; \quad i \leq j$$

The sample paths of this process are nondecreasing step-functions. Its trend function is linearly increasing: $m(t) = E(N(t)) = \lambda t$. Thus, a stationary initial distribution cannot exist. (By definition of a Poisson process, $P(N(0)=0)=1$.) □

Example 6.2 At time $t=0$, n systems start operating. Their lifetimes are independent, identically distributed exponential random variables with parameter λ. Let $X(t)$ denote the number of systems which are still operating at time t. Then $\{X(t), t \geq 0\}$ is a Markov chain with state space $\mathbf{Z} = \{0, 1, \dots, n\}$, initial distribution $P(X(0)=n)=1$, and transition probabilities

$$p_{ij}(t) = \binom{i}{i-j} (1-e^{-\lambda t})^{i-j} e^{-\lambda t j}, \quad n \geq i \geq j \geq 0$$

The structure of these transition probabilities is based on the memoryless property of the exponential distribution (see example 1.2). Of course, this Markov chain cannot be stationary. □

Example 6.3 Let $\mathbf{Z} = \{0, 1\}$ be the state space and

$$\mathbf{P}(t) = \begin{pmatrix} \frac{1}{t+1} & \frac{t}{t+1} \\ \frac{t}{t+1} & \frac{1}{t+1} \end{pmatrix}$$

the transition matrix of a stochastic process $\{X(t), t \geq 0$. It is to check whether this process can be a Markov chain. Assuming the initial distribution

$$p_0(0) = P(X(0)=0) = 1$$

and applying formula (6.7) yields the absolute probability of state 0 at time $t=3$ to be

$$p_0(3) = p_0(0)\, p_{00}(3) = 1 \cdot \frac{1}{3+1} = \frac{1}{4}$$

On the other hand, applying (6.6) with $t=2$ and $\tau=1$ yields the (wrong) result

$$p_0(3) = p_{00}(2)p_{00}(1) + p_{01}(2)p_{10}(1)$$
$$= \frac{1}{2+1} \cdot \frac{1}{1+1} + \frac{2}{2+1} \cdot \frac{1}{1+1} = \frac{1}{2}$$

Therefore, the Chapman-Kolmogorov equations are not valid so that $\{X(t),\, t \geq 0\}$ cannot be a Markov chain. □

Classification of states The classification concepts already introduced for discrete-time Markov chains can be defined analogously for continuous-time Markov chains. In what follows, some concepts are defined, but not discussed in detail.

A state set $\mathbf{C} \subseteq \mathbf{Z}$ is called *closed* if

$$p_{ij}(t) = 0 \quad \text{for all} \quad t > 0,\ i \in \mathbf{C} \ \text{and}\ j \notin \mathbf{C}$$

If, in particular, $\{i\}$ is a closed set, then i is called an *absorbing state*.

The state j is *accessible* from i if there exists a t with $p_{ij}(t) > 0$. If i and j are accessible from each other, then they are said to *communicate*. Thus, equivalence classes, essential and inessential states as well as irreducible and reducible Markov chains can be defined as in section 5.2 for discrete Markov chains.
The state i is *recurrent* (*transient*) if $\int_0^\infty p_{ii}(t)\, dt$ diverges (converges). A recurrent state i is said to be *positive recurrent* if the expected value of its recurrence time (time between two successive occurences of state i) is finite. Since it can easily be shown that $p_{ij}(t_0) > 0$ implies $p_{ij}(t) > 0$ for all $t > t_0$, introducing the concept of a period analogously to section 5.2.3 makes no sense, because each state would have period 1.

6.2 Kolmogorov's Differential Equations

This section discusses some structural properties of continuous Markov chains which are fundamental to the mathematical modeling real systems.

Theorem 1.1 Assuming (6.3), the transition probabilities $p_{ij}(t)$ are differentiable in $[0, \infty)$ for all $i, j \in \mathbf{Z}$.

Proof For any $h > 0$, the Chapman-Kolmogorov equations yield

$$p_{ij}(t+h) - p_{ij}(t) = \sum_{k \in \mathbf{Z}} p_{ik}(h) p_{kj}(t) - p_{ij}(t)$$
$$= -(1 - p_{ii}(h)) p_{ij}(t) + \sum_{\substack{k \in \mathbf{Z} \\ k \neq i}} p_{ik}(h) p_{kj}(t)$$

Thus,

$$-(1-p_{ii}(h)) \le -(1-p_{ii}(h))p_{ij}(t) \le p_{ij}(t+h)-p_{ij}(t)$$

$$\le \sum_{\substack{k\in\mathbf{Z}\\ k\neq i}} p_{ik}(h)p_{kj}(t) \le \sum_{\substack{k\in\mathbf{Z}\\ k\neq i}} p_{ik}(h)$$

$$= 1-p_{ii}(h)$$

Hence,

$$\left|p_{ij}(t+h)-p_{ij}(t)\right| \le 1-p_{ii}(h)$$

The uniform continuity of the transition probabilities and, therefore, their differentiability for all $t \ge 0$ is now a consequence of (6.3). ■

Transition rates The following limits play an important role in all future derivations. For any $i,j \in \mathbf{Z}$, let

$$q_i = \lim_{h\to 0} \frac{1-p_{ii}(h)}{h} \tag{6.11}$$

and

$$q_{ij} = \lim_{h\to 0} \frac{p_{ij}(h)}{h}, \quad i \neq j \tag{6.12}$$

By (6.3) and (6.4),

$$p'_{ii}(0) = \left.\frac{dp_{ii}(t)}{dt}\right|_{t=0} = -q_i \tag{6.13}$$

$$p'_{ij}(0) = \left.\frac{dp_{ij}(t)}{dt}\right|_{t=0} = q_{ij}, \quad i \neq j \tag{6.14}$$

Theorem 1.1 guarantees the existence of the limits q_i and q_{ij}. For $h \to 0$, their definition is equivalent to

$$p_{ii}(h) = 1 - q_i\, h + o(h) \tag{6.15}$$

and

$$p_{ij}(h) = q_{ij}\, h + o(h), \quad i \neq j \tag{6.16}$$

The parameters q_i and q_{ij} are the *transition rates* of the Markov chain. More exactly, q_i is the *unconditional transition rate* of leaving the state i for any other, and, starting from state i, q_{ij} is the *conditional transition rate* of making a transition to state j. According to (6.2),

$$\sum_{\{j,j\neq i\}} q_{ij} = q_i, \quad i \in \mathbf{Z} \tag{6.17}$$

Kolmogorov's differential equations In what follows, systems of differential equations for the transition probabilities and the absolute state probabilities of a Markov chain are derived. For this purpose, the Chapman-Kolmogorov equations are written in the form

$$p_{ij}(t+h) = \sum_{k\in\mathbf{Z}} p_{ik}(h)p_{kj}(t)$$

It follows that

$$\frac{p_{ij}(t+h)-p_{ij}(t)}{h} = \sum_{k\neq i}\frac{p_{ik}(h)}{h}p_{kj}(t) - \frac{1-p_{ii}(h)}{h}p_{ij}(t)$$

By (6.13) and (6.14), letting $h\to 0$ yields *Kolmogorov's backward equations*:

$$p'_{ij}(t) = \sum_{k\neq i} q_{ik}p_{kj}(t) - q_i p_{ij}(t), \quad t\geq 0 \tag{6.18}$$

Analogously, starting with

$$p_{ij}(t+h) = \sum_{k\in\mathbf{Z}} p_{ik}(t)p_{kj}(h)$$

yields *Kolmogorov's forward equations*:

$$p'_{ij}(t) = \sum_{k\neq j} p_{ik}(t)q_{kj} - q_j p_{ij}(t), \quad t\geq 0 \tag{6.19}$$

Let $\{p_i(0),\ i\in\mathbf{Z}\}$ be any initial distribution. Multiplying Kolmogorov's forward equations (6.19) by $p_i(0)$ and summing with respect to i yields

$$\begin{aligned}\sum_{i\in\mathbf{Z}} p_i(0)p'_{ij}(t) &= \sum_{i\in\mathbf{Z}} p_i(0)\sum_{k\neq j} p_{ik}(t)q_{kj} - \sum_{i\in\mathbf{Z}} p_i(0)q_j p_{ij}(t)\\ &= \sum_{k\neq j} q_{kj}\sum_{i\in\mathbf{Z}} p_i(0)p_{ik}(t) - q_j\sum_{i\in\mathbf{Z}} p_i(0)p_{ij}(t)\end{aligned}$$

Thus, in view of (6.7), the absolute state probabilities satisfy the system of linear differential equations

$$p'_j(t) = \sum_{k\neq j} q_{kj}p_k(t) - q_j p_j(t), \quad t\geq 0, \quad j\in\mathbf{Z} \tag{6.20}$$

Furthermore, it is assumed that the absolute state probabilities satisfy the normalizing condition

$$\sum_{i\in\mathbf{Z}} p_i(t) = 1 \tag{6.21}$$

This condition is always fulfilled if $\mathbf{Z}$ is finite.

Note: If the initial distribution has structure $p_i(0) = 1$, $p_j(0) = 0$ for $j \neq i$, then the absolute state probabilities $\{p_j(t), j \in \mathbf{Z}\}$ are equal to the transition probabilities $\{p_{ij}(t), j \in \mathbf{Z}\}$.

Transition times and transition rates It is only possible to model real systems exactly by continuous-time Markov chains if the time periods between changes of states are exponentially distributed, since in this case the "memoryless property" of exponential distribution (example 1.2) implies the Markov property. If the times between transitions have known exponential distributions, then it is no problem to determine the transition rates. For instance, if the sojourn time of a Markov chain in state 0 is exponentially distributed with parameter λ_0, then, according to (6.11), the unconditional rate of leaving this state is given by

$$q_0 = \lim_{h \to 0} \frac{1 - p_{00}(h)}{h} = \lim_{h \to 0} \frac{1 - e^{-\lambda_0 h}}{h}$$

$$= \lim_{h \to 0} \frac{\lambda_0 h + o(h)}{h}$$

$$= \lambda_0 + \lim_{h \to 0} \frac{o(h)}{h}$$

Hence,

$$q_0 = \lambda_0 \tag{6.22}$$

Now let the Markov chain stay in state 0 for the random time $Y_0 = \min(Y_{01}, Y_{02})$. If $Y_{01} < Y_{02}$, then the Markov chain makes a transition to state 1 and otherwise to state 2. If Y_{01} and Y_{02} are independent exponential random variables with parameters λ_1 and λ_2, respectively, then the conditional transition rate from state 0 to state 1 is, by (6.12),

$$q_{01} = \lim_{h \to 0} \frac{p_{01}(h)}{h}$$

$$= \lim_{h \to 0} \frac{(1 - e^{-\lambda_1 h})\, e^{-\lambda_2 h} + o(h)}{h}$$

$$= \lim_{h \to 0} \frac{\lambda_1 h (1 - \lambda_2 h)}{h} + \lim_{h \to 0} \frac{o(h)}{h}$$

Thus, since the roles of Y_{01} and Y_{02} can be interchanged,

$$q_{01} = \lambda_1, \quad q_{02} = \lambda_2, \quad q_0 = \lambda_1 + \lambda_2 \tag{6.23}$$

The results (6.22) and (6.23) are generalized in section 6.4.

Transition graphs In practice, establishing the Kolmogorov equations can be facilitated by *transition graphs.* These graphs are constructed analogously to the transition graphs of discrete-time Markov chains: The nodes of a transition graph represent the states of the Markov chain. A (directed) edge from node i to node j exists if and only if $q_{ij} > 0$. The edges are weighted by their corresponding transition rates. Thus, two sets of states (possibly empty ones) can be assigned to each node i: edges with initial node i and edges with end node i, that is, edges which leave node i and edges which end in node i. The unconditional transition rate q_i is equal to the sum of the weights of all those edges which leave node i. If there is an edge ending in state i and no edge leaving state i, then i is an absorbing state.

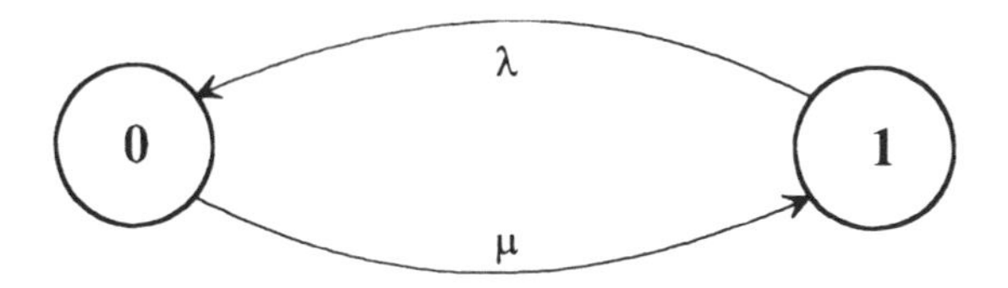

Figure 6.1 Transition graph of an alternating renewal process (example 6.4)

Example 6.4 (***system with renewal***) Let the random lifetime of a system be exponentially distributed with parameter λ. After each failure the system is replaced by an equivalent new one. Each replacement time is exponentially distributed with parameter μ. All life times and replacement times are assumed to be independent. Thus, if L and Z denote the lifetime and the repair time, respectively, an alternating renewal process (section 4.6) with "typical renewal cycle" (L, Z) is given.

Consider the Markov chain $\{X(t), t \geq 0\}$ defined by

$$X(t) = \begin{cases} 1 & \text{if the system is operating} \\ 0 & \text{if the system is being replaced} \end{cases}$$

Its state space is $\mathbf{Z} = \{0, 1\}$. The absolute state probability $p_1(t) = P(X(t) = 1)$ is the *point availability* of the system.

In this simple example, only state changes from 0 to 1 and from 1 to 0 are possible. Hence, by (6.22),

$$q_0 = q_{01} = \mu \text{ and } q_1 = q_{10} = \lambda$$

The corresponding Kolmogorov differential equations (6.20) are

$$p_0'(t) = -\mu p_0(t) + \lambda p_1(t)$$

$$p_1'(t) = +\mu p_0(t) - \lambda p_1(t)$$

These two equations are linearly dependent. (The respective sums of the left hand parts and the right hand parts are equal to 0.) Replacing $p_0(t)$ in the second equation by $1-p_1(t)$ yields a first-order inhomogeneous differential equation with constant coefficients for $p_1(t)$:

$$p_1'(t)+(\lambda+\mu)p_1(t)=\mu$$

Given the initial condition $p_1(0)=1$, the solution is

$$p_1(t)=\frac{\mu}{\lambda+\mu}+\frac{\lambda}{\lambda+\mu}e^{-(\lambda+\mu)t}, \quad t\ge 0$$

The corresponding stationary availability is

$$\pi_1=\lim_{t\to\infty}p_1(t)=\frac{\mu}{\lambda+\mu}$$

In example 4.6, the same results have been obtained by applying the Laplace transform. (However, there the notation $L=Y$, $\lambda=\lambda_1$ and $\mu=\lambda_0$ is used.) □

Example 6.5 ***(two-unit redundant system, standby redundancy)*** A system consists of two identical units. The system is available if and only if at least one of its units is available. If both units are available, then one of them is in standby redundancy (cold redundancy), that is, in this state it does not age and cannot fail. After the failure of a unit, the other one (if available) is immediately switched from the redundancy state to the operating state and the replacement of the failed unit begins. The replaced unit becomes the standby unit if the other unit is still operating. Otherwise it immediately resumes its work. The lifetimes and replacement times of the units are independent random variables, identically distributed as L and Z, respectively. L and Z are assumed to be exponentially distributed with respective parameters λ and μ. Let L_s denote the system lifetime, i.e. the random time to a system failure. (A system failure occurs when a unit fails during the replacement period of the other.)
A Markov chain $\{X(t), t\ge 0\}$ with state space $\mathbf{Z}=\{0, 1, 2\}$ is introduced in the following way: $X(t)=i$ when i units are unavailable at time t. Let Y_i be the unconditional sojourn time of the system in state i and and Y_{ij} be the conditional sojourn time of the system in state i given that the system makes a transition from state i into state j. From state 0, the system can only make a transition to state 1. Hence, $Y_0=Y_{01}=L$. According to (6.22), the corresponding transition rate is given by $q_0=q_{01}=\lambda$. If the system makes a transition from state 1 to state 2, then its conditional sojourn time in state 1 is $Y_{12}=L$, whereas in case of a transition to state 0 it spends time $Y_{10}=Z$ in state 1. The unconditional sojourn time of the system in state 1 is given by $Y_1=\min(L,Z)$. Hence, according to (6.23), the corresponding transition rates are $q_{12}=\lambda$, $q_{10}=\mu$ and $q_1=\lambda+\mu$. When the sys-

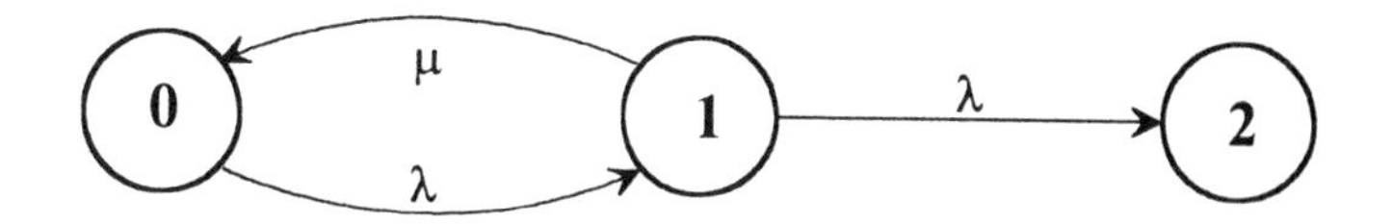

Figure 6.2 Transition graph for example 6.5 a)

tem returns from state 1 to state 0, then it again spends time L in state 0, since the operating unit is "as good as new" in view of the memoryless property of the exponential distribution.

a) *Survival probability* In this case, only the time to entering state 2 (system failure) is of interest. Hence, state 2 is considered to be absorbing (Figure 6.2) so that $q_{20} = q_{21} = 0$, and the survival probability of the system is given by

$$\bar{F}_s(t) = P(L_s > t) = p_0(t) + p_1(t)$$

The corresponding system of differential equations (6.20) is

$$\begin{aligned} p_0'(t) &= -\lambda p_0(t) + \mu p_1(t) \\ p_1'(t) &= +\lambda p_0(t) - (\lambda + \mu) p_1(t) \\ p_2'(t) &= +\lambda p_1(t) \end{aligned} \tag{6.24}$$

It is solved assuming that both units are available at time $t = 0$. Combining the first two differential equations in (6.24) yields a homogeneous differential equation of the second order with constant coefficients for $p_0(t)$:

$$p_0''(t) + (2\lambda + \mu) p_0'(t) + \lambda^2 p_0(t) = 0$$

The corresponding characteristic equation is

$$x^2 + (2\lambda + \mu)x + \lambda^2 = 0$$

Its solutions are

$$x_{1,2} = -\left(\lambda + \frac{\mu}{2}\right) \pm \sqrt{\lambda\mu + \mu^2/4}$$

Hence, since $p_0(0) = 1$,

$$p_0(t) = a \sinh \frac{c}{2} t, \quad t \geq 0,$$

where

$$c = \sqrt{4\lambda\mu + \mu^2}$$

Since $p_1(0) = 0$, the first differential equation in (6.24) yields $a = 2\lambda/c$ and

$$p_1(t) = e^{-\frac{2\lambda+\mu}{2}t}\left(\frac{\mu}{c}\sinh\frac{c}{2}t + \cosh\frac{c}{2}t\right), \quad t \geq 0$$

Thus, the survival probability of the system is

$$\overline{F}_s(t) = e^{-\frac{2\lambda+\mu}{2}}\left[\cosh\frac{c}{2}t + \frac{2\lambda+\mu}{c}\sinh\frac{c}{2}t\right], \quad t \geq 0$$

(For a definition of the hyperbolic functions sinh and cosh, see page 74.) The expected value of the system lifetime L_s is most easily obtained by applying (1.12) :

$$E(L_s) = \frac{2}{\lambda} + \frac{\mu}{\lambda^2} \tag{6.25}$$

For the sake of comparision, in case of no replacement ($\mu = 0$), L_s has an Erlang-distribution with parameters 2 and λ:

$$\overline{F}_s(t) = (1 + \lambda t)e^{-\lambda t}$$

$$E(L_s) = 2/\lambda$$

b) *Availability* If the replacement of failed units is also continued after a system failure, then the availability $A_s(t) = p_0(t) + p_1(t)$ of the system is of particular interest. In this case, the transition rate from state 2 to state 1 is positive. However, it depends on the number $r = 1$ or $r = 2$ of mechanics which are in charge of the replacement of failed units. Assuming that a mechanic cannot replace two failed units at the same time, the transition rate from state 2 to state 1 is $q_2 = q_{21} = r\mu$ (Figure 6.3). Note that, for $r = 2$, the sojourn time of the system in state 2 is given by $Y_2 = \min(Z_1, Z_2)$, where Z_1 and Z_2 are independent random variables, which are identically distributed as Z. Analogously, the sojourn time in state 1 is given by $Y_1 = \min(L, Z)$. Hence, the transition rates q_{10} and q_{12} have the same values as under a).

The corresponding system of differential equations (6.20) is, replacing the last differential equation by the normalizing condition (6.21),

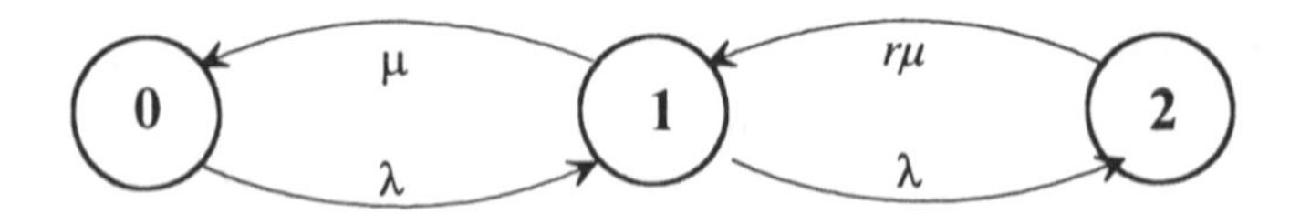

Figure 6.3 Transition graph for example 6.5 b)

$$p_0'(t) = -\lambda p_0(t) + \mu p_1(t)$$

$$p_1'(t) = +\lambda p_0(t) - (\lambda + \mu) p_1(t) + r\mu p_2(t)$$

$$1 = \quad p_0(t) + \qquad p_1(t) + \quad p_2(t)$$

The solution is left as an exercise to the reader. □

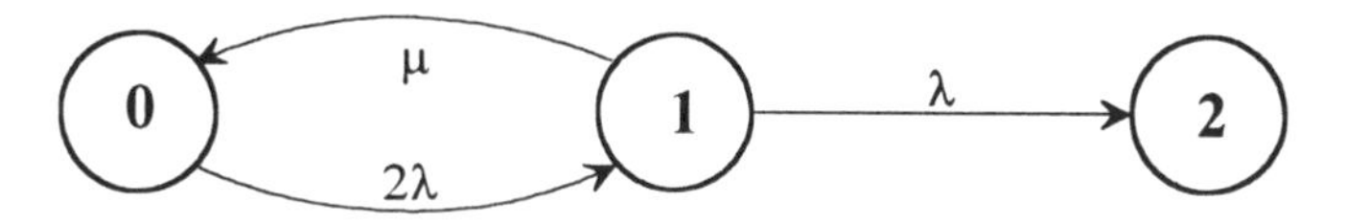

Figure 6.4 Transition graph for example 6.6 a)

Example 6.6 (*two-unit system, parallel redundancy*) Now assume that both units of the system operate at the same time when they are available. All other assumptions and the notation of the previous example hold as before. In particular, the system is available if and only if at least one unit is available. In view of the initial condition $p_0(0) = 1$ and the memoryless property of the exponential distribution, the system spends time $Y_0 = \min(L_1, L_2)$ in state 0. Since Y_0 is exponentially distributed with parameter 2λ and from state 0 only a transition to state 1 is possible, $Y_0 = Y_{01}$ and

$$q_0 = q_{01} = 2\lambda$$

When the system is in state 1, then it behaves as in example 6.5 (Figure 6.4):

$$q_{10} = \mu, \quad q_{12} = \lambda, \quad q_1 = \lambda + \mu$$

a) *Survival probability* Since $q_{20} = q_{21} = 0$, the corresponding systemof differential equations (6.20) is

$$p_0'(t) = -2\lambda p_0(t) + \mu p_1(t)$$

$$p_1'(t) = +2\lambda p_0(t) - (\lambda + \mu) p_1(t)$$

$$p_2'(t) = +\lambda p_1(t)$$

Since these three differential equations are linearly dependent, one of them has to be replaced by the normalizing condition (6.21). Combining, for instance, the first two differential equations yields a homogeneous differential equation of the second order with constant coefficients for $p_0(t)$:

$$p_0''(t) + (3\lambda + \mu) p_0'(t) + 2\lambda^2 p_0(t) = 0$$

The solution is

$$p_0(t) = e^{-\left(\frac{3\lambda+\mu}{2}\right)t}\left[\cosh\frac{c}{2}t + \frac{\mu-\lambda}{c}\sinh\frac{c}{2}t\right]$$

where $c = \sqrt{\lambda^2 + 6\lambda\mu + \mu^2}$. Furthermore,

$$p_1(t) = \frac{4\lambda}{c}e^{-\left(\frac{3\lambda+\mu}{2}\right)t}\sinh\frac{c}{2}t$$

Therefore, the survival probability of the system $\bar{F}_s(t) = P(L_s > t) = p_0(t) + p_1(t)$ is

$$\bar{F}_s(t) = e^{-\left(\frac{3\lambda+\mu}{2}\right)t}\left[\cosh\frac{c}{2}t + \frac{3\lambda+\mu}{c}\sinh\frac{c}{2}t\right], \quad t \ge 0 \qquad (6.26)$$

The expected system lifetime is

$$E(L_s) = \frac{3}{2\lambda} + \frac{\mu}{2\lambda^2}$$

For the sake of comparision, in the case without replacement ($\mu = 0$),

$$\bar{F}(t) = 2e^{-\lambda t} - e^{-2\lambda t}, \quad E(L_s) = \frac{3}{2\lambda}$$

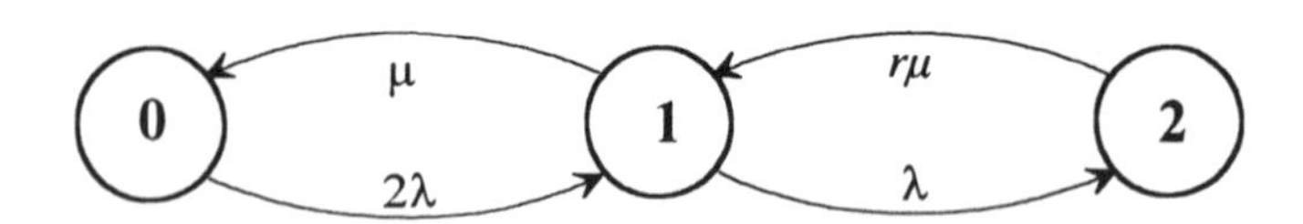

Figure 6.5 Transition graph for example 6.6 b)

b) *Availability* If r ($r = 1, 2$) mechanics replace failed units, then $q_2 = q_{21} = r\mu$. The other transition rates are the same as those under a) (Figure 6.5).

The absolute state probabilities satisfy the system of differential equations

$$\begin{aligned} p_0'(t) &= -2\lambda p_0(t) + \mu p_1(t) \\ p_1'(t) &= +2\lambda p_0(t) - (\lambda+\mu)p_1(t) + r p_2(t) \\ 1 &= p_0(t) + p_1(t) + p_2(t) \end{aligned}$$

The solution is again left as an exercise to the reader. □

6.3 Stationary State Probabilities

If $\{\pi_j, j \in \mathbf{Z}\}$ is a stationary distribution of the Markov chain $\{X(t), t \geq 0\}$, then this special absolute distribution must satisfy the Kolmogorov equations (6.20). Since the π_j are constant, all the left-hand sides of these equations are equal to 0. Hence, the system of linear differential equations (6.20) simplifies to a system of linear algebraic equations in the unknowns π_j:

$$0 = \sum_{\substack{k \in \mathbf{Z} \\ k \neq j}} q_{kj} \pi_k - q_j \pi_j, \quad j \in \mathbf{Z} \tag{6.27}$$

This system of equations is frequently written in the form

$$q_j \pi_j = \sum_{\substack{k \in \mathbf{Z} \\ k \neq j}} q_{kj} \pi_k, \quad j \in \mathbf{Z} \tag{6.28}$$

This form clearly illustrates that the stationary state probabilities refer to an equilibrium state of the Markov chain: The mean intensity per unit time of leaving state j (namely $q_j \pi_j$), is equal to the mean intensity per unit time of arriving in state j. According to assumption (6.21), only those solutions of (6.27) satisfying the normalizing condition

$$\sum_{j \in \mathbf{Z}} \pi_j = 1 \tag{6.29}$$

are of interest. It is now assumed that the Markov chain is irreducible and positive recurrent. (Note that an irreducible Markov chain with finite state space $\mathbf{Z}$ is always positive recurrent.) Then it can be shown that a unique stationary distribution $\{\pi_j, j \in \mathbf{Z}\}$ satisfying (6.27) and (6.29) exists. Moreover, in this case the limits $p_j = \lim_{t\to\infty} p_{ij}(t)$ exist, and they are independent of i. Hence, for any initial distribution, there exist also the limits of the absolute state probabilities $\lim_{t\to\infty} p_j(t)$ and they are equal to p_j:

$$p_j = \lim_{t\to\infty} p_j(t), \quad j \in \mathbf{Z} \tag{6.30}$$

Furthermore, for all $j \in \mathbf{Z}$,

$$\lim_{t\to\infty} p_j'(t) = 0$$

Otherwise, $p_j(t)$ would increase unboundedly as $t \to \infty$ contradicting $p_j(t) \leq 1$. Hence, when passing to the limit as $t \to \infty$ in (6.20) and (6.21), the limits (6.30) are seen to fulfil equations (6.27) and (6.29) and, since these equations have a unique solution, the limits p_j and the stationary probabilities π_j must coincide:

$$p_j = \pi_j, \quad j \in \mathbf{Z}$$

For a detailed discussion of the relationship between the solvability of (6.27) and the existence of a stationary distribution see *Feller* (1968).

Continuation of example 6.5 (*two-unit system, standby redundancy*) Since the system is available if and only if at least one unit is available, its stationary availability is given by $A = \pi_0 + \pi_1$. Substituting the transition rates from Figure 6.3 into (6.27) and (6.29), the stationary state probabilities π_j are seen to satisfy the following system of algebraic equations:

$$\begin{aligned} -\lambda\pi_0 + \quad \mu\pi_1 \quad &= 0 \\ +\lambda\pi_0 - (\lambda+\mu)\pi_1 + r\pi_2 &= 0 \\ \pi_0 + \quad \pi_1 + \pi_2 &= 0 \end{aligned}$$

***r* = 1**

$$\pi_0 = \frac{\mu^2}{(\lambda+\mu)^2 - \lambda\mu}, \quad \pi_1 = \frac{\lambda\mu}{(\lambda+\mu)^2 - \lambda\mu}, \quad \pi_2 = \frac{\lambda^2}{(\lambda+\mu)^2 - \lambda\mu}$$

$$A = \pi_0 + \pi_1 = \frac{\mu^2 + \lambda\mu}{(\lambda+\mu)^2 - \lambda\mu}$$

***r* = 2**

$$\pi_0 = \frac{2\mu^2}{(\lambda+\mu)^2 + \mu^2}, \quad \pi_1 = \frac{2\lambda\mu}{(\lambda+\mu)^2 + \mu^2}, \quad \pi_2 = \frac{\lambda^2}{(\lambda+\mu)^2 + \mu^2}$$

$$A = \pi_0 + \pi_1 = \frac{2\mu^2 + 2\lambda\mu}{(\lambda+\mu)^2 + \mu^2}$$

Continuation of example 6.6 (*two-unit system, parallel redundancy*) Given the transition rates in Figure 6.5, the π_j satisfy the equations

$$\begin{aligned} -2\lambda\pi_0 + \quad \mu\pi_1 \quad &= 0 \\ +2\lambda\pi_0 - (\lambda+\mu)\pi_1 + r\mu\pi_2 &= 0 \\ \pi_0 + \quad \pi_1 + \pi_2 &= 0 \end{aligned}$$

***r* = 1**

$$\pi_0 = \frac{\mu^2}{(\lambda+\mu)^2 + \lambda^2}, \quad \pi_1 = \frac{2\lambda\mu}{(\lambda+\mu)^2 + \lambda^2}, \quad \pi_2 = \frac{2\lambda^2}{(\lambda+\mu)^2 + \lambda^2}$$

$$A = \pi_0 + \pi_1 = \frac{\mu^2 + 2\lambda\mu}{(\lambda+\mu)^2 + \lambda^2}$$

***r* = 2**

$$\pi_0 = \frac{\mu^2}{(\lambda+\mu)^2}, \quad \pi_1 = \frac{2\lambda\mu}{(\lambda+\mu)^2}, \quad \pi_2 = \frac{\mu^2}{(\lambda+\mu)^2}$$

$$A = \pi_0 + \pi_1 = 1 - \left(\frac{\lambda}{\lambda+\mu}\right)^2$$

Figure 6.6 shows a) the mean lifetimes and b) the stationary availabilities of the two-unit system for $r = 1$ as functions of $\rho = \lambda/\mu$. As expected, standby redundancy yields better results if, as assumed, switching a unit from the standby redundancy state to the operating state is absolutely reliable. With parallel redundancy this switching problem does not exist since the spare unit is also operating when available. □

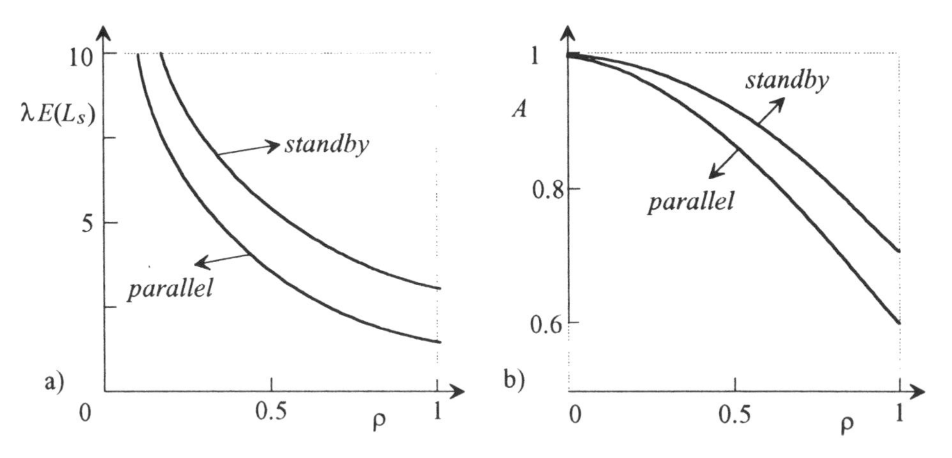

Figure 6.6 Expected lifetime a) and stationary availability b) in the case of standby and parallel redundancy (r = 1)

Example 6.7 A system has two different failure types: type 1 and type 2. After a type *i*-failure the system is said to be in failure state *i*; $i = 1, 2$. The random time to a type *i*-failure L_i is assumed to be exponentially distributed with parameter λ_i; $i = 1, 2$. L_1 and L_2 are assumed to be independent. Thus, if at time $t = 0$ a new system starts working, the time to its first failure is $Y_0 = \min(L_1, L_2)$. After a type 1-failure, the system is switched from failure state 1 into failure state 2. The time required for this is exponentially distributed with parameter ν. After entering failure state 2, the renewal of the system begins. A renewed system immediately starts working. The renewal time is exponentially distributed with parameter μ. This process continues indefinitely. All life- and renewal times as well as switching times are assumed to be independent.

This model is, for example, of importance in traffic safety engineering (*Fischer* (1984)): When the red signal in a traffic light fails (type 1-failure), then the whole traffic light is switched off (type 2-failure). That is, a *dangerous failure state* is removed by inducing a *blocking failure state.*

Consider the following system states:

0 system is operating

1 type 1-failure state

2 type 2-failure state

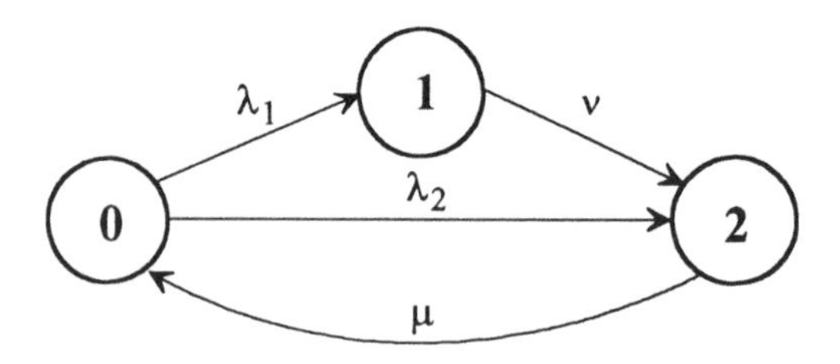

Figure 6.7 Transition graph for example 6.7

If $X(t)$ denotes the state of the system at time t, then $\{X(t), t \geq 0\}$ is a homogeneous Markov chain with state space $\mathbf{Z} = \{0, 1, 2\}$. Its transition rates are (Figure 6.7)

$$q_{01} = \lambda_1, \; q_{02} = \lambda_2, \; q_0 = \lambda_1 + \lambda_2, \; q_{12} = q_1 = \nu, \; q_{20} = q_2 = \mu$$

Hence the stationary state probabilities fulfil the system of equations

$$\begin{aligned} -(\lambda_1 + \lambda_2)\pi_0 + \qquad\qquad \mu\pi_2 &= 0 \\ \lambda_1\pi_0 - \nu\pi_1 \qquad &= 0 \\ \pi_0 + \pi_1 + \pi_2 &= 1 \end{aligned}$$

The solution is

$$\pi_0 = \frac{\mu\nu}{(\lambda_1 + \lambda_2)\nu + (\lambda_1 + \nu)\mu}$$

$$\pi_1 = \frac{\lambda_1\mu}{(\lambda_1 + \lambda_2)\nu + (\lambda_1 + \nu)\mu}$$

$$\pi_2 = \frac{(\lambda_1 + \lambda_2)\nu}{(\lambda_1 + \lambda_2)\nu + (\lambda_1 + \nu)\mu}$$

□

6.4 Construction of Markovian Systems

In a *Markovian* (*Markov*) *system* state changes are controlled by a Markov process. Markovian systems in which the underlying Markov process is a homogeneous, continuous-time Markov chain with state space **Z** are frequently special cases of the following basic model: The sojourn time of the system in state i is

$$Y_i = \min(Y_{i1}, Y_{i2}, ..., Y_{in_i}),$$

where the Y_{ij} are independent, exponentially distributed random variables with parameters λ_{ij}; $j = 1, 2, ..., n_i$; $i, j \in \mathbf{Z}$. A transition from state i to state j is made if and only if $Y_i = Y_{ij}$. If $X(t)$ denotes the state of the system at time t, then, in view of the memoryless property of the exponential distribution, $\{X(t), t \geq 0\}$ is a homogeneous Markov chain with transition rates

$$q_{ij} = \lim_{h \to 0} \frac{p_{ij}(h)}{h} = \lambda_{ij}, \quad q_i = \sum_{j=1}^{n_i} \lambda_{ij}$$

The representation of q_i results from (6.17). It reflects the fact that Y_i, as the minimum of independent, exponentially distributed random variables Y_{ij}, also has an exponential distribution whose parameter is obtained by summing the parameters of the Y_{ij}.

Example 6.8 (***repairman problem***) Let n machines with lifetimes $L_1, L_2, \ldots, L_n$ start operating at time $t = 0$. The L_i are assumed to be independent, exponential random variables with parameter λ. Failed machines are repaired. A repaired machine is "as good as new". There is one mechanic who can only handle one failed machine at a time. Thus, when there are k, $k > 1$, failed machines, $k - 1$ are waiting for repair. The repair times are assumed to be mutually independent and identically distributed as an exponential random variable Z with parameter μ. Moreover, they are independent of the lifetimes. Immediately after its repair, a machine resumes its work.

If $X(t)$ denotes the number of machines which are in the failed state at time t, then $\{X(t), t \geq 0\}$ is a Markov chain with state space $\mathbf{Z} = \{0, 1, ..., n\}$ and stationary state probabilities

$$\pi_j = \lim_{t \to \infty} P(X(t) = j), \quad j = 0, 1, 2, ..., n$$

The system stays in state 0 for a random time $Y_0 = \min(L_1, L_2, \ldots, L_n)$ and then makes a transition to state 1. The corresponding transition rate is

$$q_0 = q_{01} = n\lambda$$

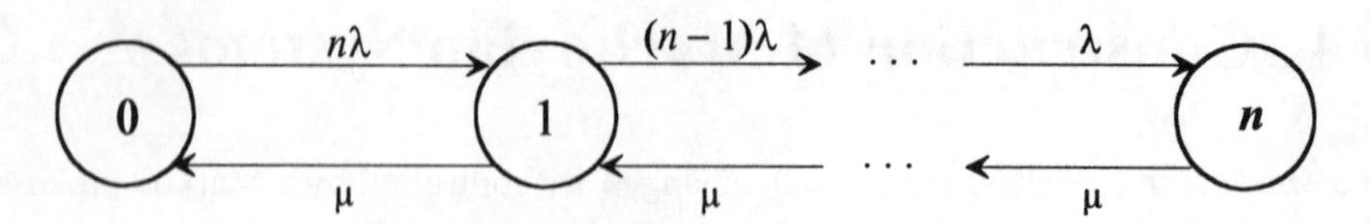

Figure 6.8 Transition graph for example 6.8 (repairman problem)

The system stays in state 1 for a random time $Y_1 = \min(L_1, L_2, \ldots, L_{n-1}, Z)$. From state 1 it makes a transition to state 2 if $Y_1 = L_k$ for $k \in (1, 2, \ldots, n-1)$, and a transition to state 0 if $Y_1 = Z$. Hence,

$$q_{10} = \mu,\ q_{12} = (n-1)\lambda \quad \text{and} \quad q_1 = (n-1)\lambda + \mu$$

In general (Figure 6.8),

$$\begin{aligned} q_{j-1,j} &= (n-j+1)\,\lambda; && j = 1, 2, \ldots, n \\ q_{j+1,j} &= \mu; && j = 0, 1, \ldots, n-1 \\ q_{ij} &= 0; && |i-j| \geq 2 \\ q_j &= (n-j)\lambda + \mu; && j = 1, 2, \ldots, n \\ q_0 &= n\lambda \end{aligned}$$

The corresponding system of equations (6.28) is

$$\mu\pi_1 = n\lambda\pi_0$$

$$(n-j+1)\lambda\pi_{j-1} + \mu\pi_{j+1} = ((n-j)\lambda + \mu)\pi_j; \quad j = 1, 2, \ldots, n-1$$

$$\mu\pi_n = \lambda\pi_{n-1}$$

By successively solving for the π_i, beginning with the first equation, one obtains

$$\pi_j = \frac{n!}{(n-j)!}\,\rho^j\,\pi_0; \quad j = 0, 1, \ldots, n;$$

where $\rho = \lambda/\mu$. From (6.29),

$$\pi_0 = \left[\sum_{i=0}^{n} \frac{n!}{(n-i)!}\,\rho^i\right]^{-1}$$

Hence the stationary state probabilities are

$$\pi_j = \frac{\frac{n!}{(n-j)!}\,\rho^j}{\sum_{i=0}^{n} \frac{n!}{(n-i)!}\,\rho^i}; \quad j = 0, 1, \ldots, n$$

□

6.5 Erlang's Phase Method

Systems with Erlang-distributed sojourn times in their states can be transformed into Markovian systems by increasing the state space by means of dummy states. This is due to the fact that a random variable which is Erlang-distributed with parameters n and μ can be represented as a sum of n independent exponential random variables with parameter μ (example 1.6). Hence, if the time interval which the system spends in a state i, say, is Erlang-distributed with parameters n_i and μ_i, then this interval is partitioned into n_i disjoint subintervals (*phases*), the lengths of which are independent, identically distributed exponential random variables with parameter μ_i. Introducing new states $j_1, j_2, ..., j_{n_i}$ to denote these phases, the original non-Markovian system becomes a Markovian system.

In what follows, a general theoretical treatment of Erlang's phase method will not be presented. Instead, the principle of the approach outlined is demonstrated by an example.

Example 6.9 ***(two-unit system, parallel redundancy)*** As in example 6.6, a two-unit system with parallel redundancy is considered. The lifetimes of the units are identically distributed as an exponential random variable L with parameter λ. The replacement times of the units are identically distributed as Z, where Z is Erlang-distributed of order $n = 2$ with parameter μ. There is only one mechanic in charge of the replacement of failed units. All other assumptions and model specifications are as in example 6.6.

The following system states are introduced:

0 both units are operating

1 one unit is operating, the replacement of the other one is in phase 1

2 one unit is operating, the replacement of the other one is in phase 2

3 no unit is operating, the replacement of the one being maintained is in phase 1

4 no unit is operating, the replacement of the one being maintained is in phase 2

The transition rates are (Figure 6.9):

$$q_{01} = 2\lambda, \qquad q_0 = 2\lambda$$

$$q_{12} = \mu, \; q_{13} = \lambda, \qquad q_1 = \lambda + \mu$$

$$q_{20} = \mu, \; q_{23} = \lambda, \qquad q_2 = \lambda + \mu$$

$$q_{34} = \mu, \qquad q_3 = \mu$$

$$q_{41} = \mu, \qquad q_4 = \mu$$

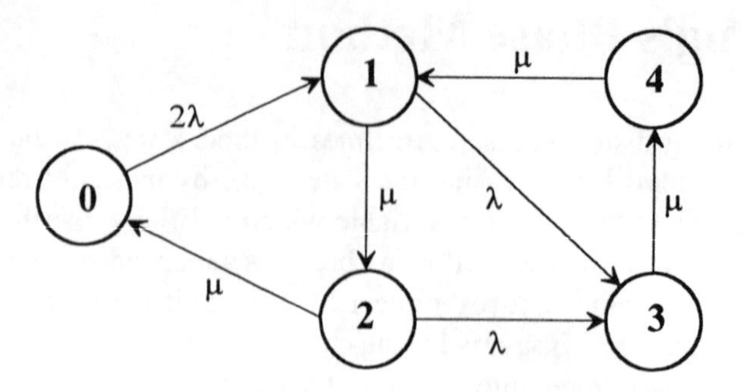

Figure 6.9 Transition graph for example 6.9

Hence the stationary state probabilities satisfy the following system of equations:

$$\mu\pi_2 = 2\lambda\pi_0$$

$$2\lambda\pi_0 + \mu\pi_4 = (\lambda+\mu)\pi_1$$

$$\mu\pi_1 = (\lambda+\mu)\pi_2$$

$$\lambda\pi_1 + \lambda\pi_2 = \mu\pi_3$$

$$\mu\pi_3 = \mu\pi_4$$

$$1 \quad = \pi_0 + \pi_1 + \pi_2 + \pi_3 + \pi_4$$

Let π_i^* denote the stationary probability that i units are failed. Then,

$$\pi_0^* = \pi_0, \quad \pi_1^* = \pi_1 + \pi_2, \quad \pi_2^* = \pi_3 + \pi_4$$

The probabilities π_i^* are the ones of interest. Letting

$$\rho = E(Z)/E(L) = 2\lambda/\mu \,,$$

they are seen to be

$$\pi_0^* = \left[1 + 2\rho + \tfrac{3}{2}\rho^2 + \tfrac{1}{4}\rho^3\right]^{-1}$$

$$\pi_1^* = \left[2\rho + \tfrac{1}{2}\rho^2\right]^{-1}\pi_0^*$$

$$\pi_2^* = \left[\rho^2 + \tfrac{1}{4}\rho^3\right]^{-1}\pi_0^*$$

The stationary system availability is given by $A = \pi_0^* + \pi_1^*$.

Unfortunately, applying Erlang's phase method to real systems may lead to rather complex Markovian systems which are not even tractable on high-speed computers. □

6.6 Birth- and Death Processes

6.6.1 Time-Dependent State Probabilities

Analogously to the particular discrete Markov chain considered in section 5.4, the case where only transitions of the Markov chain into "neighbouring" states are possible is now discussed in more detail. Most of the applications which have been considered so far belong to this category.

Definition 6.3 (*Birth- and death process*) A Markov chain with state space $\mathbf{Z} = \{0, 1, \ldots, n\}$, $n \leq \infty$, is called a *birth- and death process* if from any state i only a transition to $i-1$ or $i+1$ is possible, provided that $i-1 \in \mathbf{Z}$ and $i+1 \in \mathbf{Z}$. ●

Therefore, the transition rates of a birth- and death process have properties

$$q_{i,i+1} > 0 \quad \text{for } i = 1, 2, \ldots$$

$$q_{i,i-1} > 0 \quad \text{for } i = 1, 2, \ldots$$

$$q_{ij} = 0 \quad \text{for } |i-j| > 1$$

The transition rates $\lambda_i = q_{i,i+1}$ and $\mu_i = q_{i,i-1}$ are called *birth rates* and *death rates*, respectively. According to the given state space, $\lambda_n = 0$ for $n < \infty$ and $\mu_0 = 0$ (Figure 6.10).

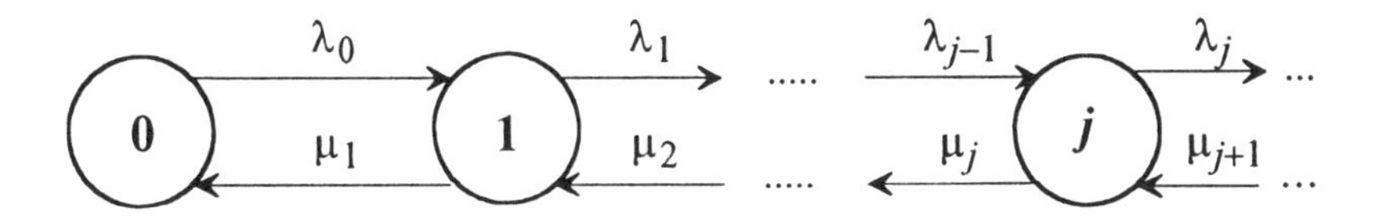

Figure 6.10 Transition graph of a birth- and death process

The investigation of birth- and death processes was initiated by attempts to model the quantitative development of biological populations. In this special application, when arriving in state 0, the population is extinguished. Thus, without the possibility of immigration from outside, state 0 is absorbing ($\lambda_0 = 0$).

Birth- and death processes have proved to be an important tool for modeling queuing and reliability systems (section 6.8). In the economical sciences, birth- and death processes are, for instance, used to describe the development of the number of enterprises in a particular area. In physics they are used to model flows of radioactive, cosmic and other particles.

If $\{X(t), t \geq 0\}$ is a birth- and death process, then, according to (6.20), its absolute state probabilities

$$p_j(t) = P(X(t) = j), \quad j \in \mathbf{Z},$$

satisfy the system of linear differential equations

$$\begin{aligned}
p_0'(t) &= -\lambda_0 p_0(t) + \mu_1 p_1(t) \\
p_j'(t) &= +\lambda_{j-1} p_{j-1}(t) - (\lambda_j + \mu_j) p_j(t) + \mu_{j+1} p_{j+1}(t), \quad j = 1, 2, ... \qquad (6.31) \\
&\ldots\ldots\ldots\ldots \\
p_n'(t) &= +\lambda_{n-1} p_{n-1}(t) - \mu_n p_n(t), \quad n < \infty
\end{aligned}$$

In the sequel, this system is solved for some important special cases.

Pure birth processes A birth- and death process is a *pure birth process* if all its death rates are equal to 0. Hence, the sample paths of pure birth processes are nondecreasing step functions. The homogeneous Poisson process with intensity λ is the simplest example of a pure birth process. In this case, $\lambda_j = \lambda,\ j = 0, 1, ...$

Given the initial distribution $p_m(0) = P(X(0) = m) = 1$ (i.e. in the beginning the population consists of m individuals), the absolute state probabilities $p_j(t)$ are equal to the transition probabilities $p_{mj}(t)$. The $p_j(t)$ are identically 0 for $j < m$ and, according to (6.31), for $j \geq m$ they satisfy the system of linear differential equations

$$\begin{aligned}
p_m'(t) &= -\lambda_m p_m(t) \\
p_j'(t) &= +\lambda_{j-1} p_{j-1}(t) - \lambda_j p_j(t); \quad j = m+1, m+2, ... \qquad (6.32) \\
&\ldots\ldots\ldots\ldots \\
p_n'(t) &= +\lambda_{n-1} p_{n-1}(t), \quad n < \infty
\end{aligned}$$

Note that, in case of a finite state space $\mathbf{Z} = \{m, m+1, ..., n\}$, state n is absorbing, $0 \leq m < n$.

From the first differential equation in (6.32),

$$p_m(t) = e^{-\lambda_m t}, \quad t \geq 0 \qquad (6.33)$$

For $j = m+1, m+2, ...$, the differential equations are equivalent to

$$e^{\lambda_j t}\left(p_j'(t) + \lambda_j p_j(t)\right) = \lambda_{j-1} e^{\lambda_j t} p_{j-1}(t)$$

or

$$\frac{d}{dt}\left(e^{\lambda_j t} p_j(t)\right) = \lambda_{j-1} e^{\lambda_j t} p_{j-1}(t)$$

Integrating,

$$p_j(t) = \lambda_{j-1} e^{-\lambda_j t} \int_0^t e^{\lambda_j x} p_{j-1}(x)\, dx \qquad (6.34)$$

Formulas (6.33) and (6.34) allow the successive determination of the absolute state probabilities $p_j(t)$ for $j = m+1, m+2, \ldots$

In particular, if the initial distribution is $p_0(0) = 1\,(m = 0)$, then, for $\lambda_0 \neq \lambda_1$ and $\lambda_0 > 0$,

$$p_1(t) = \lambda_0\, e^{-\lambda_1 t} \int_0^t e^{\lambda_1 x}\, e^{-\lambda_0 x}\, dx$$

$$= \lambda_0\, e^{-\lambda_1 t} \int_0^t e^{-(\lambda_0 - \lambda_1)x} dx$$

$$= \frac{\lambda_0}{\lambda_0 - \lambda_1} \left(e^{-\lambda_1 t} - e^{-\lambda_0 t} \right), \quad t \geq 0$$

If all birth rates are different from each other, then one obtains from (6.34) by induction

$$p_j(t) = \sum_{i=0}^{j} C_{ij} \lambda_i e^{-\lambda_i t}, \quad j = 0, 1, \ldots \qquad (6.35)$$

with

$$C_{ij} = \frac{1}{\lambda_i} \prod_{\substack{k=0 \\ k \neq i}}^{j} \frac{\lambda_k}{\lambda_k - \lambda_i}; \quad 0 \leq i \leq j; \quad C_{00} = 1$$

Example 6.10 ***(linear birth process)*** A pure birth process is called a *linear birth process* or *a Yule-Furry-process* if its birth rates are given by

$$\lambda_j = j\lambda; \; j = 1, 2, \ldots$$

Linear birth processes occur, for instance, if, in the interval $[t,\, t+h]$, each member of a population (bacterium, physical particle) splits with probability $\lambda h + o(h)$ as $h \to 0$. Assuming $X(0) = 1$, equations (6.32) specialize to

$$p_j'(t) = -\lambda\left[j p_j(t) - (j-1) p_{j-1}(t) \right]; \quad j = 1, 2, \ldots ;$$

with

$$p_1(0) = 1, \quad p_j(0) = 0; \quad j = 2, 3, \ldots$$

From(6.35),

$$p_j(t) = e^{-\lambda t}(1 - e^{-\lambda t})^{j-1}; \quad j = 1, 2, \ldots$$

Thus, $X(t)$ has a geometric distribution with parameter $e^{-\lambda t}$. Hence the trend function of this process is

$$m(t) = e^{\lambda t}, \quad t \geq 0$$

□

If $\mathbf{Z}$ is finite, then there always exists a solution to (6.32) satisfying

$$\sum_{i \in \mathbf{Z}} p_i(t) = 1 \tag{6.36}$$

In the case of an infinite state space $\mathbf{Z} = \{0, 1, \ldots\}$, the following theorem gives a necessary and sufficient condition for the existence of a solution of (6.32) satisfying (6.36). Without loss of generality, this theorem is proved on condition that $p_0(0) = 1$.

Theorem 6.2 (*Feller-Lundberg*) A solution $\{p_0(t), p_1(t), \ldots\}$ of (6.32) satisfies condition (6.36) if and only if the series

$$\sum_{j=0}^{\infty} \frac{1}{\lambda_j} \tag{6.37}$$

diverges.

Proof Let

$$s_k(t) = p_0(t) + p_1(t) + \ldots + p_k(t)$$

Summing the middle equation of (6.32) from $j = 1$ to k yields

$$s_k'(t) = -\lambda_k p_k(t)$$

Integrating, taking into account $s_k(0) = 1$,

$$1 - s_k(t) = \lambda_k \int_0^t p_k(x)\, dx \tag{6.38}$$

Since $s_k(t)$ is monotonically increasing as $k \to \infty$, the limit $\lim_{k \to \infty} (1 - s_k(t))$ exists:

$$r(t) = \lim_{k \to \infty} (1 - s_k(t))$$

From (6.38),

$$\lambda_k \int_0^t p_k(x)\, dx \ge r(t)$$

Summing these inequalities from 0 to k,

$$\int_0^t s_k(x)\, dx \ge r(t)\left(\frac{1}{\lambda_0} + \frac{1}{\lambda_1} + \ldots + \frac{1}{\lambda_k}\right)$$

Since $s_k(t) \le 1$ for all $t \ge 0$,

$$t \ge r(t)\left(\frac{1}{\lambda_0} + \frac{1}{\lambda_1} + \ldots + \frac{1}{\lambda_k}\right)$$

If the series (6.37) diverges, then this inequality implies that $r(t) \equiv 0$. But this identity is equivalent to (6.36).

Conversely, from (6.38),

$$\lambda_k \int_0^t p_k(x)\, dx \le 1$$

so that

$$\int_0^t s_k(x)\,dx \le \frac{1}{\lambda_0} + \frac{1}{\lambda_1} + \ldots + \frac{1}{\lambda_k}$$

Passing to the limit as $k \to \infty$ yields

$$\int_0^t (1 - r(t))dt \le \sum_{j=0}^{\infty} \frac{1}{\lambda_j}$$

On condition that $r(t) \equiv 0$, the left-hand side of this inequality is equal to t. Since t can be arbitrarily large, the series (6.37) must diverge. This result completes the proof of the theorem. ■

According to the theorem of Feller-Lundberg it is theoretically possible that within a finite interval $[0, t]$ the population grows beyond all finite bounds. This occurs with positive probability

$$1 - \sum_{i=0}^{\infty} p_i(t)$$

if the birth rates grow sufficiently fast. In such a case an *explosive growth* of the population takes place. It is remarkable that an explosive growth, when it is happening, then it happens in an arbitrarily small interval. This is due to the fact that the convergence of the series (6.37) does not depend on t. An explosive growth would occur, for example, if $\lambda_j = j^2 \lambda$; $j = 1, 2, \ldots$, since

$$\sum_{j=1}^{\infty} \frac{1}{\lambda_j} = \frac{1}{\lambda} \sum_{j=1}^{\infty} \frac{1}{j^2} = \frac{\pi^2}{6\lambda} < \infty$$

Pure death processes A birth- and death process is called a *pure death process* if all its birth rates are zero. The sample paths of such processes are nonincreasing step functions. For pure death processes, the system of linear differential equation (6.31) simplifies to (assuming $p_n(0) = 1$)

$$p_n'(t) = -\mu_n p_n(t)$$

$$p_j'(t) = -\mu_j p_j(t) + \mu_{j+1} p_{j+1}(t); \quad j = 0, 1, \ldots, n-1$$

From the first differential equation,

$$p_n(t) = e^{-\mu_n t}, \quad t \ge 0$$

Starting with $p_n(t)$ the complete solution of this system of differential equations can be determined recursively by the relationship

$$p_j(t) = \mu_{j+1} e^{-\mu_j t} \int_0^t e^{\mu_j x} p_{j+1}(x)\,dx; \quad j = n-1, \ldots, 1, 0 \tag{6.39}$$

For instance, assuming $\mu_n \ne \mu_{n-1}$,

$$p_{n-1}(t) = \mu_n e^{-\mu_{n-1} t} \int_0^t e^{-(\mu_n - \mu_{n-1})x} dx$$

$$= \frac{\mu_n}{\mu_n - \mu_{n-1}} \left(e^{-\mu_{n-1} t} - e^{-\mu_n t} \right)$$

If the death rates are different from each other, the complete solution is inductively seen to be

$$p_j(t) = \sum_{i=j}^{n} D_{ij} \mu_i e^{-\mu_i t}, \quad 0 \le j \le n, \tag{6.40}$$

where

$$D_{ij} = \frac{1}{\mu_j} \prod_{\substack{k=j \\ k \neq i}}^{n} \frac{\mu_k}{\mu_k - \mu_i}, \quad j \le i \le n, \quad D_{nn} = \frac{1}{\mu_n}$$

Example 6.11 (***linear death process***) A system consisting of n subsystems starts operating at time $t = 0$. The lifetimes of the subsystems are independent, exponential random variables with parameter λ. If $X(t)$ denotes the number of subsystems still working at time t, then $\{X(t), t \ge 0\}$ is a pure death process with death rates

$$\mu_j = j\lambda; \; j = 0, 1, ..., n$$

$\{X(t), t \ge 0\}$ is called a *linear death process.* Since $p_n(0) = 1$,

$$p_n(t) = e^{-n\lambda t}, \quad t \ge 0$$

Starting with $p_n(t)$, one obtains inductively,

$$p_j(t) = \binom{n}{j} e^{-j\lambda t} (1 - e^{-\lambda t})^{n-j}; \quad j = 0, 1, ..., n$$

Hence, $X(t)$ is binomially distributed with parameters $e^{-\lambda t}$ and n so that the trend function of a linear death process is $m(t) = n e^{-\lambda t}, t \ge 0$. □

Special birth- and death processes In the following two examples, the state probabilities $\{p_0(t), p_1(t), ...\}$ of two important birth- and death processes are determined via their z-transforms (section 1.5.1):

$$M(t, z) = \sum_{i=0}^{\infty} p_i(t) z^i$$

The initial conditions assumed are

$$p_0(0) = 1 \quad \text{or} \quad p_1(0) = 1$$

In terms of the z-transform, these initial conditions are equivalent to

$$M(0, z) \equiv 1 \quad \text{and} \quad M(0, z) \equiv z,$$

respectively. (In general, the initial condition $p_i(0) = P(X(0) = i) = 1$ is equivalent to $M(0, z) \equiv z^i$; $i = 0, 1, ...$) Furthermore, the partial derivatives

$$\frac{\partial M(t,z)}{\partial t} = \sum_{i=0}^{\infty} p_i'(t) z^i \quad \text{and} \quad \frac{\partial M(t,z)}{\partial z} = \sum_{i=1}^{\infty} i p_i(t) z^{i-1} \tag{6.41}$$

will be needed.

Example 6.12* (***linear birth- and death process***) A birth- and death process $\{X(t), t \ge 0\}$ with transition rates

$$\lambda_j = j\lambda, \quad \mu_j = j\mu; \quad j = 0, 1, ...$$

is called a *linear birth- and death process.* In what follows, this process is analyzed on condition that $p_1(0) = P(X(0) = 1) = 1$. (Assuming $p_0(0) = 1$ would make no sense since state 0 is absorbing.)

The system of differential equations (6.31) is

$$p_0'(t) = \mu p_1(t)$$

$$p_j'(t) = (j-1)\lambda p_{j-1}(t) - j(\lambda+\mu) p_j(t) + (j+1)\mu p_{j+1}(t); \quad j = 1, 2, ... \tag{6.42}$$

Multiplying the j-th differential equation by z^j and summing from $j = 0$ to $j = \infty$, taking into account (6.41), yields the following linear homogeneous partial differential equation in $M(t,z)$:

$$\frac{\partial M(t,z)}{\partial t} - (z-1)(\lambda z - \mu)\frac{\partial M(t,z)}{\partial z} = 0 \tag{6.43}$$

The corresponding (ordinary) characteristic differential equation is

$$\frac{dz}{dt} = -(z-1)(\lambda z - \mu)$$

After separation of variables,

$$\frac{dz}{(z-1)(\lambda z - \mu)} = -dt \tag{6.44}$$

a) $\lambda \neq \mu$ Integration on both sides of (6.44) yields

$$-\frac{1}{\lambda - \mu} \ln\left(\frac{\lambda z - \mu}{z-1}\right) = -t + C$$

The general solution $z = z(t)$ of the characteristic differential equation in implicit form is, therefore, given by

$$c = (\lambda - \mu)t - \ln\left(\frac{\lambda z - \mu}{z-1}\right),$$

where c is an arbitrary constant. Thus, the general solution $M(t,z)$ of (6.43) has structure

$$M(t,z) = f\left((\lambda - \mu)t - \ln\left(\frac{\lambda z - \mu}{z-1}\right)\right),$$

where f is a continuously differentiable function. f can be determined by making use of the initial condition $p_1(0) = 1$ or, equivalently, $M(0, z) = z$. Since

$$M(0,z) = f\left(-\ln\left(\frac{\lambda z-\mu}{z-1}\right)\right)$$

$$= f\left(\ln\left(\frac{z-1}{\lambda z-\mu}\right)\right) = z,$$

f is given by

$$f(x) = \frac{\mu e^x - 1}{\lambda e^x - 1}$$

Hence,

$$M(t,z) = \frac{\mu\exp\left\{(\lambda-\mu)t - \ln\left(\frac{\lambda z-\mu}{z-1}\right)\right\} - 1}{\lambda\exp\left\{(\lambda-\mu)t - \ln\left(\frac{\lambda z-\mu}{z-1}\right)\right\} - 1}$$

After simplification, $M(t,z)$ becomes

$$M(t,z) = \frac{\mu\left[1-e^{(\lambda-\mu)t}\right] - \left[\lambda-\mu e^{(\lambda-\mu)t}\right]z}{\left[\mu-\lambda e^{(\lambda-\mu)t}\right] - \lambda\left[1-\mu e^{(\lambda-\mu)t}\right]z}$$

This representation of $M(t,z)$ allows its expansion as a power series in z. The coefficient of z^j is the desired absolute state probability $p_j(t)$. Letting $\rho = \lambda/\mu$, the $p_j(t)$ are seen to be

$$p_0(t) = \frac{1-e^{(\lambda-\mu)t}}{1-\rho e^{(\lambda-\mu)t}},$$

$$p_j(t) = (1-\rho)\rho^{j-1}\frac{\left[1-e^{(\lambda-\mu)t}\right]^{j-1}}{\left[1-\rho e^{(\lambda-\mu)t}\right]^{j+1}}e^{(\lambda-\mu)t}, \quad j = 1,2,...$$

Since 0 is an absorbing state, $p_0(t)$ is the probability that the population is extinguished at time t. Moreover,

$$\lim_{t\to\infty} p_0(t) = \begin{cases} 1 & \text{for } \lambda < \mu \\ \frac{\mu}{\lambda} & \text{for } \lambda > \mu \end{cases}$$

Thus, if $\lambda > \mu$, the population will survive infinitely with positive probability μ/λ. If $\lambda < \mu$, the population will certainly be extinguished sooner or later. In the latter case it makes sense to ask for the lifetime L of the population. L is a random variable with distribution function

$$P(L \le t) = p_0(t) = \frac{1 - e^{(\lambda-\mu)t}}{1 - \rho e^{(\lambda-\mu)t}}, \quad t \ge 0$$

Hence,

$$P(L > t) = 1 - p_0(t)$$

is the *survival probability* of the population. Using (1.12), the expected lifetime of the population turns out to be

$$E(L) = \frac{1}{\mu - \lambda} \ln\left(2 - \frac{\lambda}{\mu}\right)$$

The trend function $m(t) = E(X(t))$ of the linear birth- and death process is

$$m(t) = \sum_{j=0}^{\infty} j p_j(t)$$

According to (1.54), it can be also obtained from the moment generating function:

$$m(t) = \left. \frac{\partial M(t,z)}{\partial z} \right|_{z=1}$$

If $M(t,z)$ is not known, then it is more convenient to determine $m(t)$ from the system of differential equations (6.42) by multiplying the j-th differential equation by j and summing from $j=0$ to ∞ to obtain the following first-order differential equation in $m(t)$:

$$m'(t) = (\lambda - \mu) m(t) \tag{6.45}$$

Taking into account the initial condition $p_1(0) = 1$, its solution is

$$m(t) = e^{(\lambda-\mu)t}$$

Analogously, by multiplying the j-th differential equation of (6.42) by j^2 and summing from $j = 0$ to ∞, a second order differential equation in $Var(X(t))$ is obtained, which has solution

$$Var(X(t)) = \frac{\lambda + \mu}{\lambda - \mu}\left[1 - e^{-(\lambda-\mu)t}\right] e^{2(\lambda-\mu)t}$$

Of course, since $M(t,z)$ is known, $Var(X(t))$ can be also obtained from (1.54).

If the linear birth- and death process starts in state i, no principal additional problems arise up to the determination of $M(t, z)$. But it will be more complicated to expand $M(t,z)$ as a power series in z. The corresponding trend function, however, is immediately obtained from (6.45):

$$m(t) = i\, e^{(\lambda-\mu)t}, \quad t \ge 0$$

b) $\lambda = \mu$ In this case the characteristic differential equation (6.44) is

$$\frac{dz}{\lambda(z-1)^2} = -dt$$

Integration yields

$$c = \lambda t - \frac{1}{z-1},$$

where c is an arbitrary constant. Therefore, $M(t, z)$ has structure

$$M(t,z) = f\left(\lambda t - \frac{1}{z-1}\right),$$

where f is a continuously differentiable function. Since $p_1(0) = 1$, f satisfies condition

$$f\left(-\frac{1}{z-1}\right) = z$$

Hence,

$$f(x) = 1 - \frac{1}{x}$$

so that the z-transform is

$$M(t,z) = \frac{\lambda t + (1-\lambda t)z}{1 + \lambda t - \lambda t z}$$

Expanding $M(t, z)$ as a power series in z yields the absolute state probabilities:

$$p_0(t) = \frac{\lambda t}{1+\lambda t}, \qquad p_j(t) = \frac{(\lambda t)^{j-1}}{(1+\lambda t)^{j+1}}; \quad j = 1, 2, \ldots$$

An equivalent form of the absolute state probabilities is

$$p_0(t) = \frac{\lambda t}{1+\lambda t}, \qquad p_j(t) = [1 - p_0(t)]^2\, [p_0(t)]^{j-1}; \quad j = 1, 2, \ldots$$

Expected value and variance of $X(t)$ are

$$E(X(t)) = 1, \quad Var(X(t)) = 2\lambda t$$

This example shows that the analysis of apparently simple birth- and death processes requires some effort. □

Example 6.13 Let $\lambda_j = \lambda$, $\mu_j = j\mu$; $j = 0, 1, \ldots$ and $p_0(0) = P(X(0) = 0) = 1$. The corresponding system of linear differential equations (6.31) becomes

$$p_0'(t) = \mu p_1(t) - \lambda p_0(t)$$

$$p_j'(t) = \lambda p_{j-1}(t) - (\lambda + \mu j) p_j(t) + (j+1)\mu p_{j+1}(t); \quad j = 1, 2, \ldots \tag{6.46}$$

Multiplying the j-th equation by z^j and summing from $j = 0$ to ∞ yields a homogeneous linear partial differential equation for the moment generating function $M(t, z) = \Sigma_{j=0}^{\infty} p_j(t) z^j$:

$$\frac{\partial M(t,z)}{\partial t} + \mu(z-1)\frac{\partial M(t,z)}{\partial z} = \lambda(z-1)\,M(t,z) \tag{6.47}$$

The corresponding system of characteristic differential equations is

$$\frac{dz}{dt} = \mu(z-1)$$

$$\frac{dM(t,z)}{dt} = \lambda(z-1)\,M(t,z)$$

After separation of variables and subsequent integration, the first differential equation yields

$$c_1 = \ln(z-1) - \mu t\,,$$

where c_1 is an arbitrary constant. By combining both differential equations,

$$\frac{dM(t,z)}{M(t,z)} = \frac{\lambda}{\mu}\,dz$$

Integration yields

$$c_2 = \ln M(t,z) - \frac{\lambda}{\mu} z\,,$$

where c_2 is an arbitrary constant. As a solution of (6.47), $M(t,z)$ satisfies the condition $c_2 = f(c_1)$, where f is an arbitrary continuous function, or, equivalently,

$$\ln M(t,z) - \frac{\lambda}{\mu} z = f(\ln(z-1) - \mu t)$$

Therefore,

$$M(t,z) = \exp\left\{f(\ln(z-1) - \mu t) + \frac{\lambda}{\mu} z\right\}$$

The initial condition $p_0(0) = 1$ is equivalent to

$$M(0,z) \equiv 1$$

Hence, f must satisfy

$$f(\ln(z-1)) = -\frac{\lambda}{\mu} z$$

Thus,

$$f(x) = -\frac{\lambda}{\mu}(e^x + 1)$$

so that the moment generating function becomes

$$M(t,z) = \exp\left\{-\frac{\lambda}{\mu}\left(e^{\ln(z-1) - \mu t} + 1\right) + \frac{\lambda}{\mu} z\right\}$$

or, equivalently,

$$M(t,z) = e^{-\frac{\lambda}{\mu}(1-e^{-\mu t})} \cdot e^{+\frac{\lambda}{\mu}(1-e^{-\mu t})z}$$

In this form, it is easy to expand $M(t, z)$ as a power series in z. The coefficients of z^j are

$$p_j(t) = \frac{\left(\frac{\lambda}{\mu}(1-e^{-\mu t})\right)^j}{j!} e^{-\frac{\lambda}{\mu}(1-e^{-\mu t})}; \quad j = 0, 1, ... \tag{6.48}$$

This is a Poisson distribution with intensity function $\frac{\lambda}{\mu}(1-e^{-\mu t})$. Therefore, the birth- and death process has trend function

$$m(t) = \frac{\lambda}{\mu}(1-e^{-\mu t})$$

If the process starts in state i, $i > 0$, then the absolute state probabilities are not Poisson distributed. They have a rather complicated structure which will not be presented here. Instead, the system of linear differential equations (6.46) can be used to establish ordinary differential equations for the trend function $m(t)$ and the variance of $X(t)$. Given the initial distribution $p_i(0) = 1$, their respective solutions are

$$m(t) = \frac{\lambda}{\mu}(1-e^{-\mu t}) + i e^{-\mu t},$$

$$Var(X(t)) = (1-e^{-\mu t})\left(\frac{\lambda}{\mu} + i e^{-\mu t}\right)$$

The special birth- and death process considered in this example is of some importance in queueing theory (section 6.7). □

6.6.2 Stationary State Probabilities

According to (6.27), the stationary state probabilities of a birth- and death process

$$\pi_j = \lim_{t\to\infty} p_j(t)$$

satisfy the following system of linear algebraic equations

$$\begin{aligned} &\lambda_0\pi_0 - \mu_1\pi_1 = 0 \\ &\lambda_{j-1}\pi_{j-1} - (\lambda_j + \mu_j)\pi_j + \mu_{j+1}\pi_{j+1} = 0; \quad j = 1, 2, ... \\ &\lambda_{n-1}\pi_{n-1} - \mu_n\pi_n = 0, \quad n < \infty \end{aligned} \tag{6.49}$$

This system is equivalent to the following one:

$$\begin{aligned}
&\mu_1\pi_1 = \lambda_0\pi_0 \\
&\mu_{j+1}\pi_{j+1} + \lambda_{j-1}\pi_{j-1} = (\lambda_j + \mu_j)\pi_j\,; \quad j = 1, 2, \ldots \\
&\mu_n\pi_n = \lambda_{n-1}\pi_{n-1}\,, \quad n < \infty
\end{aligned} \tag{6.50}$$

In contrast to the system of linear differential equations in the time-dependent state probabilities (6.31), it is possible to obtain the general solution of (6.49) or (6.50). Letting

$$h_j = -\lambda_j\pi_j + \mu_{j+1}\pi_{j+1}\,; \quad j = 0, 1, \ldots$$

the system (6.49) simplifies to

$$\begin{aligned}
&h_0 = 0 \\
&h_j - h_{j-1} = 0 \\
&\ldots\ldots\ldots \\
&h_{n-1} = 0, \quad n < \infty
\end{aligned}$$

Starting with $j = 0$, one successively obtains

$$\begin{aligned}
&\pi_1 = \frac{\lambda_0}{\mu_1}\pi_0 \\
&\pi_2 = \frac{\lambda_0}{\mu_1}\frac{\lambda_1}{\mu_2}\pi_0 \\
&\ldots\ldots\ldots \\
&\pi_j = \prod_{i=1}^{j}\frac{\lambda_{i-1}}{\mu_i}\pi_0\,; \quad j = 1, 2, \ldots, n
\end{aligned} \tag{6.51}$$

If $n < \infty$, then the stationary state probabilities satisfy the normalizing condition

$$\sum_{i=0}^{n}\pi_i = 1 \tag{6.52}$$

Solving for π_0 yields

$$\pi_0 = \left[1 + \sum_{j=1}^{n}\prod_{i=1}^{j}\frac{\lambda_{i-1}}{\mu_i}\right]^{-1} \tag{6.53}$$

If $n = \infty$, then equation (6.53) shows that the convergence of the series

$$\sum_{j=1}^{\infty}\prod_{i=1}^{j}\frac{\lambda_{i-1}}{\mu_i} \tag{6.54}$$

is necessary for the existence of a stationary distribution. A sufficient condition for the convergence of this series is the existence of a positive integer N such that

$$\frac{\lambda_{i-1}}{\mu_i} \le \alpha < 1 \quad \text{for all } i > N \tag{6.55}$$

Intuitively this condition seems to be obvious: If the birth rates are greater than the death rates, then the process will drift to infinity with probability 1. The following theorem is given without proof.

Theorem 6.3 The convergence of the series (6.54) and the divergence of the *Karlin-Gregor-series*

$$\sum_{j=1}^{\infty} \prod_{i=1}^{j} \frac{\mu_i}{\lambda_i}$$

are sufficient for the existence of a stationary state distribution. The divergence of the Karlin-Gregor-series is, moreover, sufficient for the existence of a time-dependent solution $\{p_0(t), p_1(t). \dots\}$ of (6.31) which satisfies

$$\sum_{j=0}^{\infty} p_j(t) = 1, \quad t \geq 0$$

■

Example 6.14 (*repairman problem*) The repairman problem, introduced in example 6.8, is considered once more. However, now it is assumed that there are r mechanics to repair the n machines, $1 \leq r \leq n$. All the other assumptions as well as the notation are as in example 6.8 (Figure 6.11).

If $X(t)$ denotes the number of failed machines at time t, then $\{X(t), t \geq 0\}$ is a birth- and death process with state space $\mathbf{Z} = \{0, 1, \dots, n\}$. Its transition rates are

$$\lambda_j = (n-j)\lambda, \quad 0 \leq j \leq n$$

and

$$\mu_j = \begin{cases} j\mu, & 0 \leq j \leq r \\ r\mu, & r < j \leq n \end{cases},$$

respectively. (In this excmple, the terminology of birth- and death rates does not reflect the technological situation.) If the *service rate* $\rho = \lambda/\mu$ is introduced, formulas (6.51) and (6.53) yield the stationary state probabilities

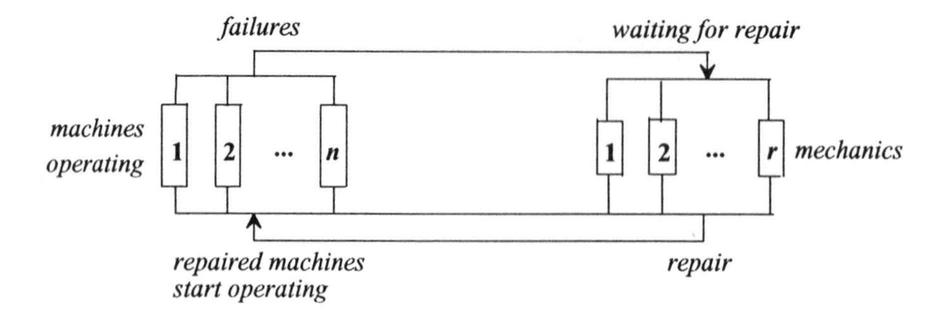

Figure 6.11 Repairman problem

$$\pi_j = \begin{cases} \binom{n}{j} \rho^j \pi_0 ; & 1 \le j \le r \\ \dfrac{n!}{r^{j-r}\, r!\,(n-j)!} \rho^j \pi_0 ; & r \le j \le n \end{cases} \tag{6.56}$$

$$\pi_0 = \left[\sum_{j=0}^{r} \binom{n}{j} \rho^j + \sum_{j=r+1}^{n} \frac{n!}{r^{j-r}\, r!\,(n-j)!} \rho^j \right]^{-1}$$

A practical application of the stationary state probabilities (6.56) is illustrated by a numerical example: Let be $n = 10$, $\rho = 0.3$ and $r = 2$. The efficiency of two maintenance policies will be compared:

1) Both mechanics are in charge of the repair of any of the 10 machines.

2) The mechanics are each assigned 5 machines for whose repair they alone are responsible.

Let $X_{n,r}$ and $Z_{n,r}$ be the random number of failed machines and the random number of mechanics, which are busy with repairing failed machines, respectively; dependent on the number n of machines and the number r of available mechanics. From table 6.1:

$$E(X_{10,2}) = \Sigma_{j=1}^{10} j \pi_j = 3.902$$

$$E(Z_{10,2}) = 1 \cdot \pi_1 + 2 \Sigma_{j=2}^{10} \pi_j = 1.8296$$

policy 1: $n = 10$, $r = 2$		*policy* 2: $n = 5$, $r = 1$	
j	π_j	j	π_j
0	0.0341	0	0.1450
1	0.1022	1	0.2175
2	0.1379	2	0.2611
3	0.1655	3	0.2350
4	0.1737	4	0.1410
5	0.1564	5	0.0004
6	0.1173		
7	0.0704		
8	0.0316		
9	0.0095		
10	0.0014		

Table 6.1 Numerical results for example 6.14 (j = number of failed machines)

$$E(X_{5,1}) = \Sigma_{j=1}^{5} j\pi_j = 2.011$$

$$E(Z_{5,1}) = \Sigma_{j=1}^{5} \pi_j = 0.855$$

When applying policy 2, the average number of failed machines out of the 10 and the average number of busy mechanics out of the 2 are

$$2E(X_{5,1}) = 4.022 \quad \text{and} \quad 2E(Z_{5,1}) = 1.71,$$

respectively. Thus, on the one hand, the expected number of failed machines under policy 1 is smaller than under policy 2, and, on the other hand, the mechanics are less busy under policy 2 than under policy 1. Hence, policy 1 should be preferred if there are no other relevant performance criteria. □

Example 6.15 The repairman problem of example 6.14 is modified in the following way: The available maintenance capacity of r units (which needs not necessarily be human) is always fully used for repairing failed machines. Thus, if only one machine has failed, then all r units are busy with repairing this machine. If several machines are down, the full maintenance capacity of r units is uniformly distributed to the failed machines. This adaptation is repeated after each failure of a machine and after each completion of a repair. In this case, no machines have to wait for repair.

If j machines have failed, then the repair rate of each failed machine is $(r/j)\,\mu$. Therefore, the death rates of the corresponding birth- and death process are constant:

$$\mu_j = j \cdot \frac{r}{j}\mu = r\mu; \quad j = 1, 2, ...$$

The birth rates are as in example 6.14

$$\lambda_j = (n-j)\,\lambda; \quad j = 0, 1, ...$$

Thus, the stationary state probabilities are according to (6.51)

$$\pi_j = \frac{n!}{(n-j)!}\left(\frac{\lambda}{r\mu}\right)^j \pi_0; \quad j = 1, 2, ...$$

$$\pi_0 = \left[\sum_{j=1}^{n} \frac{n!}{(n-j)!}\left(\frac{\lambda}{r\mu}\right)^j\right]^{-1}$$

Comparing this result with the stationary state probabilities (6.56), it is apparent that in the case of $r = 1$ the uniform distribution of the repair capacity to the failed machines has no influence on the stationary state probabilities. This fact is not surprising, since in this case the available maintenance capacity of one unit, if required, is always fully used. □

6.7 Sojourn Times*

So far the fact has been used that independent, exponentially distributed times between state changes allow modeling by homogeneous Markov chains. Conversely, by making use of the relationships (6.8) and (6.15), it can be shown that, for any $i \in \mathbf{Z}$, the sojourn time Y_i of a homogeneous Markov chain $\{X(t), t \geq 0\}$ in state i also has an exponential distribution:

$$P(Y_i > t | X(0) = i)$$

$$= P(X(s) = i,\ 0 < s \leq t | X(0) = i)$$

$$= \lim_{n \to \infty} P\left(X\left(\tfrac{k}{n}t\right) = i;\ k = 1, 2, \ldots, n \,\middle|\, X(0) = i\right)$$

$$= \lim_{n \to \infty} \left[p_{ii}\left(\tfrac{1}{n}t\right)\right]^n$$

$$= \lim_{n \to \infty} \left[1 - q_i \tfrac{t}{n} + o\left(\tfrac{1}{n}\right)\right]^n$$

Since e can be represented as the limit

$$e = \lim_{x \to \infty} \left(1 + \tfrac{1}{x}\right)^x, \tag{6.57}$$

it follows that

$$P(Y_i > t | X(0) = i) = e^{-q_i t}, \quad t \geq 0 \tag{6.58}$$

Thus, Y_i is exponentially distributed with parameter q_i.

Given that $X(0) = i$, $X(Y_i)$ is the state to which the Markov chain makes a transition on leaving state i. Let $m(nt)$ denote the greatest integer m satisfying $m/n \leq t$ or, equivalently, $nt - 1 < m(nt) \leq nt$, then the joint probability distribution of Y_i and $X(Y_i)$, $i \neq j$, can be obtained as follows:

$$P(X(Y_i) = j,\ Y_i > t | X(0) = i)$$

$$= P(X(Y_i) = j,\ X(s) = i \text{ for } 0 < s \leq t | X(0) = i)$$

$$= \lim_{n \to \infty} \sum_{m=m(nt)}^{\infty} P\left(X\left(\tfrac{m+1}{n}\right) = j,\ Y_i \in \left[\tfrac{m}{n}, \tfrac{m+1}{n}\right) \,\middle|\, X(0) = i\right)$$

$$= \lim_{n \to \infty} \sum_{m=m(nt)}^{\infty} P\left(X\left(\tfrac{m+1}{n}\right) = j,\ X\left(\tfrac{k}{n}\right) = i \text{ for } 1 \leq k \leq m \,\middle|\, X(0) = i\right)$$

$$= \lim_{n\to\infty} \sum_{m=m(nt)}^{\infty} \left[1 - q_i \frac{1}{n} + o\left(\frac{1}{n}\right)\right]^m \left[q_{ij} \frac{1}{n} + o\left(\frac{1}{n}\right)\right]$$

$$= \lim_{n\to\infty} \frac{\left[1 - q_i \frac{1}{n} + o\left(\frac{1}{n}\right)\right]^{m(nt)}}{q_i \frac{1}{n} + o\left(\frac{1}{n}\right)} \left[q_{ij} \frac{1}{n} + o\left(\frac{1}{n}\right)\right]$$

Hence, by (6.57),

$$P(X(Y_i) = j,\ Y_i > t | X(0) = i) = \frac{q_{ij}}{q_i} e^{-q_i t}; \quad i \neq j;\ i,j \in \mathbf{Z}. \tag{6.59}$$

Passing to the marginal distribution of Y_i (i.e. summing the equations (6.59) with respect to $j \in \mathbf{Z}$) again proves the result (6.58). Further important conclusions are:

1) The one-step transition probability from state i into state j is given by

$$p_{ij} = P(X(Y_i) = j | X(0) = i) = \frac{q_{ij}}{q_i}$$

This formula results from (6.59) by letting there $t = 0$. Hence, the probability distribution of the state following state i is

$$\left\{\frac{q_{ij}}{q_i};\ j \in \mathbf{Z}\right\}$$

2) The state following state i is independent of Y_i (and, of course, independent of the history of the Markov chain before arriving in state i).

Knowledge of the transition probabilities p_{ij} suggests to observe the continuous-time Markov chain $\{X(t),\ t \geq 0\}$ only at those discrete time points where state changes take place. Let X_n be the state of the Markov chain immediately after the n-th change of state, $X_0 = X(0)$. Then $\{X_0, X_1, ...\}$ is a discrete-time homogeneous Markov chain with transition probabilities $p_{ij};\ i,j \in \mathbf{Z}$, which can be written in the form

$$p_{ij} = P(X_n = j | X_{n-1} = i); \quad n = 1, 2, ...$$

In this sense, the discrete-time Markov chain $\{X_0, X_1, ...\}$ is *embedded* in the continuous-time Markov chain $\{X(t),\ t \geq 0\}$. Embedded Markov chains can also be found in non-Markov processes (see also example 5.5). In these cases, they may facilitate the investigation of non-Markov processes. Embedded discrete-time Markov chains are frequently the main tool for analyzing non-Markovian, continuous-time stochastic processes. Section 6.9 deals with semi-Markov processes whose framework is an embedded Markov chain.

6.8 Applications in Queueing Theory

6.8.1 Introduction

One of the most important applications of continuous-time Markov chains is the stochastic modeling of service facilities. The basic situation is the following: Customers arrive at a service system (= queueing system) at random time points. If all servers are busy, an arriving customer either waits for service a random time or leaves the system without having been served. Otherwise an available server takes care of the customer. After a random service time the customer leaves the system. The arriving customers constitute the *input* (the *traffic*, the *flow of demands*) and the leaving customers the *output* of the queueing system. A queueing system is called a *loss system* if it has no waiting capacity for customers which do not find an available server on arriving at the system. These customers leave the system immediately after arrival and are said to be *lost.* A *waiting system* has an infinite waiting capacity for customers who do not immediately find an available server and are willing to wait any length of time for service. A *waiting-loss system* has a bounded waiting capacity for customers. An arriving customer is lost if he finds all servers busy and the waiting capacity occupied. A *multi-server queueing system* has s, $s > 1$, servers. Correspondingly, a *single-server queueing system* has only one server.

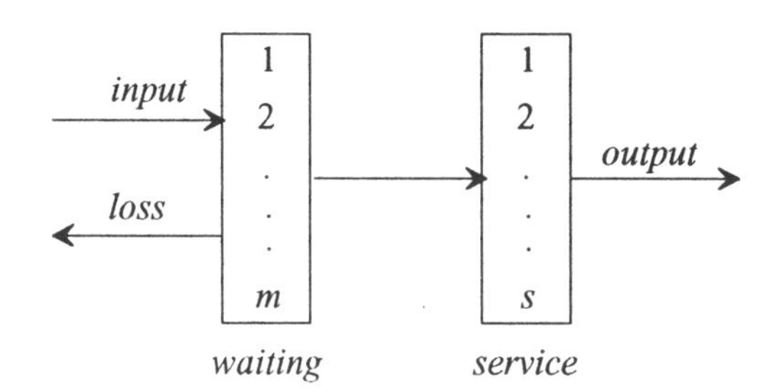

Figure 6.12 Scheme of a standard queueing system

Examples Supermarkets are simple examples of queueing systems. Its input (flow of customers) is served at checkout counters. Filling stations can also be modeled by queueing systems. The petrol pumps are the servers. Filling stations are typical waiting-loss systems. Even a car park has the typical features of a standard queueing system. Each single car space can be considered to be a server. In this case, the service time only depends on the customer himself. An anti-aircraft battery is a queueing system in the sense that it "serves" the enemy aircraft. During recent years the stochastic modeling of communication systems, in particular computer networks, has stimulated the application of known queueing models and the creation of new, more sophisticated ones. However, the investigation of queueing systems goes back to the Danish engineer *A. K. Erlang* in the early 1900s when he had to plan telephone exchanges to meet criteria such as "what is the mean waiting time of a customer before being connected" or "how many lines (servers) are necessary to guarantee that, with a given probability, a customer can immediately be connected"? The repairman problem considered in example 6.14 also fits into the framework of a queueing system. The failed machines constitute the input and the mechanics are the servers. This example is distinguished by a particular feature: each demand ("customer") is generated by one of a finite number n of different sources, namely by the failure of a specific machine. Classes of queueing systems having this particular feature are called *closed service systems.*

The objective of queueing theory is to enable the appropriate design of service systems for specific situations. The global criterion is to provide the required service at minimal expense. In particular, managers of service systems do not want to "employ" more servers than are necessary for meeting given performance criteria. Important criteria are:

1) The probability that an arriving customer finds an available server.

2) The expected waiting time of a customer for service.

It is common practice to characterize the structure of standard queueing systems by *Kendall's notation* $A/B/s/m$. In this code, A and B refer to the input and service time distributions, respectively, s is the number of servers, and waiting capacity is available for m customers.

$A = M$ (*Markov*): Customers arrive in accordance with a homogeneous Poisson process (*Poisson input*)

$A = GI$ (*general independent*): Customers arrive in accordance with an ordinary renewal process (*recurrent input*).

$A = D$: The distances between the arrivals of neighbouring customers are constant (*deterministic input*).

$B = M$: The service times are independent, identically distributed exponential random variables.

$B = G$: The service times are independent, identically distributed random variables with arbitrary distribution.

For instance, $M/M/1/0$ is a loss system with Poisson input, 1 server, and exponential service times. $GI/M/3/\infty$ is a waiting system with recurrent input, exponential service times, and 3 servers. For queueing systems with an infinite number of servers no waiting capacity is necessary. Hence their code is $A/B/\infty$.

In waiting systems and waiting-loss systems there are several possible ways of choosing waiting customers for service. These possibilities are called *service disciplines* (*queueing disciplines*). The most important ones are:

1) *FCFS* (*first come - first served*) Waiting customers are served in accordance with their order of arrival. This discipline is also called *FIFO* (*first in-first out*)

2) *LCFS* (*last come-first served*) The customer which arrived last is served first. This discipline is also called *LIFO* (*last in-first out*).

3) *SIRO* (*service in random order*) A server, after having finished with a customer, randomly picks one of the waiting customers for service.

There is a close relationship between service disciplines and priority systems. In a *priority system*, arriving customers have different *priorities* of being served. A customer with higher priority is served before a customer with lower priority (*head of the line priority discipline*), but no interruption of service takes place. When a customer with *absolute priority* arrives and finds all servers busy, then the service of a customer with lower priority has to be interrupted (*preemptive priority discipline*).

Notation This section deals essentially with *M/M/s/m*-systems where all servers have identically distributed service times. The parameter λ of the Poisson input is said to be the *arrival intensity* (*arrival rate*). The parameter μ of the exponential service time is said to be the *service intensity* (*service rate*). If Y and Z denote the random times between the arrival of two neighbouring customers and the random service time of a customer, respectively, then $E(Y) = 1/\lambda$ and $E(Z) = 1/\mu$. The ratio

$$\rho = \lambda/\mu$$

is called the *traffic intensity* of the queueing system. The *degree of server utilisation* η is defined as

$$\eta = E(S)/s,$$

where S is the number of busy servers in the steady state. Thus, in the steady state, the coefficient η can be interpreted as the proportion of time any single server is busy. (Here and in what follows *in the steady state* refers to the stationary phase. More precisely, the underlying Markov chain $\{X(t), t \geq 0$, which describes the behaviour of the system, is stationary.) In sections 6.8.2 to 6.8.4, $X(t)$ denotes the total number of customers in the queueing system (either waiting or being served) at time t. If X denotes the corresponding number in the steady state, then the stationary probability π_j of state j satisfies

$$\begin{aligned}\pi_j &= \lim_{t\to\infty} p_j(t) \\ &= \lim_{t\to\infty} P(X(t) = j) \\ &= P(X = j); \quad j = 0, 1, \ldots, s+m; \quad s, m \leq \infty\end{aligned}$$

6.8.2 Loss Systems

***M/M/∞*- system** Strictly speaking, this system is neither a loss nor a waiting system. $\{X(t),\ t \geq 0\}$ is a birth-and death process with state space $\mathbf{Z} = \{0, 1, ...\}$ and transition rates (example 6.13)

$$\lambda_j = \lambda; \quad \mu = j\mu; \quad j = 0, 1, ...$$

Its time-dependent state probabilities $p_j(t)$ are given by (6.48). The stationary state probabilities may be obtained by passing to the limit as $t \to \infty$ in $p_j(t)$ or by solving the corresponding system of equations (6.49) :

$$\pi_j = \frac{\rho^j}{j!} e^{-\rho}; \quad j = 0, 1, ... \tag{6.60}$$

This is a Poisson distribution with parameter ρ. Hence, in the steady state, the expected number of busy servers is equal to the traffic intensity of the system:

$$E(X) = \rho$$

Note that in loss systems the number of customers in the system equals the number of busy servers. □

***M/M/s/0*- system** In this case, $\{X(t), t \geq 0\}$ is a birth- and death process with state space $\mathbf{Z} = \{0, 1, ... , s\}$ and transition rates

$$\lambda_j = \lambda; \quad j = 0, 1, ..., s-1$$

$$\lambda_j = 0 \text{ for } j \geq s$$

$$\mu_j = j\mu; \quad j = 0, 1, ..., s$$

The system of equations for the stationary state probabilities (6.50) is

$$\mu\pi_1 = \lambda\pi_0$$

$$\lambda\pi_{j-1} + (j+1)\mu\pi_{j+1} = (\lambda + j\mu)\pi_j; \quad j = 1, 2, ..., s-1$$

$$s\mu\pi_s = \lambda\pi_{s-1}$$

By successively solving.these equations

$$\pi_j = \frac{1}{j!}\rho^j \pi_0; \quad j = 0, 1, ..., s$$

The still missing probability π_0 that there is no customer in the system is determined by the normalizing condition

$$\Sigma_{j=0}^{s} \pi_i = 1 \tag{6.61}$$

Solving for π_0 yields

$$\pi_0 = \left[\sum_{i=0}^{s} \frac{1}{i!} \rho^i \right]^{-1}$$

Hence the stationary state probabilities are given by

$$\pi_j = \frac{\frac{1}{j!} \rho^j}{\sum_{i=0}^{s} \frac{1}{i!} \rho^i} \; ; \quad j = 0, 1, \ldots, s \tag{6.62}$$

In particular, the *loss probability* is

$$\pi_s = \frac{\frac{1}{s!} \rho^s}{\sum_{i=0}^{s} \frac{1}{i!} \rho^i} \tag{6.63}$$

π_s is the probability that an arriving customer does not find an idle server so that he has to leave the system. (6.63) is the famous *Erlang's loss formula.* The following recursive formula for the loss probability as a function of s can be easily verified:

$$\pi_0 = 1 \text{ for } s = 0; \; \frac{1}{\pi_s} = \frac{s}{\rho} \frac{1}{\pi_{s-1}} + 1; \quad s = 1, 2, \ldots$$

The expected number of busy servers is

$$E(X) = \sum_{i=1}^{s} i\pi_i = \sum_{i=1}^{s} i \frac{\rho^i}{i!} \pi_0$$

$$= \rho \sum_{i=1}^{s} \frac{\rho^{i-1}}{(i-1)!} \pi_0 = \rho \sum_{i=0}^{s-1} \frac{\rho^i}{i!} \pi_0$$

In view of (6.61),

$$E(X) = \rho(1 - \pi_s)$$

Hence, the *degree of server utilization* is

$$\eta = \frac{\rho}{s}(1 - \pi_s)$$

Special case $s = 1$ In a single-server system, the vacant and loss probabilities are given by

$$\pi_0 = \frac{1}{1+\rho} \quad \text{and} \quad \pi_1 = \frac{\rho}{1+\rho},$$

respectively. Since $\rho = E(Z)/E(Y)$,

$$\pi_0 = \frac{E(Y)}{E(Y) + E(Z)} \quad \text{and} \quad \pi_1 = \frac{E(Z)}{E(Y) + E(Z)}$$

Hence, π_0 (π_1) is formally equal to the stationary availability (nonavailability) of a system with expected lifetime $E(Y)$ and expected renewal time $E(Z)$, whose operation is governed by an alternating renewal process (section 4.6, formula (4.43)).

Example 6.16 It is assumed that a telephone exchange can be modeled by an $M/M/s/0$-queueing system. The input (calls of subscribers wishing to be connected) has intensity $\lambda = 2$ $[min^{-1}]$. Thus, the expected time between successive calls is equal to $E(Y) = 1/\lambda = 0.5$ $[min]$. On average, each subscriber occupies a line for $E(Z) = 3$ *min.*

1) What is the loss probability in the case of $s = 7$ lines?

The traffic intensity is $\rho = E(Z)/E(Y) = 6$. By (6.63), the loss probability equals

$$\pi_7 = \frac{\frac{1}{7!}6^7}{1+6+\frac{6^2}{2!}+\frac{63}{3!}+\frac{6^4}{4!}+\frac{6^5}{5!}+\frac{6^6}{6!}+\frac{6^7}{7!}} = 0.185$$

Hence, the expected number of occupied lines is

$$E(X) = \rho(1-\pi_7) = 6(1-0.185) = 4.89$$

and the degree of server (line) utilization is

$$\eta = 4.89/7 = 0.70$$

2) How many lines at least have to be provided in order to guarantee that no less than 95% of the desired connections can be made?

The loss probabilities for $s = 9$ and $s = 10$ are

$$\pi_9 = 0.075 \quad \text{and} \quad \pi_{10} = 0.043,$$

resepectively. Hence the minimum number of lines required is $s_{\min} = 10$. However, in this case the degree of server utilization is smaller than with $s = 7$ lines

$$\eta = 0.574$$ □

It is interesting and practically important that the stationary state probabilities of the queueing system $M/G/s/0$ also have the structure (6.62). That is, if the respective arrival intensities and the service rates of the systems $M/M/s/0$ and $M/G/s/0$ are equal, then their traffic intensities coincide and so do their stationary state probabilities: for both systems they are given by (6.62). A corresponding result holds also for the queueing systems $M/M/\infty$ and $M/G/\infty$. (Compare the stationary state probabilities (3.22) and (6.60).) Queueing systems having this property are said to be *insensitive* with respect to the probability distribution of the service times. An analogous property can be defined with respect to the input. A comprehensive treatment of the *insensitivity* of queueing systems is given in *Gnedenko* and *König* (1984).

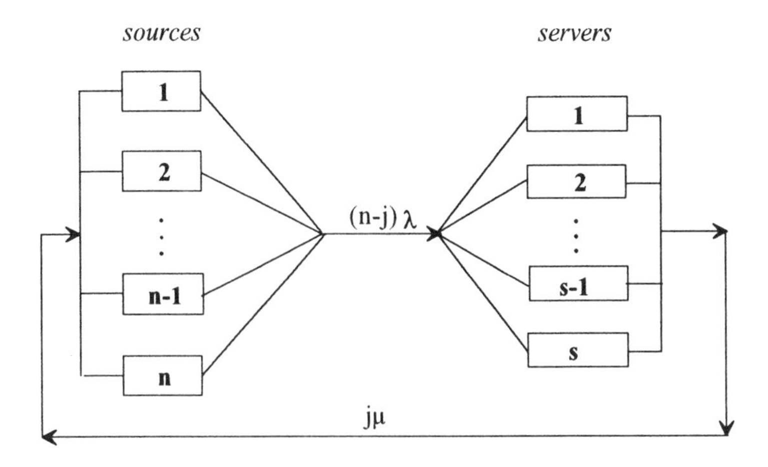

Figure 6.13 Engset's loss system in state $X(t) = j$

Engset's loss system n sources generate n independent Poisson inputs with common intensity λ, which are served by s servers, $s \leq n$. The service times are independent exponential random variables with parameter μ. As long as a customer from a particular source is being served, this source cannot produce another customer (Compare with the repairman problem considered in example 6.14: during the repair of a machine, this machine cannot produce another demand for repair.) A customer who does not find an available server is lost. If $X(t)$ denotes the number of customers being served at time t, then $\{X(t), t \geq 0\}$ is a birth- and death process with state space $\mathbf{Z} = \{0, 1, \dots, s\}$. If $X(t) = j$, only $n - j$ sources are active, i.e. they are able to generate customers. Therefore, its positive transition rates are

$$\lambda_j = (n-j)\lambda; \quad j = 0, 1, 2, \dots, s-1$$

$$\mu_j = j\mu; \quad j = 1, 2, \dots, s$$

Hence, the system of equations in the stationary state probabilities (6.50) is

$$n\lambda\pi_0 = \mu\pi_1$$

$$(n-j+1)\lambda\pi_{j-1} + (j+1)\mu\pi_{j+1} = ((n-j)\lambda + j\mu)\pi_j; \quad j = 1, 2, \dots, s-1$$

$$s\mu\pi_s = (n-s+1)\lambda\pi_{s-1}$$

The solution is

$$\pi_j = \frac{\binom{n}{j}\rho^j}{\sum\limits_{i=0}^{s}\binom{n}{i}\rho^i}; \quad j = 0, 1, \dots, s$$

In particular, π_0 and the loss probability π_s are

$$\pi_0 = \frac{1}{\sum\limits_{i=0}^{s} \binom{n}{i} \rho^i}, \qquad \pi_s = \frac{\binom{n}{s} \rho^s}{\sum\limits_{i=0}^{s} \binom{n}{i} \rho^i}$$

Engset's loss system is as the repairman problem a closed queueing system. □

6.8.3 Waiting Systems

Note that, if $X(t) = j > s$, then $j - s$ is the number of waiting customers, i.e. the length of the *queue*, at time t. (By convention, $X(t)$ is the total number of customers in the queueing system at time t.)

***M/M/s/∞*-system** $\{X(t),\ t \geq 0\}$ is a birth- and death process with countably infinite state space $\mathbf{Z} = \{0, 1, ...\}$ and transition rates

$$\lambda_j = \lambda; \quad j = 0, 1, ...$$
$$\mu_j = j\mu; \quad j = 0, 1, ..., s$$
$$\mu_j = s\mu; \quad j > s$$

Providing their existence, the stationary state probabilities satisfy the following system of equations:

$$\lambda\pi_0 = \mu\pi_1$$
$$\lambda\pi_{j-1} + (j+1)\mu\pi_{j+1} = (\lambda + j\mu)\pi_j; \quad j = 1, 2, ..., s-1$$
$$\lambda\pi_{j-1} + s\mu\pi_{j+1} = (\lambda + s\mu)\pi_j; \quad j \geq s$$

According to (6.51), the solution has structure

$$\pi_j = \frac{\rho^j}{j!}\pi_0 \qquad \text{for } j = 0, 1, ..., s-1$$

$$\pi_j = \frac{\rho^j}{s!\, s^{j-s}}\pi_0 \quad \text{for } j \geq s \tag{6.64}$$

In what follows, it is assumed that $\rho = \lambda/\mu < s$. In this case the arrival intensity λ of customers is smaller than the maximum service rate $s\mu$ of the system. Hence, the existence of a stationary distribution can be expected. If $\rho \geq s$, a stationary distribution does not exist since, at least in the long-run, the servers cannot cope with the input. In this case, the length of the waiting queue will tend to infinity as $t \to \infty$. Hence, no equilibrium state between arriving and leaving customers is possible.

By making use of the normalizing condition (6.29) and the geometric series, the probability π_0 that there is no customer in the system is seen to be

$$\pi_0 = \left[\sum_{i=0}^{s-1} \frac{1}{i!} \rho^i + \frac{\rho^s}{(s-1)!\,(s-\rho)} \right]^{-1}$$

The probability π_w that an arriving customer finds all servers busy and, therefore, must wait for service, is

$$\pi_w = \sum_{i=s}^{\infty} \pi_i$$

π_w is called *waiting probability*. Making use of the geometrical series again yields a simple formula for π_w:

$$\pi_w = \frac{\pi_s}{1-\rho/s}$$

The expected value of the number S of busy servers is

$$E(S) = \sum_{i=0}^{s-1} i\pi_i + s\pi_w$$

$$= \rho$$

(The details are left as an exercise to the reader.) Hence the degree of server utilization is $\eta = \rho/s$.

The expected value of the total number of customers in the system is

$$E(X) = \sum_{i=1}^{\infty} i\pi_i = \rho\left[1 + \frac{s}{(s-\rho)^2}\pi_s \right]$$

Waiting time distribution In the steady state, let W be be the random time a customer has to wait for service if on his arrival at the system all servers are busy and the service discipline *FCFS* is in effect. By the total probability rule,

$$P(W>t) = \sum_{i=s}^{\infty} P(W>t|X=i)\,\pi_i$$

Assuming that a customer enters the system when it is in state $X=i \ge s$, then $W>t$ if within t time units after his arrival the service of at most i-s customers has been finished. The probability that the service of precisely k customers will be finished in this time interval, $0 \le k \le i-s$, is equal to

$$\frac{(s\mu t)^k}{k!} e^{-s\mu t},$$

since all servers are busy and, therefore, the output constitutes a Poisson process with intensity $s\mu$. Hence,

$$P(W > t | X = i) = e^{-s\mu t} \sum_{k=0}^{i-s} \frac{(s\mu t)^k}{k!}$$

Thus,

$$P(W > t) = e^{-s\mu t} \sum_{i=s}^{\infty} \pi_i \sum_{k=0}^{i-s} \frac{(s\mu t)^k}{k!}$$

$$= \pi_0 e^{-s\mu t} \sum_{i=s}^{\infty} \frac{\rho^i}{s!s^{i-s}} \sum_{k=0}^{i-s} \frac{(s\mu t)^k}{k!}$$

Performing the index transformation $j = i - s$, changing the order of summation according to formula (1.55), applying (6.64) with $j = s$, and making use of both the power series of e^x and the geometric series yields

$$P(W > t) = \pi_0 \frac{\rho^s}{s!} e^{-s\mu t} \sum_{j=0}^{\infty} \left(\frac{\rho}{s}\right)^j \sum_{k=0}^{j} \frac{(s\mu t)^k}{k!}$$

$$= \pi_s e^{-s\mu t} \sum_{k=0}^{\infty} \frac{(s\mu t)^k}{k!} \sum_{j=k}^{\infty} \left(\frac{\rho}{s}\right)^j$$

$$= \pi_s e^{-s\mu t} \sum_{k=0}^{\infty} \frac{(\lambda t)^k}{k!} \sum_{i=0}^{\infty} \left(\frac{\rho}{s}\right)^i$$

$$= \pi_s e^{-s\mu t} e^{\lambda t} \frac{1}{1 - \rho/s}$$

From this the distribution function of the steady state waiting time is seen to be

$$F_W(t) = P(W \le t)$$

$$= 1 - \frac{s\mu}{s\mu - \lambda} \pi_s e^{-(s\mu - \lambda)t}, \quad t \ge 0 \tag{6.65}$$

Note that $P(W > 0)$ is the waiting probability which has already been computed directly:

$$1 - F_W(0) = P(W > 0) = \frac{\pi_s}{1 - \rho/s} = \pi_w$$

According to (1.12), the expected waiting time of a customer is

$$E(W) = \int_0^{\infty} P(W > t)\, dt$$

It follows

$$E(W) = \frac{s\mu}{(s\mu - \lambda)^2} \pi_s$$

Single-server system ($s = 1$) In this case, for $\rho < 1$,

$$\pi_j = (1 - \rho)\rho^j; \quad j = 0, 1, \dots$$

Thus, in the steady state, the number of customers X in the system has a geometric distribution with expected value $E(X) = \rho/(1 - \rho)$. The waiting probability is $\pi_W = \rho$.

The waiting time of a customer has distribution function

$$F_W(t) = 1 - \rho\, e^{-(\mu - \lambda)t}, \quad t \geq 0,$$

and the expected waiting time of a customer is

$$E(W) = \frac{\rho}{\mu(1 - \rho)}$$

This result verifies the intuitive relationship

$$E(W) = E(Z)\, E(X)$$

Engset's waiting system n sources generate n independent Poisson inputs with common intensity λ to be served by s servers, $s \leq n$. The service times are independent exponential random variables with parameter μ. As long as a customer from a particular source is being served, this source cannot produce another customer. A customer waits for service, if, on arrival, he does not find an available server. There is waiting capacity for all arriving customers. If $X(t)$ denotes the number of customers being served at time t, then $\{X(t), t \geq 0\}$ is a birth- and death process with state space $\mathbf{Z} = \{0, 1, \dots, n\}$. (Note that $n - s$ is the maximum possible length of the queue.) Given that $X(t) = j$, only $n - j$ sources are active. Formally, this queueing system is only a slightly more general formulation of the repairman problem considered in example 6.14. Hence, its stationary state probabilities are given by (6.56) with r replaced by s.

6.8.4 Waiting-Loss Systems

***M/M/s/m*-system** This system has s servers and waiting capacity for m customers. A customer which on arrival finds no idle server and the waiting capacity occupied is lost. The number of customers $X(t)$ in the system at time t constitutes a birth- and death process $\{X(t), t \geq 0\}$ with state space $\mathbf{Z} = \{0, 1, \dots, s+m\}$ and positive transition rates

$$\lambda_j = \lambda, \quad 0 \leq j \leq s + m - 1,$$

$$\mu_j = \begin{cases} j\mu & \text{for} \quad 1 \leq j \leq s \\ s\mu & \text{for} \quad s < j \leq s + m \end{cases}$$

According to (6.51) and (6.53), the stationary state probabilities are

$$\pi_j = \begin{cases} \frac{1}{j!}\rho^j \pi_0 & \text{for } 1 \le j \le s-1 \\ \frac{1}{s!\, s^{j-s}}\rho^j \pi_0 & \text{for } s \le j \le s+m \end{cases}$$

$$\pi_0 = \left[\sum_{j=0}^{s-1} \frac{1}{j!}\rho^j + \sum_{j=s}^{s+m} \frac{1}{s!\, s^{j-s}}\rho^j\right]^{-1}$$

The second series in π_0 can be summed up to obtain

$$\pi_0 = \begin{cases} \left[\sum_{j=0}^{s-1} \frac{1}{j!}\rho^j + \frac{1}{s!}\rho^s \frac{1-(\rho/s)^{m+1}}{1-\rho/s}\right]^{-1} & \text{for } \rho \ne s \\ \left[\sum_{j=0}^{s-1} \frac{1}{j!}\rho^j + (m+1)\frac{s^s}{s!}\right]^{-1} & \text{for } \rho = s \end{cases}$$

π_0 is the probability that there is no customer in the system and π_{s+m} is the *loss probability*, i.e. the probability that an arriving customer is lost (rejected). The respective probabilities π_f and π_w that an arriving customer finds an idle server or waits for service are

$$\pi_f = \Sigma_{i=0}^{s-1}\, \pi_i\,, \qquad \pi_w = \Sigma_{i=s}^{s+m-1}\, \pi_i$$

Analogously to the loss system *M/M/s/0*, the expected number of busy servers is given by

$$E(S) = \rho\,(1 - \pi_{s+m})$$

Thus, the degree of server utilisation is

$$\eta = \frac{\rho}{s}(1 - \pi_{s+m})$$

In the following example, the probabilities π_0 and π_{s+m} which refer to a queueing system with s servers and waiting capacity for m customers are denoted by $\pi_0(s, m)$ and $\pi_{s+m}(s, m)$, respectively.

Example 6.17 A filling station has $s = 8$ petrol pumps and waiting capacity for $m = 6$ cars. On average, 1.2 cars arrive at the filling station per minute. The mean time a car occupies a petrol pump is 5 minutes. It is assumed that the filling station behaves like an *M/M/s/m*-queueing system. Since $\lambda = 1.2$ and $\mu = 0.2$, the traffic intensity is $\rho = 6$. Hence, the loss probability $\pi_{14}(8, 6)$ becomes

$$\pi_{14}(8, 6) = \frac{1}{8!\, 8^6}\, 6^{14}\pi_0 = 0.0167\,,$$

where

$$\pi_0(8,6) = \left[\sum_{j=0}^{7} \frac{1}{j!} 6^j + \frac{1}{8!} 6^8 \frac{1-(6/8)^7}{1-6/8}\right]^{-1} = 0.00225$$

Consequently, on average

$$E(S) = 6 \cdot (1 - 0.0167) = 5.9$$

of the petrol pumps are occupied. Having obtained this result, the owner considers 2 of the 8 petrol pumps to be superfluous and has them pulled down. It is assumed that this change does not influence the input so that cars continue to arrive with traffic intensity $\rho = 6$. Hence, the corresponding loss probability $\pi_{12} = \pi_{12}(6,6)$ becomes

$$\pi_{12}(6,6) = \frac{6^6}{6!} \pi_0(6,6) = 0.1023$$

Thus, about 10% of all arriving cars leave the station without having filled up. In order to counter this drop, the owner provides waiting capacity for another 4 cars so that $m = 10$. The corresponding loss probability $\pi_{16} = \pi_{16}(6,10)$ is

$$\pi_{16}(6,10) = \frac{6^6}{6!} \pi_0(6,10) = 0.0726$$

Since

$$\pi_{6+m}(6,m) = \frac{6^6}{6!} \left[\sum_{j=0}^{5} \frac{1}{j!} 6^j + (m+1) \frac{6^6}{6!}\right]^{-1},$$

it can easily be calculated that additional waiting capacity for 51 cars has to be provided to equalize the loss caused by reducing the number of pumps from 8 to 6. (For obvious reasons, the owner does not take into consideration this variant; see next model with impatient customers.) □

***M/M/s/∞*- system with impatient customers** Even if there is waiting capacity for arbitrarily many customers, some customers might leave the system without having been served. This happens when customers can only spend a finite time, their *patience times*, in the queue. If the service of a customer does not begin before his patience time expires, he leaves the system. For example, if somebody, whose long-distance train will depart in 10 minutes, has to wait 15 minutes to buy a ticket, then he will leave the counter without one. Real time monitoring and control systems have memories for data to be processed. But these data "wait" only as long as they are up to date. Bounded waiting times are also typical for packed switching systems, for instance in computer-aided booking systems. Generally one expects that "intelligent" customers adapt their behaviour to the actual state of the queueing system. Of the many available models dealing with such situations the following one is considered in some detail:

The patience times of customers arriving at a $M/M/s/\infty$- system are assumed to be independent, identically distributed exponential random variables with parameter ν. If $X(t)$ denotes the number of customers in the system at time t, then $\{X(t), t \geq 0\}$ is a birth- and death process with transition rates

$$\lambda_j = \lambda; \quad j = 0, 1, \ldots$$

$$\mu_j = \begin{cases} j\mu & \text{for } j = 1, 2, \ldots, s \\ s\mu + (j-s)\nu & \text{for } j = s, s+1, \ldots \end{cases}$$

If $j \to \infty$, then $\mu_j \to \infty$, whereas the birth rate remains constant. Hence the sufficient condition for the existence of a stationary distribution stated in theorem 6.2 is fulfilled. Thus, given a sufficiently long queue, the number of customers leaving the system is on average greater than the number of arriving customers per unit time. That is, the system is self-regulating, aiming at achieving the equilibrium state. From (6.51) and (6.53), the stationary state probabilities are given by

$$\pi_j = \begin{cases} \frac{1}{j!}\rho^j \pi_0 & \text{for } j = 1, 2, \ldots, s \\ \frac{\rho^s}{s!} \frac{\lambda^{j-s}}{\prod_{i=1}^{j-s}(s\mu + i\nu)} \pi_0 & \text{for } j = s+1, s+2, \ldots \end{cases}$$

and

$$\pi_0 = \left[\sum_{j=0}^{s} \frac{1}{j!}\rho^j + \frac{\rho^s}{s!} \sum_{j=s+1}^{\infty} \frac{\lambda^{j-s}}{\prod_{i=1}^{j-s}(s\mu + i\nu)} \right]^{-1}$$

Let L denote the random length of the queue in the steady state. Then,

$$E(L) = \sum_{j=s+1}^{\infty} (j-s)\pi_j$$

Inserting the π_j yields after some transformation

$$E(L) = \pi_s \sum_{j=1}^{\infty} j\lambda^j \left[\prod_{i=1}^{j} (s\mu + i\nu) \right]^{-1}$$

In this model, the *loss probability* π_v is not strictly associated with the number of customers in the system. It is the probability that a customer leaves the system without having been served because his patience time has expired. Therefore, $1 - \pi_v$ is the probability that a customer leaves the system after having been served. By applying the total probability rule with the exhaustive and mutually exclusive set of events "$X = j$"; $j = s, s+1, \ldots$; one obtains

$$\pi_v = \frac{\nu}{\lambda} E(L)$$

Hence, the loss probability is directly proportional to the expected length of the queue.

Variable input intensity Finite waiting capacities and patience times imply in the end that only a "thinned flow of potential customers" is served. Thus, it seems to be appropriate to investigate queueing systems whose input intensities depend on the state of the system. However, those customers, which actually enter the system, do not leave it without service. Since the tendency of customers to leave the system immediately after arrival increases with the number of customers in the system, the birth rates λ_j should be decreasing for $j \geq s$ as j tends to infinity. This property have, for example, the birth rates

$$\lambda_j = \begin{cases} \lambda & \text{for } j = 0, 1, \ldots, s-1 \\ \frac{s}{j+a}\lambda & \text{for } j = s, s+1, \ldots \end{cases}, \quad a > 0$$

6.8.5 Special Single-Server Queueing Systems

This section deals with single-server queueing systems with priorities and with unreliable server, respectively.

System with priorities A single-server queueing system with waiting capacity for $m = 1$ customer is subject to two independent Poisson inputs **1** and **2**, with respective intensities λ_1 and λ_2. The corresponding customers are called type 1- and type 2- customers. Type 1-customers have absolute (preemptive) priority, i.e. when both a type 1- and a type 2-customer are in the system, the type 1-customer is being served. Thus, the service of a type 2-customer is interrupted as soon as a type 1-customer arrives. The displaced customer will occupy the waiting facility if it is empty. Otherwise he leaves the system. A waiting type 2-customer also has to leave the system when a type 1- customer arrives, since the newcomer will occupy the waiting facility. (Such a situation can only happen when a type 1-customer is being served.) An arriving type 1-customer is only lost when both server and waiting facility are already occupied by other type 1-customers. The service times of type 1- and type 2-customers are exponentially distributed with respective parameters μ_1 and μ_2. The state space of the system is given by

$$\mathbf{Z} = \{(i,j);\ i,j = 0, 1, 2\}$$

where i denotes the number of type 1-customers and j the number of type 2-customers in the system. Note that if $X(t)$ denotes the system state at time t, then the stochastic process $\{X(t), t \geq 0\}$ can be treated as a one-dimensional Markov chain, since scalars can be assigned to the 6 possible system states given as two-component vectors. However, $\{X(t), t \geq 0\}$ is not a birth- and death process. Figure 6.14 shows the transition graph of this Markov chain.

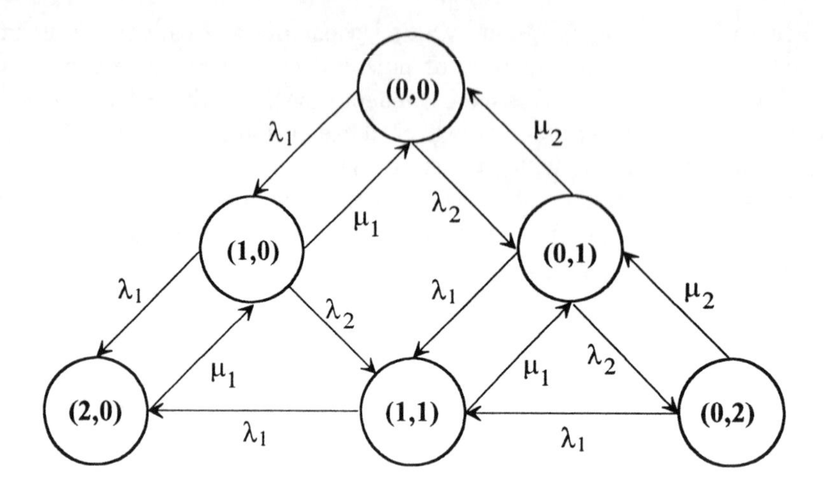

Figure 6.14 Transition graph of a single-server priority queueing system with waiting capacity for one customer

Together with the normalizing condition, the stationary state probabilities satisfy the system of equations

$$(\lambda_1 + \lambda_2)\pi_{(0,0)} = \mu_1 \pi_{(1,0)} + \mu_2 \pi_{(0,1)}$$

$$(\lambda_1 + \lambda_2 + \mu_1)\pi_{(1,0)} = \lambda_1 \pi_{(0,0)} + \mu_1 \pi_{(2,0)}$$

$$(\lambda_1 + \lambda_2 + \mu_2)\pi_{(0,1)} = \lambda_2 \pi_{(0,0)} + \mu_1 \pi_{(1,1)} + \mu_2 \pi_{(0,2)}$$

$$(\lambda_1 + \mu_1)\pi_{(1,1)} = \lambda_2 \pi_{(1,0)} + \lambda_1 \pi_{(0,1)} + \lambda_1 \pi_{(0,2)}$$

$$\mu_1 \pi_{(2,0)} = \lambda_1 \pi_{(1,0)} + \lambda_1 \pi_{(1,1)}$$

$$(\lambda_1 + \mu_2)\pi_{(0,2)} = \lambda_2 \pi_{(0,1)}$$

$$\pi_{(0,0)} + \pi_{(1,0)} + \pi_{(0,1)} + \pi_{(1,1)} + \pi_{(2,0)} + \pi_{(0,2)} = 1$$

The general solution of this system of equations is so complicated that it will not be presented here. Instead, a numerical example is considered.

Let

$$\lambda_1 = 0.1; \quad \lambda_2 = 0.2, \qquad \mu_1 = \mu_2 = 0.2 \tag{6.66}$$

Then the stationary state probabilities are

$$\pi_{(0,0)} = 0.2105, \quad \pi_{(0,1)} = 0.3073, \quad \pi_{(1,0)} = 0.0085$$

$$\pi_{(1,1)} = 0.1765, \quad \pi_{(0,2)} = 0.2048, \quad \pi_{(2,0)} = 0.0924$$

If only the number of type 1-customers in the system is of interest, then the priority queueing system becomes the waiting-loss-system *M/M/s/*1 with $s = 1$ considered in section 6.8.4, since type 2-customers do not influence the service of type 1-customers at all.

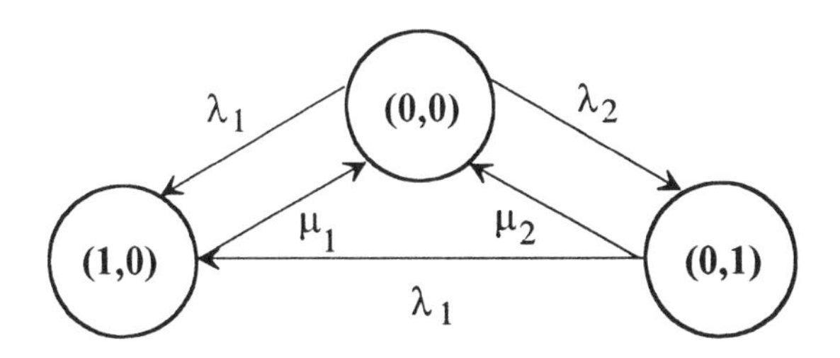

Figure 6.15 Transition graph of a 1-server priority loss system

m = 0 Since there is no waiting capacity, each customer, notwithstanding its type, is lost if the server is busy with a type 1-customer. In addition, a type 2-customer is lost if a type 1-customer arrives while it is being served. The state space now reduces to $\mathbf{Z} = \{(0,0), (0,1), (1,0)\}$. The stationary state probabilities satisfy the following system of equations (Figure 6.15):

$$(\lambda_1 + \lambda_2)\pi_{(0,0)} = \mu_1 \pi_{(1,0)} + \mu_2 \pi_{(0,1)}$$

$$\mu_1 \pi_{(1,0)} = \lambda_1 \pi_{(0,0)} + \lambda_1 \pi_{(0,1)}$$

$$1 = \pi_{(0,0)} + \pi_{(1,0)} + \pi_{(0,1)}$$

The solution is

$$\pi_{(0,0)} = \frac{\mu_1(\lambda_1 + \mu_2)}{(\lambda_1 + \mu_1)(\lambda_1 + \lambda_2 + \mu_2)}$$

$$\pi_{(0,1)} = \frac{\lambda_2 \mu_1}{(\lambda_1 + \mu_1)(\lambda_1 + \lambda_2 + \mu_2)}$$

$$\pi_{(1,0)} = \frac{\lambda_1}{\lambda_1 + \mu_1}$$

$\pi_{(1,0)}$ is the loss probability for type 1-customers. This probability does not depend on type 2-customers. It is simply the probability that the service time of type 1-customers is greater than their interarrival time.

Given that at the arrival time of a type 2-customer the server is idle, this customer is lost if and only if during its service a type 1-customer arrives. The conditional probability of this event is

$$\int_0^\infty e^{-\mu_2 t} \lambda_1 e^{-\lambda_1 t} dt = \lambda_1 \int_0^\infty e^{-(\lambda_1+\mu_2)t} dt = \frac{\lambda_1}{\lambda_1+\mu_2}$$

Therefore, the (total) loss probability for type 2-customers is

$$\pi_v = \frac{\lambda_1}{\lambda_1+\mu_2}\,\pi_{(0,0)} + \pi_{(0,1)} + \pi_{(1,0)}$$

or

$$\pi_v = 1 - \frac{\mu_1 \mu_2}{(\lambda_1+\mu_1)(\lambda_1+\lambda_2+\mu_2)}$$

For example, given the transition rates (6.66), the state probabilities are

$$\pi_{(0,0)} = \frac{6}{15}, \quad \pi_{(1,0)} = \frac{5}{15}, \quad \pi_{(0,1)} = \frac{4}{15}$$

The loss probability for type 2-customers is $\pi_v = 11/15$.

***M/M/1/m*-system with unreliable server** Failures (downtimes) of servers can strongly influence the efficiency of queueing systems. Hence, if they are not negligible, they have to be taken into account in the mathematical model. The principal approach is now demonstrated for a single-server queueing system with waiting capacity for m customers, Poisson input, and independent, identically distributed exponential service times with parameter μ. The lifetime of the server is assumed to be exponentially distributed with parameter α, both in its busy phase and in its idle phase, and the subsequent renewal time of the server is assumed to be exponentially distributed with parameter β. It is further assumed that the sequence of life- and renewal times of the server can be described by an alternating renewal process. When the server fails, all customers leave the system, i.e. the customer being served and the waiting customers, if there are any, are lost. Customers arriving during a renewal phase of the server are rejected, i.e. they are lost, too.

The stochastic process $\{X(t), t \geq 0\}$ describing the behaviour of the system is characterized as follows:

$$X(t) = \begin{cases} j & \text{if there are } j \text{ customers in the system at time } t;\ j = 0, 1, \ldots, m+1 \\ m+2 & \text{if the server is being renewed at time } t \end{cases}$$

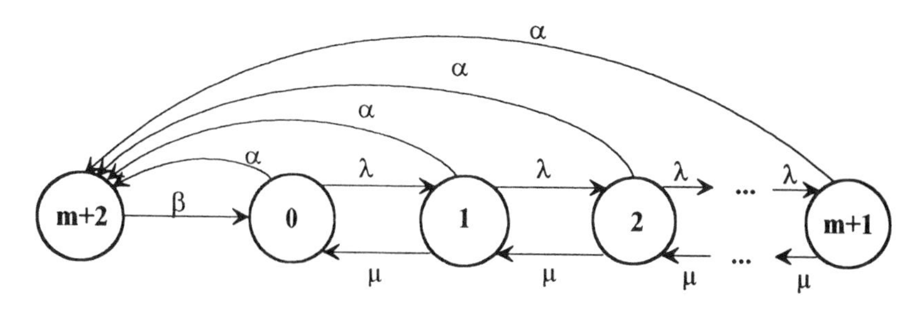

Figure 6.16 Transition graph of a queueing system with unreliable server

Its transition rates are (Figure 6.16):

$$\begin{aligned} q_{j,j+1} &= \lambda; \quad j = 0, 1, \dots, m \\ q_{j,j-1} &= \mu; \quad j = 1, 2, \dots, m+1 \\ q_{j,m+2} &= \alpha; \quad j = 0, 1, \dots m+1 \\ q_{m+2,0} &= \beta \end{aligned} \tag{6.67}$$

Comment It is interesting that this queueing system with unreliable server can formally be modeled as a queueing system with priorities and absolutely reliable server. To see this, a failure of the server has to be interpreted as a "customer" with absolute priority, whose service consists in the renewal of the server. Such a "customer" pushes away any other customer from the server, in this model even from the waiting facility. Hence it is not surprising that in many cases the theory of queueing systems with priorities also provides solutions for more complicated queueing systems with unreliable servers than the one considered here.

According to (6.28), the stationary state probabilities satisfy the system of equations

$$\begin{aligned} (\alpha + \lambda)\pi_0 &= \mu\pi_1 + \beta\pi_{m+2} \\ (\alpha + \lambda + \mu)\pi_j &= \lambda\pi_{j-1} + \mu\pi_{j+1}; \quad j = 1, 2, \dots, m \\ (\alpha + \mu)\pi_{m+1} &= \lambda\pi_m \\ \beta\pi_{m+2} &= \alpha\pi_0 + \alpha\pi_1 + \cdots + \alpha\pi_{m+1} \end{aligned} \tag{6.68}$$

The last equation is equivalent to

$$\beta\pi_{m+2} = \alpha(1 - \pi_{m+2})$$

Hence,

$$\pi_{m+2} = \frac{\alpha}{\alpha + \beta}$$

Now, starting with the first equation in (6.68), the stationary state probabilities of the system $\pi_1, \pi_2, \ldots, \pi_{m+1}$ can be successively determined. The probability π_0 is, as usual, obtained from the normalizing condition

$$\Sigma_{i=0}^{m+2} \pi_i = 1 \tag{6.69}$$

For example, in case of the corresponding loss system ($m = 0$), the stationary state probabilities are

$$\pi_0 = \frac{\beta(\alpha + \mu)}{(\alpha + \beta)(\alpha + \lambda + \mu)}$$

$$\pi_1 = \frac{\beta \lambda}{(\alpha + \beta)(\alpha + \lambda + \mu)}$$

$$\pi_2 = \frac{\alpha}{\alpha + \beta}$$

Modification of the model It makes sense to assume that the server can only fail when it is busy. In this case,

$$q_{j,m+2} = \alpha \quad \text{for } j = 1, 2, \ldots, m+1$$

The other transition rates given by (6.67) remain valid. Thus, the corresponding transition graph is again given by Figure 6.16 with the arrow from node 0 to node $m+2$ deleted. The stationary state probabilities now satisfy the system of equations

$$\begin{aligned}
&\lambda \pi_0 = \mu \pi_1 + \beta \pi_{m+2} \\
&(\alpha + \lambda + \mu) \pi_j = \lambda \pi_{j-1} + \mu \pi_{j+1}\,; \quad j = 1, 2, \ldots, m \\
&(\alpha + \mu) \pi_{m+1} = \lambda \pi_m \\
&\beta \pi_{m+2} = \alpha \pi_1 + \alpha \pi_2 + \cdots + \alpha \pi_{m+1}
\end{aligned} \tag{6.70}$$

The last equation is equivalent to

$$\beta \pi_{m+2} = \alpha(1 - \pi_0 - \pi_{m+2})$$

It follows

$$\pi_{m+2} = \frac{\alpha}{\alpha + \beta}(1 - \pi_0)$$

Starting with the first equation in (6.70), the solution $\pi_0, \pi_1, \pi_2, \ldots, \pi_{m+1}$ can now be obtained as above. For example, in case $m = 0$ the stationary state probabilities are

$$\pi_0 = \frac{\beta(\alpha+\mu)}{\beta(\alpha+\mu)+\lambda(\alpha+\beta)}$$

$$\pi_1 = \frac{\lambda\beta}{\beta(\alpha+\mu)+\lambda(\alpha+\beta)}$$

$$\pi_2 = \frac{\alpha\lambda}{\beta(\alpha+\mu)+\lambda(\alpha+\beta)}$$

6.8.6 Networks of Queueing Systems

Customers frequently need several kinds of service so that, after leaving one service station, they have to visit one or more other service stations in a fixed or random order. Each of these service stations is assumed to behave like the basic queueing system sketched in Figure 6.12. A set of queueing systems together with their interactions is called a *network of queueing systems* or a *queueing network.* Typical examples are technological processes for manufacturing (semi-) finished products. In such a case the order of service is usually prescribed. Queueing systems are frequently subject to several inputs, i.e. customers with different service requirements have to be attended. In this case they may visit the service stations in different orders. Examples of such situations are computer- and communication networks. Depending on whether the data which are to be provided, processed, or transmitted, the terminals (service stations) will be used in different orders. If technical systems have to be repaired, then, depending on the nature and the extent of the damage, service of different production departments in the maintenance firm is needed. Transport and loading systems (goods stations) also fit into the scheme of queueing networks.

Using a concept from graph theory, the service stations of a queueing network are called *nodes.* In an *open queueing network* customers arrive from "outside" at the system (external input). Each node may have its own external input. Once inside the system, the customers visit other nodes in a deterministic or random order before leaving the network. Thus, in an open network, each node may have to serve *external* and *internal customers*, where internal customers are the ones which arrive from other nodes. In *closed queueing network* there are no external inputs in the nodes and the number of customers in the network is constant. Consequently, no customer departs from the network. Queueing networks may be illustrated by directed graphs. The directed edges between the nodes symbolize the possible transitions of customers from one node to another.

In what follows, the n nodes in the network are denoted by **1**, **2**, ... , $\boldsymbol{n}$. Node **i** is assumed to have s_i servers; $1 \le s_i \le \infty$.

Open queueing networks The analytical treatment of queueing systems becomes difficult when dropping the assumptions of Poisson input and exponential service times. But the mathematical modeling and evaluation of queueing networks in all their generality is an almost impossible task. Hence, this section is restricted to a rather simple class of queueing networks, the *Jackson-queueing networks*. They are characterized by four properties:

1) Each node has an unbounded waiting capacity. The number of servers at each node is arbitrary.

2) The service times of all servers at node ***i*** are independent, identically distributed exponential random variables with parameter (intensity) μ_i. They are also independent of the service times at other nodes.

3) External customers arrive at node ***i*** in accordance with a homogeneous Poisson process with intensity λ_i. All external inputs are independent of each other and of all service times.

4) When the service of a customer at node ***i*** has been finished, he makes a transition to node ***j*** with probability p_{ij} or leaves the network with probability a_i. The *transition* or *routing matrix* $\mathbf{P} = ((p_{ij}))$ is assumed to be independent of the current state of the network and of its past. If **I** denotes the identity matrix, then the matrix **I** - **P** is assumed to be nonsingular, i.e. there exists the inverse matrix $(\mathbf{I}-\mathbf{P})^{-1}$.

According to the definition of the a_i and p_{ij},

$$a_i + \sum_{j=1}^{n} p_{ij} = 1 \tag{6.71}$$

In a Jackson-queueing network, each node is generally subject to two inputs, the external input consisting of customers arriving from outside and the internal input consisting of customers arriving from other nodes. Let α_j denote the total arrival intensity at node ***j***. In the steady state, the output intensity from node ***j*** must be equal to α_j. Otherwise, the system could not be in an equilibrium state. Hence, in the steady state, the portion of internal input intensity to node ***j*** which is due to customer from node ***i*** is given by $\alpha_i p_{ij}$. Thus, $\Sigma_{i=1}^{n} \alpha_i \, p_{ij}$ is the total internal input intensity to node ***j***. Consequently, in the steady state,

$$\alpha_j = \lambda_j + \sum_{i=1}^{n} \alpha_i p_{ij}; \quad j = 1, 2, \ldots, n \tag{6.72}$$

By introducing the vectors $\alpha = (\alpha_1, \alpha_2, \ldots, \alpha_n)$ and $\lambda = (\lambda_1, \lambda_2, \ldots, \lambda_n)$, the relationship (6.72) can be written as

$$\alpha(\mathbf{I}-\mathbf{P}) = \lambda$$

Since $\mathbf{I}-\mathbf{P}$ is assumed to be nonsingular, the vector of the total input intensities $\boldsymbol{\alpha}$ is

$$\boldsymbol{\alpha} = \boldsymbol{\lambda}(\mathbf{I}-\mathbf{P})^{-1} \tag{6.73}$$

Even under the assumptions stated, the total inputs at the nodes and the outputs from the nodes are generally not homogeneous Poisson processes.

Let $X_i(t)$ be the random number of customers at node $\boldsymbol{i}$ at time t. Its realizations are denoted by x_i; $x_i = 0, 1, \dots$ The state of the network at time t is is characterized by the vector

$$\mathbf{X}(t) = (X_1(t), X_2(t), \dots, X_n(t))$$

with realizations $\mathbf{x} = (x_1, x_2, \dots, x_n)$. The set of all these vectors $\mathbf{x}$ forms the state space of the Markov chain $\{\mathbf{X}(t),\ t \geq 0\}$. Using set-theory notation, the state space is denoted by $\mathbf{Z} = \{0, 1, \dots\}^n$. Since $\mathbf{Z}$ is countably infinite, this n-dimensional Markov chain becomes one-dimensional by arranging the states as a sequence. To determine the transition rates, the n-dimensional vector $\mathbf{e}_i$ is introduced. Its i-th component is a 1 and the other components are zeros:

$$\mathbf{e}_i = (\underset{1}{0}, \underset{2}{0}, \underset{\dots}{\dots}, 0, \underset{i}{1}, 0, \underset{\dots}{\dots}, \underset{n}{0}). \tag{6.74}$$

Thus, $\mathbf{e}_i$ is the i-th row of the identity matrix $\mathbf{I}$. Since the components of any state vector $\mathbf{x}$ are nonnegative integers, every $\mathbf{x}$ can be represented as a linear combination of all or some of the $\mathbf{e}_1, \mathbf{e}_2, \dots, \mathbf{e}_n$. In particular, $\mathbf{x}+\mathbf{e}_i$ $(\mathbf{x}-\mathbf{e}_i)$ is the vector which arises from $\mathbf{x}$ by increasing (decreasing) the i-th component by 1. Starting from state $\mathbf{x}$, the Markov chain $\{X(t),\ t \geq 0\}$ can make the following one-step transitions:

1) When a customer arrives at node i, the Markov chain makes a transition to state $\mathbf{x}+\mathbf{e}_i$.

2) When a service at node i is finished, $x_i > 0$, and the served customer leaves the network, the Markov chain makes a transition to state $\mathbf{x}-\mathbf{e}_i$.

3) When a service at node i is finished, $x_i > 0$, and the served customer leaves node i for node j, the Markov chain makes a transition to state $\mathbf{x}-\mathbf{e}_i+\mathbf{e}_j$.

Therefore, starting from state $\mathbf{x} = (x_1, x_2, \dots, x_n)$, the transition rates are

$$q_{\mathbf{x},\mathbf{x}+\mathbf{e}_i} = \lambda_i$$

$$q_{\mathbf{x},\mathbf{x}-\mathbf{e}_i} = \min(x_i, s_i)\, \mu_i\, a_i$$

$$q_{\mathbf{x},\mathbf{x}-\mathbf{e}_i+\mathbf{e}_j} = \min(x_i, s_i)\, \mu_i\, p_{ij}, \quad i \neq j$$

In view of (6.71), $\sum_{j,j\neq i} p_{ij} = 1 - p_{ii} - a_i$. Hence the rate of leaving state $\mathbf{x}$ is

$$q_{\mathbf{x}} = \sum_{i=1}^{n} \lambda_i + \sum_{i=1}^{n} \mu_i (1 - p_{ii}) \min(x_i, s_i)$$

If they exist, the stationary state probabilities

$$\pi_{\mathbf{x}} = \lim_{t \to \infty} P(\mathbf{X}(t) = \mathbf{x}), \quad \mathbf{x} \in \mathbf{Z}$$

satisfy according to (6.28) the system of equations

$$q_{\mathbf{x}} \pi_{\mathbf{x}} = \sum_{i=1}^{n} \lambda_i \pi_{\mathbf{x}-\mathbf{e}_i} + \sum_{i=1}^{n} a_i \mu_i \min(x_i + 1, s_i) \pi_{\mathbf{x}+\mathbf{e}_i}$$

$$+ \sum_{j=1}^{n} \sum_{\substack{i=1 \\ i \neq j}}^{n} a_i \mu_i \min(x_i + 1, s_i) p_{ij} \pi_{\mathbf{x}+\mathbf{e}_i - \mathbf{e}_j} \tag{6.75}$$

$$\sum_{\mathbf{x} \in \mathbf{Z}} \pi_{\mathbf{x}} = 1$$

In order to be able to present the solution in a convenient form, let us recall the stationary state probabilities of the waiting system $M/M/s_i/\infty$ with parameters α_i, μ_i, and $\rho_i = \alpha_i/\mu_i$ denoting the intensity of the Poisson input, the service intensities of all s_i servers, and the traffic intensity of the system, respectively. Then, assuming $\rho_i < s_i$, the stationary state probabilities of this system are (see 6.62)

$$\phi_i(j) = \begin{cases} \frac{1}{j!} \rho_i^j \phi_i(0) & \text{for } j = 1, 2, \ldots, s_i - 1 \\ \frac{1}{s_i! \, s_i^{j-s_i}} \rho_i^j \phi_i(0) & \text{for } j = s_i, s_i + 1, \ldots \end{cases}$$

$$\phi_i(0) = \left[\sum_{j=0}^{s_i - 1} \frac{1}{j!} \rho_i^j + \frac{\rho_i^{s_i}}{(s_i - 1)! \, (s_i - \rho_i)} \right]^{-1}$$

(In this context, the notation $\phi(j)$ for the stationary state probabilities is common practice.) The stationary state probabilities of the queueing network are simply obtain by multiplying the corresponding stationary state probabilities of the queueing systems $M/M/s_i/\infty$; $i = 1,2,\ldots n$:

> Provided that the vector of the total input intensities $\alpha = (\alpha_1, \alpha_2, \ldots, \alpha_n)$ given by (6.73) satisfies the conditions
>
> $$\alpha_i < s_i \mu_i; \quad i = 1, 2, \ldots, n;$$
>
> the stationary probability of state $\mathbf{x} = (x_1, x_2, \ldots, x_n)$ is
>
> $$\pi_{\mathbf{x}} = \prod_{i=1}^{n} \phi_i(x_i), \quad \mathbf{x} \in \mathbf{Z} \tag{6.76}$$

Thus, the stationary state distribution of a Jackson-queueing system is given in *product form*. From this one concludes that each node of the network behaves like an $M/M/s/\infty$-system. However, it is generally not a queueing system of this type, because $\{X_i(t), t \geq 0\}$ is usually not a birth- and death process. In particular, the total input into a node need not be a homogeneous Poisson process. But the product form (6.76) of the stationary state probabilities proves that the queue lengths at the nodes are independent random variables. (There is a vast amount of literature dealing with assumptions under which the stationary distribution of a queueing network has the product form (see, for instance, *van Dijk, N.*, 1993).) To verify that the stationary state distribution actually has the product form (6.76), one has to substitute (6.76) into the system of equations (6.75). Using (6.71) and (6.72), one obtains an identity after some tedious algebra.

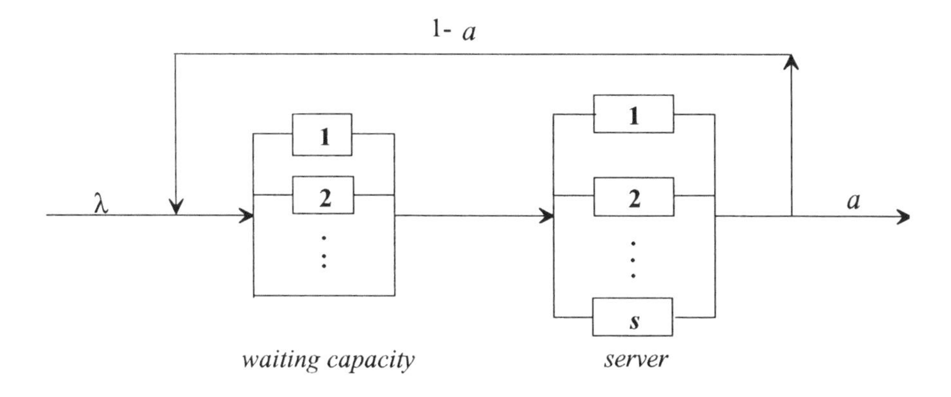

Figure 6.17 Queueing system with feedback

Example 6.18 The simplest case of a Jackson-queueing network is when $n = 1$. The only difference from the queueing system $M/M/s/\infty$ is that now a positve proportion of customers who have departed from the network after having been served will return and require further service. This is a case of a queueing system with *feedback* (Figure 6.17). For instance, when servers have done a bad job, then the corresponding customers will soon return to exercise possible guarantee claims. Formally, these customers remain in the network. Roughly speaking, a single-node Jackson queueing network is a mixture between an open and a closed waiting system.

A customer leaves the system with probability a or joins the queue of the system again with probability $p_{11} = 1 - a$. When there is an idle server, then, clearly, the service of such a customer starts immediately. From (6.71) and (6.72), the total input rate α into the system satisfies the condition

$$\alpha = \lambda + \alpha(1 - a)$$

(Here and in what follows, the index 1 of all system parameters is dropped.) Thus,

$$\alpha = \lambda/a$$

Hence there exists a stationary distribution if $\lambda/a < s\mu$ or, equivalently, if $\rho < as$ with $\rho = \lambda/\mu$. In this case the stationary state probabilities are

$$\pi_j = \begin{cases} \frac{1}{j!}\left(\frac{\rho}{a}\right)^j \pi_0 & \text{for } j = 1, 2, \dots, s-1 \\ \frac{1}{s!\, s^{j-s}}\left(\frac{\rho}{a}\right)^j \pi_0 & \text{for } j = s, s+1, \dots \end{cases}$$

where

$$\pi_0 = \left[\sum_{j=1}^{s-1} \frac{1}{j!}\left(\frac{\rho}{a}\right)^j + \frac{\left(\frac{\rho}{a}\right)^s}{(s-1)!\left(s - \frac{\rho}{a}\right)}\right]^{-1}$$

This is the stationary state distribution of the queueing system $M/M/s/\infty$ (without feedback) the input of which has intensity λ/a. □

Example 6.19 In technological processes, the sequence of service is usually fixed.

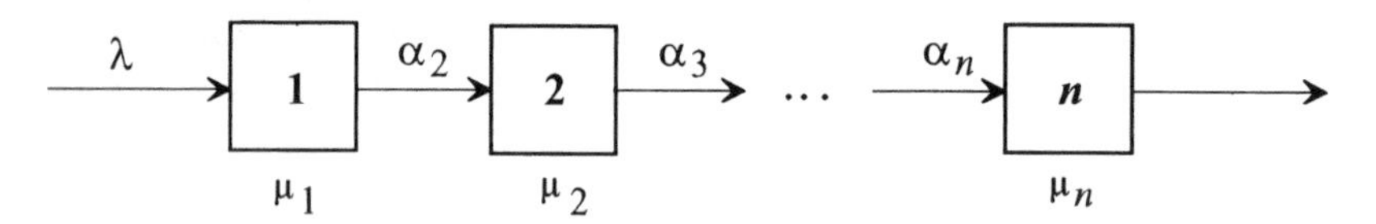

Figure 6.18 Sequential queueing network

(For example, a "customer" may be a car being manufactured on an assembly line.) Hence series connections of queueing systems, i.e. *sequential* or *tandem queueing networks* (Figure 6.18) are of considerable practical interest: External customers arrive only at node **1** (arrival intensity: $\lambda_1 = \lambda$). They subsequently visit the nodes **1**, **2**, ... , ***n*** and leave then the network. The corresponding parameters are

$$\lambda_i = 0; \qquad i = 2, 3, \dots, n$$
$$p_{i,i+1} = 1; \quad i = 1, 2, \dots, n-1$$
$$a_n = 1, \qquad a_1 = a_2 = \cdots = a_{n-1} = 0$$

According to (6.72), the total input intensities of all nodes in the steady state must be the same:

$$\alpha_1 = \alpha_2 = \cdots = \alpha_n = \lambda$$

Hence, given single-server nodes ($s_1 = s_2 = \cdots = s_n = 1$), a stationary state distribution exists if

$$\rho_i = \lambda/\mu_i < 1 ; \quad i = 1, 2, \ldots, n$$

or, equivalently,

$$\lambda < \min(\mu_1, \mu_2, \ldots, \mu_n)$$

Thus, it is the slowest server which determines the efficiency of a sequential network. The stationary probability of state $\mathbf{x} = (x_1, x_2, \cdots, x_n)$ is given by

$$\pi_{\mathbf{x}} = \prod_{i=1}^{n} \rho_i^{x_i}(1-\rho_i); \quad \mathbf{x} \in \mathbf{Z}$$

Of course, the sequential network can be generalized by taking feedback into account. This is left as an exercise to the reader. □

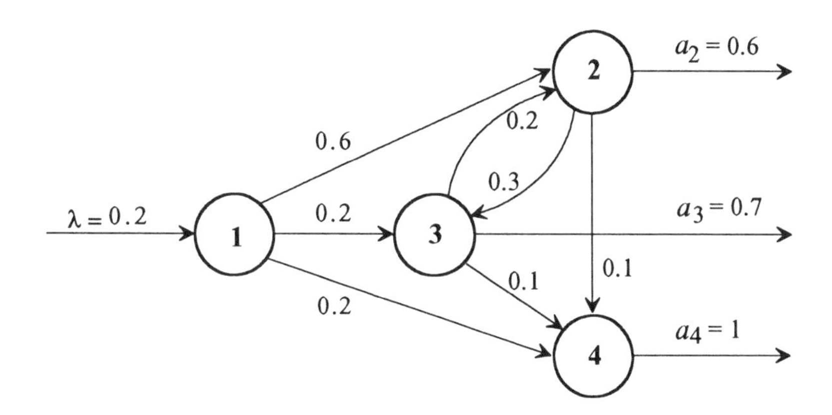

Figure 6.19 Maintenance workshop as a queueing network

Example 6.20 Defective robots arrive at the admissions department of a maintenance workshop in accordance with a homogeneous Poisson process with intensity $\lambda = 0.2\ [h^{-1}]$. In the admissions department (denoted by (**1**)) a first failure diagnosis is carried out. Depending on the result, the robots will have to visit other departments of the workshop. These are departments for checking and repairing the mechanics (**2**), electronics (**3**), and software (**4**) of the robots, respectively. The failure diagnosis in the admissions department results in 60% of the arriving robots being sent to department (**2**) and 20% each to the departments (**3**) and (**4**). After having being maintained in department (**2**), 60% of the robots leave the workshop, 30% are sent to department (**3**), and 10% to department (**4**). After having being served by department (**3**), 70% of the robots leave the workshop, 20% are sent to department (**2**), and 10% are sent to department (**4**). After elimination of possible software failures all robots leave the workshop (Figure 6.19). Naturally, a robot can be sent several times to one and the same department.

The following transition probabilities are a consequence of the transfer of robots between the departments:

$$p_{12}=0.6,\quad p_{13}=0.2,\quad p_{14}=0.2,\quad p_{23}=0.3$$

$$p_{24}=0.1,\quad p_{32}=0.2,\quad p_{34}=0.1$$

$$a_1=0,\qquad a_2=0.6,\qquad a_3=0.7,\qquad a_4=1$$

The service intensities are are assumed to be

$$\mu_1=1,\ \mu_2=0.45,\ \mu_3=0.4, \text{ and } \mu_4=0.1\quad [\text{in } h^{-1}]$$

The graph depicted in Figure 6.19 illustrates the possible transitions between the departments. The edges of the graph are weighted by the corresponding transition probabilities. The system of equations (6.72) in the total input intensities is

$$\alpha_1=0.2$$

$$\alpha_2=0.6\,\alpha_1 \qquad\qquad +0.2\,\alpha_3$$

$$\alpha_3=0.2\,\alpha_1+0.3\,\alpha_2$$

$$\alpha_4=0.2\,\alpha_1+0.1\,\alpha_2+0.1\,\alpha_3$$

The solution is (after rounding)

$$\alpha_1=0.20,\quad \alpha_2=0.135,\quad \alpha_3=0.08,\quad \alpha_4=0.06$$

The corresponding traffic intensities $\rho_i=\alpha_i/\mu_i$ are

$$\rho_1=0.2,\quad \rho_2=0.3,\quad \rho_3=0.2,\quad \rho_4=0.6$$

From (6.76), the stationary probability of state $\mathbf{x}=(x_1,x_2,x_3,x_4)$ in case of single-server nodes is

$$\pi_{\mathbf{x}}=\prod_{i=1}^{4}\rho^{x_i}(1-\rho_i)$$

or

$$\pi_{\mathbf{x}}=0.1792\,(0.2)^{x_1}\,(0.3)^{x_2}\,(0.2)^{x_3}\,(0.6)^{x_4}\,;\quad \mathbf{x}\in\mathbf{Z}=\{0,1,\ldots\}^4$$

In particular, $\pi_{\mathbf{x}_0}=0.1792$ with $\mathbf{x}_0=(0,0,0,0)$ is the stationary probability that there is no robot in the workshop. The stationary probability that there is at least one robot in the admissions department is

$$P(X_1>0)=0.8\sum_{i=1}^{\infty}(0.2)^i=0.2$$

Analogously,

$$P(X_2 > 0) = 0.3, \quad P(X_3 > 0) = 0.2, \quad \text{and} \quad P(X_4 > 0) = 0.6$$

(X_i is the random number of robots at node $\boldsymbol{i}$ in the steady state.) Thus, when there is a delay in checking and repairing defective robots, the cause is most probably department **4** in view of the comparatively high amount of time necessary for finding and removing software failures. □

Closed queueing networks Analogously to the closed queueing system, no customers arrive at a *closed queueing network* "from outside". Customers which have been served at a node, do not leave the network, but move to another node for service. Hence, the number of customers in a closed queueing network is constant, say, N. Practical examples of closed queueing networks are multiprogrammed computer and communication systems.

When the service of a customer at node $\boldsymbol{i}$ is finished, then the customer moves to node $\boldsymbol{j}$ for further service with probability p_{ij}. Since the customer does not leave the network,

$$\sum_{j=1}^{n} p_{ij} = 1; \quad i = 1, 2, \ldots, n, \tag{6.77}$$

where n is again the number of nodes. Provided the discrete Markov chain given by transition matrix $\mathbf{P} = ((p_{ij}))$ and state space $\mathbf{Z} = \{1, 2, \ldots, n\}$ is irreducible, it has a stationary state distribution $\{\pi_1, \pi_2, \ldots, \pi_n\}$ which, according to (5.7), is the unique solution of the system of equations

$$\pi_j = \sum_{i=1}^{n} p_{ij}\pi_i; \quad j = 1, 2, \ldots, n \tag{6.78}$$

$$1 = \sum_{i=1}^{n} \pi_i$$

Let $X_i(t)$ be the random number of customers at node $\boldsymbol{i}$ at time t and

$$\mathbf{X}(t) = (X_1(t), X_2(t), \ldots, X_n(t))$$

The state space of the Markov chain $\{\mathbf{X}(t), t \geq 0\}$ is

$$\mathbf{Z} = \left\{ \mathbf{x} = (x_1, x_2, \ldots, x_n) \text{ with } \sum_{i=1}^{n} x_i = N \text{ and } 0 \leq x_i \leq N \right\} \tag{6.79}$$

Clearly, $\mathbf{Z}$ contains

$$\binom{n+N-1}{N}$$

elements (states).

Let $\mu_i(x_i)$ be the common service intensity of all servers at node $\boldsymbol{i}$ if there are x_i customers at this node, $\mu_i(0) = 0$. Then $\{\mathbf{X}(t),\ t \geq 0\}$ has the positive transition rates

$$q_{\mathbf{x},\mathbf{x}-\mathbf{e}_i+\mathbf{e}_j} = \mu_i(x_i)p_{ij}; \quad x_i \geq 1,\ i \neq j,$$

and

$$q_{\mathbf{x}-\mathbf{e}_i+\mathbf{e}_j,\mathbf{x}} = \mu_j(x_j+1)p_{ji}; \quad i \neq j,\ \mathbf{x}-\mathbf{e}_i+\mathbf{e}_j \in \mathbf{Z},$$

where the vectors $\mathbf{e}_i$ are defined in (6.74). From (6.77), the rate of leaving state $\mathbf{x} = (x_1, x_2, \dots, x_n)$ is

$$q_{\mathbf{x}} = \sum_{i=1}^{n} \mu_i(x_i)(1-p_{ii})$$

Hence, according to (6.28), the stationary distribution $\{\pi_{\mathbf{x}},\ \mathbf{x} \in \mathbf{Z}\}$ of the Markov chain $\{X(t),\ t \geq 0\}$ satisfies the system of equations

$$\sum_{i=1}^{n} \mu_i(x_i)(1-p_{ii})\pi_{\mathbf{x}} = \sum_{\substack{i,j=1 \\ i \neq j}}^{n} \mu_j(x_j+1)p_{ji}\pi_{\mathbf{x}-\mathbf{e}_i+\mathbf{e}_j}, \tag{6.80}$$

where $\mathbf{x} = (x_1, x_2, \dots, x_n) \in \mathbf{Z}$. In these equations, all $\pi_{\mathbf{x}-\mathbf{e}_i+\mathbf{e}_j}$ with $\mathbf{x}-\mathbf{e}_i+\mathbf{e}_j \notin \mathbf{Z}$ are equal to 0. Let $\phi_i(0) = 1$ and

$$\phi_i(j) = \prod_{m=1}^{j} \left(\frac{\pi_i}{\mu_i(m)}\right); \quad i = 1, 2, \dots, n;\ j = 1, 2, \dots, N$$

The stationary probability of state $\mathbf{x} = (x_1, x_2, \dots, x_n) \in \mathbf{Z}$ is

$$\pi_{\mathbf{x}} = h \prod_{i=1}^{n} \phi_i(x_i) \tag{6.81}$$

where

$$h = \left[\sum_{\mathbf{y} \in \mathbf{Z}} \prod_{i=1}^{n} \phi_i(y_i)\right]^{-1}, \quad \mathbf{y} = (y_1, y_2, \dots, y_n)$$

By substituting (6.81) into (6.80) one readily verifies that $\{\pi_{\mathbf{x}},\ \mathbf{x} \in \mathbf{Z}\}$ is indeed a stationary distribution of the Markov chain $\{X(t),\ t \geq 0\}$.

Example 6.21 A closed sequential queueing network has a single server at each of its n nodes. There is only $N = 1$ customer in the system. This customer is being served by exactly one node so that the other nodes are empty. Hence, with the vectors $\mathbf{e}_i$ as defined by (6.74), the state space of the corresponding Markov chain $\{\mathbf{X}(t),\ t \geq 0\}$ is $\mathbf{Z} = \{\mathbf{e}_1, \mathbf{e}_2, \dots, \mathbf{e}_n\}$.

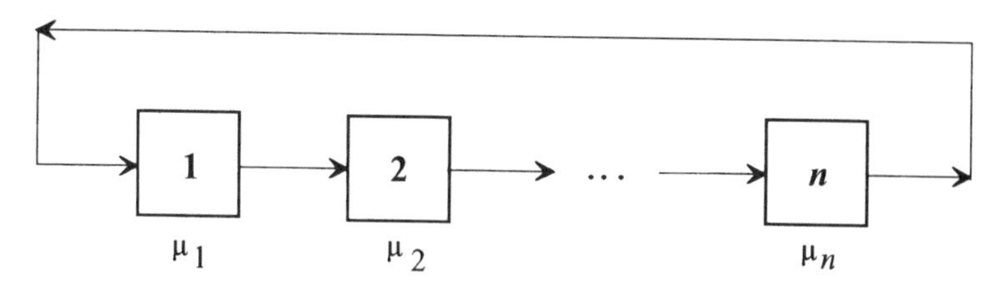

Figure 6.20 Closed sequential queueing network

Let $\mu_i = \mu_i(1)$ be the service intensity at node $\boldsymbol{i}$; $i = 1, 2, \ldots, n$ (Figure 6.20). The transition probabilities are

$$p_{i,i+1} = 1; \quad i = 1, 2, \ldots, n-1; \quad p_{n,1} = 1$$

The solution of the system of equations (6.78) has a simple structure:

$$\pi_i = 1/n; \quad i = 1, 2, \ldots, n$$

Therefore,

$$\phi_i(0) = 1 \quad \text{and} \quad \phi_i(1) = \frac{1}{n\mu_i}; \quad i = 1, 2, \ldots, n;$$

so that

$$h = n\left[\sum_{i=1}^{n} \frac{1}{\mu_i}\right]^{-1}$$

The stationary state probabilities are

$$\pi_{\mathbf{e}_i} = \frac{1/\mu_i}{\sum_{i=1}^{n} \frac{1}{\mu_i}}; \quad i = 1, 2, \ldots, n$$

If $\mu_i = \mu$; $i = 1, 2, \ldots, n$; then the states are uniformly distributed:

$$\pi_{\mathbf{e}_i} = 1/n; \quad i = 1, 2, \ldots, n$$

If there are $N \geq 1$ customers in the system and the μ_i do not depend on x_i, then the stationary state probabilities are

$$\pi_{\mathbf{x}} = \frac{(1/\mu_1)^{x_1}(1/\mu_2)^{x_2}\cdots(1/\mu_n)^{x_n}}{\sum_{\mathbf{y}\in\mathbf{Z}}\prod_{i=1}^{n}\left(\frac{1}{\mu_i}\right)^{y_i}},$$

where $\mathbf{x} = (x_1, x_2, \ldots, x_n) \in \mathbf{Z}$. Given $\mu_i = \mu$; $i = 1, 2, \ldots, n$; the states are again uniformly distributed:

$$\pi_{\mathbf{x}} = \frac{1}{\binom{n+N-1}{N}}, \quad \mathbf{x} \in \mathbf{Z}$$

□

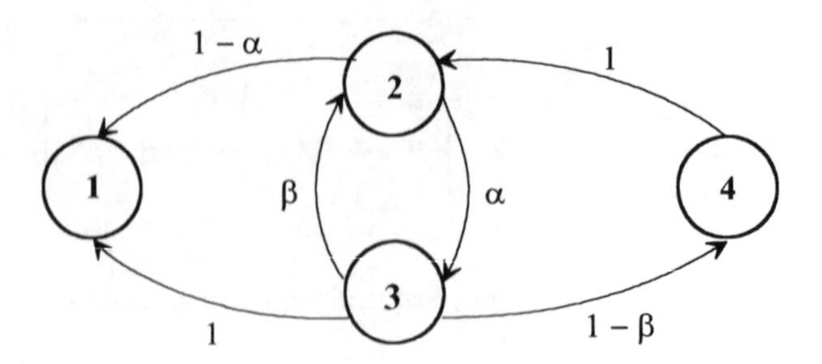

Figure 6.21 Computer system as a closed queueing network

Example 6.22 A computer system consists of two central processors **2** and **3**, a disc drive **1**, and a printer **4**. A new program starts in central processor **2**. When this processor has finished its computing job, the computing phase continues in central processor **3** with probability α or the program goes to the disc drive with probability $1-\alpha$ From the disc drive the program goes to central processor **3** with probability 1. From central processor **3** it goes to the central processor **2** with probability β or to the printer with probability $1-\beta$. Here it terminates or goes back to central processor **2**. When a program terminates, then another program (from outside) immediately joins the queue of central processor **2** so that there is always a fixed number of programs in the system. Hence, from the printer a program formally goes to the central processor **2** with probability 1. If N denotes the constant number of programs in the system, this situation represents a simple case of *multiprogramming* with N as the *level of multiprogramming.*

The state space $\mathbf{Z}$ of this system and the matrix $\mathbf{P}$ of the transition probabilities p_{ij} are given by

$$\mathbf{Z} = \{\mathbf{y} = (y_1, y_2, y_3, y_4);\ y_i = 0, 1, \ldots, N;\ y_1 + y_2 + y_3 + y_4 = N\}$$

and (Figure 6.21)

$$\mathbf{P} = \begin{pmatrix} 0 & 0 & 1 & 0 \\ 1-\alpha & 0 & \alpha & 0 \\ 0 & \beta & 0 & 1-\beta \\ 0 & 1 & 0 & 0 \end{pmatrix},$$

respectively. The corresponding solution of the system of equations (6.78) is

$$\pi_1 = \frac{1-\alpha}{4-\alpha-\beta}, \quad \pi_2 = \pi_3 = \frac{1}{4-\alpha-\beta}, \quad \pi_4 = \frac{1-\beta}{4-\alpha-\beta}$$

Let the service intensities of the nodes μ_1, μ_2, μ_3 and μ_4 be independent of the number of programs at the nodes. Then,

$$\phi_i(x_i) = \left(\frac{\pi_i}{\mu_i}\right)^{x_i}, \quad i = 1, 2, ..., n$$

Hence, the stationary probability of state

$$\mathbf{x} = (x_1, x_2, x_3, x_4) \text{ with } x_1 + x_2 + x_3 + x_4 = N$$

is given by

$$\pi_{\mathbf{x}} = \frac{h}{(4-\alpha-\beta)^N} \left(\frac{1-\alpha}{\mu_1}\right)^{x_1} \left(\frac{1}{\mu_2}\right)^{x_2} \left(\frac{1}{\mu_3}\right)^{x_3} \left(\frac{1-\beta}{\mu_4}\right)^{x_4},$$

where

$$h = \frac{(4-\alpha-\beta)^N}{\sum\limits_{\mathbf{y} \in \mathbf{Z}} \left(\frac{1-\alpha}{\mu_1}\right)^{y_1} \left(\frac{1}{\mu_2}\right)^{y_2} \left(\frac{1}{\mu_3}\right)^{y_3} \left(\frac{1-\beta}{\mu_4}\right)^{y_4}}$$

□

Application-oriented treatments of queueing networks are, for instance, *Gelenbe* and *Pujolle* (1987), *Walrand* (1988).

6.9 Semi-Markov Processes

Transitions between the states of a homogeneous Markov chain are controlled by the transition matrix $\mathbf{P} = ((p_{ij}))$. According to section 6.7, the sojourn time in a state is exponentially distributed and depends only on the current state, but not on the subsequent one. Since in the most practical applications the sojourn times in the system states are non-exponential random variables, an obvious generalization is to admit arbitrarily distributed sojourn times while retaining the transition mechanism between the states. This approach leads to the *semi-Markov processes.*

A semi-Markov process $\{X(t), t \geq 0\}$ with state space $\mathbf{Z} = \{0, 1, ...\}$ evolves in the following way: The transitions between the states are governed by a discrete-time homogeneous Markov chain $\{X_0, X_1, ...\}$ with state space $\mathbf{Z}$ and transition matrix $\mathbf{P} = ((p_{ij}))$. If the process starts in state i_0 at time $t = 0$, then the subsequent state i_1 is determined according to the transition matrix $\mathbf{P}$, while the process stays in state i_0 a random time $Y_{i_0 i_1}$. Thereafter the state i_2 following state i_1 is determined. The process stays in state i_1 a random time $Y_{i_1 i_2}$ and so on. The random variables Y_{ij} are the *conditional sojourn times* of the process in state i given that the process makes a transition from i to j. They are assumed to be independent. Hence, immediately after entering a state at time t, say, the further evolvement of a semi-Markov process depends only on this state, but not on the evolvement of the process up till t. A transition from a state to the same state may happen. The sample paths of a semi-Markov process are piecewise constant functions which, by convention, are continuous on the right. (Note that, in contrast to homogeneous continuous-time Markov chains, for a complete description of the state of a semi-Markov process at time t it is also necessary to know the "age" of the current state.)

The discrete-time stochastic process $\{X_0, X_1, ...\}$ is said to be *embedded* in the continuous-time stochastic process $\{X(t), t \geq 0\}$, or, more precisely, $\{X_0, X_1, ...\}$ is a discrete-time Markov chain embedded in the continuous-time semi-Markov process $\{X(t), t \geq 0\}$. Let $T_0, T_1, ...$ denote the sequence of time points at which the semi-Markov process makes a transition from one state to another (or to the same state). Then

$$X_n = X(T_n), \; n = 0, 1, ... ,$$

where $X_0 = X(0)$ is the initial state. Hence, the transition probabilities can be written in the following form:

$$p_{ij} = P(X(T_{n+1}) = j | X(T_n) = i); \quad n = 0, 1, ...$$

As already pointed out, the evolvement of the semi-Markov process from a *jump point* T_n is independent of the "history" of the process up till $T_n - 0$, $n = 1, 2, ...$

Special cases 1) The most simple special case of a semi-Markov process is the ordinary renewal process. In this case there is only one state, namely "system is operating". When denoting this state by "1", then $p_{11} = 1$.

2) An alternating renewal process is a semi-Markov process with state space $\mathbf{Z} = \{0, 1\}$ and transition probabilities $p_{00} = 0$, $p_{01} = 1$, $p_{10} = 1$ and $p_{11} = 0$, where state 0 (1) indicates that the system is being renewed (is operating).

Let $F_{ij}(t) = P(Y_{ij} \leq t)$ denote the distribution function of the conditional sojourn time Y_{ij} of a semi-Markov process in state i if the subsequent state is j. By the total probability rule, the *unconditional sojourn time* Y_i of the process in state i is

$$F_i(t) = P(Y_i \leq t) = \sum_{j \in \mathbf{Z}} p_{ij} F_{ij}(t), \quad i \in \mathbf{Z}$$

In what follows, semi-Markov processes are considered under the following three assumptions:

1) The embedded homogeneous Markov chain $\{X_0, X_1, ...\}$ has a unique stationary state distribution $\{\pi_0, \pi_1, ...\}$. By (5.7), this is the solution of the system of equations

$$\pi_j = \sum_{i \in \mathbf{Z}} p_{ij} \pi_i \, , \qquad \sum_{i \in \mathbf{Z}} \pi_i = 1 \tag{6.82}$$

As pointed out in section 5.3, a unique stationary state distribution exists if $\{X_0, X_1, ...\}$ is aperiodic, irreducible and positively recurrent.

2) The distribution functions $F_i(t) = P(Y_i \le t)$ are non-arithmetic (definition 4.4).

3) The expected unconditional sojourn times of the process in all states are finite:

$$\mu_i = E(Y_i) = \int_0^\infty (1 - F_i(t))\, dt < \infty, \quad i \in \mathbf{Z}$$

(Note that μ_i is no longer an intensity but a mean sojourn time.)

In what follows, a transition of the semi-Markov process into state i is called an *i-transition*. Let $N_i(t)$ be the random number of i-transitions occuring in $[0, t]$ and $H_i(t) = E(N_i(t))$. Then, for any $\tau > 0$ (see *Matthes* (1962))

$$\lim_{t \to \infty} (H_i(t+\tau) - H_i(t)) = \frac{\tau \pi_i}{\sum_{j \in \mathbf{Z}} \pi_j \mu_j}, \quad i \in \mathbf{Z} \tag{6.83}$$

This relationship implies that after a sufficiently long time period the number of i-transitions in a given time interval does no longer depend on the position of this interval, but only on its length. Thus, after a sufficiently long time, the semi-Markov process becomes stationary. (However, in what follows, the definition and properties of stationary semi-Markov processes are not discussed in detail.) The following theoretical results and applications of semi-Markov processes refer to the steady state phase (stationary phase) of the process. From (6.83), the expected number of i-transitions per unit time is

$$U_i = \frac{\pi_i}{\sum_{j \in \mathbf{Z}} \pi_j \mu_j}$$

Hence the portion of time the process is in state i is given by

$$A_i = \frac{\pi_i \mu_i}{\sum_{j \in \mathbf{Z}} \pi_j \mu_j} \tag{6.84}$$

Consequently, the fraction of time the process is in a set of states $\mathbf{Z}_0$, $\mathbf{Z}_0 \subseteq \mathbf{Z}$, is

$$A_{\mathbf{Z}_0} = \frac{\sum_{j \in \mathbf{Z}_0} \pi_j \mu_j}{\sum_{j \in \mathbf{Z}} \pi_j \mu_j} \tag{6.85}$$

In other words, $A_{\mathbf{Z}_0}$ is the probability that a visitor from outside finds the process in a state belonging to $\mathbf{Z}_0$.

Let c_i denote the cost which is caused by an i-transition of the system. Then the expected total (transition) cost per unit time amounts to

$$C = \frac{\sum_{j \in \mathbf{Z}} \pi_j c_j}{\sum_{j \in \mathbf{Z}} \pi_j \mu_j} \tag{6.86}$$

Note that the formulas (6.83) to (6.86) depend only on the unconditional sojourn times of a semi-Markov process in its states, which facilitates their practical application. The following examples, which have already been partially analysed by other methods, illustrate this fact.

Example 6.24 (***alternating renewal process***) Let 0 denote the renewal phase and 1 the operating phase of a system. Then the system of equations (6.82) becomes (one equation is superfluous and will be omitted)

$$\pi_0 = 0 \cdot \pi_0 + 1 \cdot \pi_1$$

$$1 = \pi_0 + \pi_1$$

The stationary state distribution is therefore given by $\pi_0 = \pi_1 = 1/2$. From (6.84), the probability that the system is in its operating phase, is

$$A_1 = \frac{\mu_1}{\mu_0 + \mu_1}$$

This formula has been already obtained in section 4.6 (see formula 4.43). □

Example 6.25 (***age renewal policy***) The system is renewed upon failure by an *emergency renewal* or at age τ by a *preventive renewal*, whichever occurs first.

The aim is to determine the stationary availability of the system. (This problem has been solved in section 4.8, example 4.11, by modeling the system behaviour by a regenerative stochastic process.)

The system can be in one of the following states:

0 operating

1 emergency renewal

2 preventive renewal

Let L be the random lifetime of the system, $F(t) = P(L \leq t)$ its distribution function, and $\overline{F}(t) = 1 - F(t) = P(L > t)$ its survival probability. Then the non-negative transition probabilities between the states are given by (Figure 6.22)

$$p_{01} = F(\tau), \quad p_{02} = \overline{F}(\tau), \quad p_{10} = p_{20} = 1$$

Let Z_e and Z_p be the random times for emergency renewals and preventive renewals, respectively. Then the conditional sojourn times of the system in the states are given by

$$Y_{01} = L, \quad Y_{02} = \tau, \quad Y_{10} = Z_e, \quad Y_{20} = Z_p$$

The unconditional sojourn times are

$$Y_0 = \min(L, \tau), \quad Y_1 = Z_e, \quad Y_2 = Z_p$$

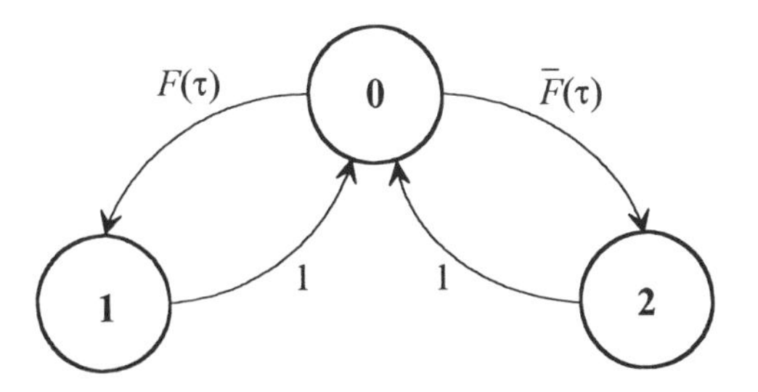

Figure 6.22 Transition graph for example 6.25

The system behaviour can be described by a semi-Markov process $\{X(t), t \geq 0\}$ with state space $\mathbf{Z} = \{0, 1, 2\}$ and the given transition probabilities and sojourn times. The corresponding system of equations (6.82) in the stationary probabilities of the embedded Markov chain is

$$\begin{aligned} \pi_0 &= \pi_1 + \pi_2 \\ \pi_1 &= F(\tau)\,\pi_0 \\ 1 &= \pi_0 + \pi_1 + \pi_2 \end{aligned}$$

The solution is

$$\pi_0 = 1/2, \quad \pi_1 = F(\tau)/2, \quad \pi_2 = \bar{F}(\tau)/2$$

Let $\mu_i = E(Y_i),\ i = 0, 1, 2$. Then,

$$\mu_0 = E(Y_0) = \int_0^\tau \bar{F}(t)\,dt$$

According to (6.84), the stationary availability $A_0 = A(\tau)$ of the system is

$$A(\tau) = \frac{\mu_0 \pi_0}{\mu_0 \pi_0 + \mu_1 \pi_1 + \mu_2 \pi_2}$$

or

$$A(\tau) = \frac{\int_0^\tau \bar{F}(t)\,dt}{\int_0^\tau \bar{F}(t)\,dt + d_e F(\tau) + d_p \bar{F}(\tau)}, \tag{6.87}$$

where $d_e = E(Z_e)$ and $d_p = E(Z_p)$. It is practical important that this result does not depend on the probability distributions of $Y_1 = Z_e$ and $Y_2 = Z_p$, but only on their expected values. Assuming constant renewal times d_e and d_p for emergency and preventive renewals, respectively, equation (6.87) implies the result obtained in example 4.11 by using regenerative stochastic processes.

If the renewal times are negligibly small, but the respective expected costs c_e and c_p for emergency and preventive renewals are relevant, then, from (6.86), the expected renewal cost per unit time in the steady state is given by

$$K(\tau) = \frac{c_e \pi_1 + c_p \pi_2}{\mu_0 \pi_0}$$

or,

$$K(\tau) = \frac{c_e F(\tau) + c_p \bar{F}(\tau)}{\int_0^\tau \bar{F}(t)\, dt}$$

Analogously to the corresponding renewal times, c_e and c_p can be the expected values of arbitrarily distributed renewal costs. Denoting the failure rate of the system by $\lambda(t)$, a cost-optimal renewal interval $\tau = \tau*$ must satisfy the necessary condition $dK(\tau)/d\tau = 0$ or

$$\lambda(\tau) \int_0^\tau \bar{F}(t)\, dt - F(\tau) = \frac{c}{1-c}, \tag{6.88}$$

where $c = c_p/c_e$. If $c < 1$ and $\lambda(t)$ is unboundedly and strictly increasing, then there exists a unique solution $\tau = \tau*$ of this equation.

Comparing the equations (4.54) and (6.88) proves that both minimizing the expected renewal cost per unit time and maximizing the stationary availability lead to the same analytical type of equations for determining the corresponding optimal renewal intervals. □

Example 6.26 A series system consists of n subsystems $e_1, e_2, \dots, e_n$. The lifetimes of the subsystems $L_1, L_2, \dots, L_n$ are independent exponential random variables with parameters $\lambda_1, \lambda_2, \dots, \lambda_n$, respectively. Let

$$G_i(t) = P(L_i \le t) = 1 - e^{-\lambda_i t}, \quad g_i(t) = \lambda_i e^{-\lambda_i t}, \quad t \ge 0;\ i = 1, 2, \dots, n$$

When a subsystem fails, the system interrupts its work. As soon as the renewal of the failed subsystem is finished, the system continues operating. Let μ_i be the average renewal time of subsystem e_i. As long as a subsystem is being renewed the other subsystems cannot fail, i.e. during such a time period they are in the cold-standby mode. Letting $X(t) = 0$ if the system is operating and $X(t) = i$ if e_i is being renewed, then $\{X(t), t \ge 0\}$ is a semi-Markov process with state space

$$\mathbf{Z} = \{0, 1, \dots, n\}.$$

(The process $\{X(t), t \ge 0\}$ would be a Markov chain if the renewal times are exponentially distributed, too.) The conditional sojourn times in state 0 of this semi-Markov process are $Y_{0i} = L_i$; $i = 1, 2, \dots, n$. Its unconditional sojourn time in state 0 is

$$Y_0 = \min\{L_1, L_2, \dots, L_n\}$$

Thus, Y_0 has distribution function

$$F_0(t) = 1 - \bar{G}_1(t) \cdot \bar{G}_2(t) \cdots \bar{G}_n(t)$$

Letting

$$\lambda = \lambda_1 + \lambda_2 + \ldots + \lambda_n$$

implies

$$F_0(t) = 1 - e^{-\lambda t}, \quad t \geq 0$$

$$\mu_0 = E(Y_0) = 1/\lambda$$

The system makes a transition from state 0 into state i with probability

$$\begin{aligned} p_{0i} &= P(Y_0 = L_i) \\ &= \int_0^\infty \bar{G}_1(x) \cdot \bar{G}_2(x) \cdots \bar{G}_{i-1}(x) \cdot \bar{G}_{i+1}(x) \cdots \bar{G}_n(x) g_i(x)\, dx \\ &= \int_0^\infty e^{-(\lambda_1 + \lambda_2 + \ldots + \lambda_{i-1} + \lambda_{i+1} + \ldots + \lambda_n)x} \lambda_i e^{-\lambda_i x}\, dx \\ &= \int_0^\infty e^{-\lambda x} \lambda_i\, dx \end{aligned}$$

Hence,

$$p_{0i} = \frac{\lambda_i}{\lambda}, \quad p_{i0} = 1; \quad i = 1, 2, \ldots, n$$

Thus, the system of equations (6.82) becomes

$$\pi_0 = \pi_1 + \pi_2 + \cdots + \pi_n$$

$$\pi_i = \frac{\lambda_i}{\lambda} \pi_0; \quad i = 1, 2, \ldots, n$$

In view of $\pi_1 + \pi_2 + \cdots + \pi_n = 1 - \pi_0$, the solution is easily seen to be

$$\pi_0 = \frac{1}{2}; \quad \pi_i = \frac{\lambda_i}{2\lambda}; \quad i = 1, 2, \ldots, n$$

In view of (6.84), the stationary availability of the system becomes

$$A_0 = \frac{1}{1 + \sum_{i=1}^{n} \lambda_i / \mu_i}$$

Thus, the stationary availability of the system increases (decreases) when the expected lifetimes of the subsystems (expected renewal times of the subsystems) increase. □

Example 6.27 The same model as in example 6.7 is considered, however, under more general assumptions on the underlying probability distributions: A system has two different types of failures 1 and 2 (see example 6.7 for motivation). After a type 1-failure, the system is switched from failure state 1 into failure state 2. On entering failure state 2, the renewal of the system begins. The renewed system im-

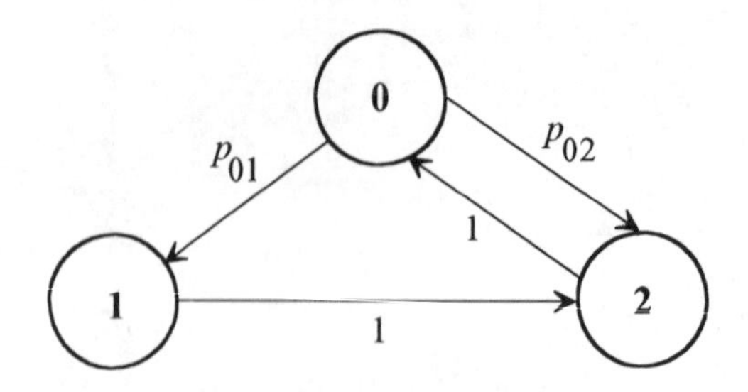

Figure 6.23 Transition graph of example 6.27

mediately starts working. This process is continued indefinitely. All life times and renewal times as well as the switching times are assumed to be independent.

The system can be in one of the following states (Figure 6.23):

0 system is available
1 type 1-failure state
2 type 2-failure state

The random times to type 1- and type 2-failures L_1 and L_2 are assumed to be exponentially distributed with parameters λ_1 and λ_2, respectively. Hence, the distribution function of the unconditional sojourn time in state 0, namely

$$Y_0 = \min(L_1, L_2),$$

and the corresponding expected value are

$$F_0(t) = 1 - e^{-(\lambda_1+\lambda_2)t}, \quad t \geq 0, \qquad \mu_0 = \frac{1}{\lambda_1 + \lambda_2}$$

The positive transition probabilities are

$$p_{01} = P(L_1 < L_2) = \frac{\lambda_1}{\lambda_1 + \lambda_2}$$

$$p_{02} = P(L_1 > L_2) = \frac{\lambda_2}{\lambda_1 + \lambda_2}$$

$$p_{12} = 1, \quad p_{20} = 1$$

Therefore, the stationary state probabilities of the embedded Markov chain satisfy the following system of equations:

$$\begin{aligned}
\pi_0 &= \pi_2 \\
\pi_1 &= p_{01}\,\pi_0 \\
1 &= \pi_0 + \pi_1 + \pi_2
\end{aligned}$$

The solution is

$$\pi_0 = \pi_2 = \frac{\lambda_1 + \lambda_2}{3\lambda_1 + 2\lambda_2}$$

$$\pi_1 = \frac{\lambda_1}{3\lambda_1 + 2\lambda_2}$$

Thus, the stationary state probabilities of the system are given by

$$A_0 = \frac{1}{1 + \lambda_1\mu_1 + (\lambda_1 + \lambda_2)\mu_2} \qquad \textit{(stationary availability)}$$

$$A_1 = \frac{\lambda_1\mu_1}{1 + \lambda_1\mu_1 + (\lambda_1 + \lambda_2)\mu_2}$$

$$A_2 = \frac{(\lambda_1 + \lambda_2)\mu_2}{1 + \lambda_1\mu_1 + (\lambda_1 + \lambda_2)\mu_2}$$

Note that μ_1 and μ_2 are the expected values of arbitrarily distributed sojourn times in the states 1 and 2, respectively. Letting

$$\mu_1 = 1/\nu \quad \text{and} \quad \mu_2 = 1/\mu,$$

one obtains the stationary state probabilities derived in example 6.7 if the sojourn times in states 1 and 2 are exponentially distributed with parameters ν and μ, respectively. (The probabilities π_i determined in example 6.7 correspond to the probabilities A_i in this example.) The special application of this model to a problem of the traffic safety pointed out in example 6.7 assumes Y_1 to be a switching time so that the assumption $\mu_1 = 0$ might be justified.

Computing the stationary state probabilities of this system by (6.84) is also possible if L_1 and L_2 have arbitrary distribution functions $G_1(t)$ and $G_2(t)$ with densities $g_1(t)$ and $g_2(t)$, respectively (provided the other model assumptions remain valid.) In this case,

$$\mu_0 = E(Y_0) = \int_0^\infty \bar{G}_1(t)\bar{G}_2(t)\,dt$$

$$p_{01} = P(L_1 < L_2) = \int_0^\infty \bar{G}_2(t)\,g_1(t)\,dt$$

$$p_{02} = P(L_1 > L_2) = 1 - p_{01} = \int_0^\infty \bar{G}_1(t)\,g_2(t)\,dt$$

The fact that the formulas (6.84) to (6.86) hold for any probability distributions of the sojourn times is a great advantage of semi-Markov processes and allows their application to modeling a much broader class of practical problems than Markov processes. □

The time-dependent behaviour of semi-Markov processes is discussed, for example, in *Gaede* (1977) and *Kulkarni* (1995).

Exercises

6.1) Let $\mathbf{Z} = \{0, 1\}$ be the state space and

$$\mathbf{P}(t) = \begin{pmatrix} e^{-t} & 1-e^{-t} \\ 1-e^{-t} & e^{-t} \end{pmatrix}$$

the transition matrix of a continuous-time stochastic process $\{X(t), t \geq 0\}$.
Check whether $\{X(t), t \geq 0\}$ is a homogeneous Markov chain.

6.2) A system fails after a random lifetime L. Then it waits a random time W for renewal. A renewal takes another random time Z. The random variables L, W and Z have exponential distributions with parameters λ, ν and μ, respectively. On completion of the renewal, the system immediately resumes its work. This process continues indefinitely. All life, waiting, and renewal times are assumed to be independent. Let the system be in states **0**, **1** and **2** when it is operating, waiting or being renewed, respectively.
(1) Draw the transition graph of the corresponding Markov chain $\{X(t), t \geq 0\}$.
(2) Determine the point and the stationary availability of the system, $P(X(0) = 0) = 1$.

6.3) Consider a 1-out-of-2-system, i.e. the system is operating when at least one of its two subsystems is operating. When a subsystem fails, the other one continues working to its failure. Then the common renewal of both subsystems begins. On its completion both subsystems resume their work at the same time. The lifetimes of the subsystems are identically exponential with parameter λ and the (common) renewal time is exponential with with parameter μ. All life and renewal times are independent of each other. Let $X(t)$ be the number of subsystems operating at time t.
(1) Draw the transition graph of the corresponding Markov chain $\{X(t), t \geq 0\}$.
(2) Determine the time-dependent state probabilities $p_i(t) = P(X(t) = i)$; $i = 0, 1, 2$; given that $P(X(0) = 2) = 1$.
3) Determine the stationary state distribution.

Hint Consider the cases $(\lambda + \mu + \nu)^2 (=)(<)(>)\ 4(\lambda\mu + \lambda\nu + \mu\nu)$ separately.

6.4) A launderette has 10 washing machines which are in constant use. The times between two successive failures of a washing machine have an exponential distribution with expected value 100 *hours*. There are two mechanics who repair failed machines. A defective machine is repaired by only one mechanic. During this time, the second mechanic is busy repairing another failed machine, if there is any, or he is idle. All repair times have an exponential distribution with expected value 4 *hours*. Times between failures and repair times are independent. Consider the steady state.
1) What is the average percentage of operating machines?
2) What is the average percentage of idle mechanics?

6.5) Consider the two-unit system with standby redundancy discussed in example 6.6a) on condition that the lifetimes of the units are exponential with respective parameters λ_1 and λ_2. The other model assumptions listed in example 6.6 remain valid.
Describe the behaviour of the system by a Markov chain and develop the corresponding transition graph.

6.6) Consider the two-unit system with parallel redundancy discussed in example 6.7 on condition that the lifetimes of the units are exponential with parameters λ_1 and λ_2, respectively. The other model assumptions listed in example 6.7 remain valid.

Describe the behaviour of the system by a Markov chain and develop the corresponding transition graph.

6.7) The system considered in example 6.7 is generalized as follows: If the system makes a direct transition from state 0 to the blocking state 2, then the subsequent renewal time is exponential with parameter μ_0. If the system makes a transition from state 1 to state 2, then the subsequent renewal time is exponential with parameter μ_1.

(1) Describe the behaviour of the system by a Markov chain and develop the corresponding transition graph.
(2) What is the stationary probability that the system is blocked?

6.8) Consider a two-unit system with standby redundancy and one mechanic. The repair times of failed units each have an Erlang distribution with parameters $n = 2$ and μ. Apart from this, the other model assumptions listed in example 6.6 remain valid.

(1) Describe the behaviour of the system by a Markov chain and develop the corresponding transition graph.
(2) Determine the stationary state probabilities of the system.
(3) Sketch the stationary availability of the system as a function of $\rho = \lambda/\mu$.

6.9) A pure birth process $\{X(t), t \geq 0\}$ with state space $\{0, 1, 2, ...\}$ has positive birth rates $\lambda_0 = 2$, $\lambda_1 = 3$, and $\lambda_2 = 1$.

Given that $X(0) = 0$, determine the time-dependent state probabilities $p_i(t) = P(X(t) = i)$ for $i = 0, 1, 2$.

6.10) A linear pure birth process with birth rates $\lambda_j = j\lambda$; $j = 0, 1, ...$ has state space $\mathbf{Z} = \{0, 1, 2, ...\}$.

(1) Given that $X(0) = 1$ determine the distribution function of the random time point T_3 at which the process enters state 3.
(2) Given that $X(0) = 1$, compute the expected value of the random time point T_n at which the process enters state n, $n > 1$.

6.11) The number of physical particles of a particular type in a closed container evolves as follows: There is one particle at time $t = 0$. It splits into two particles of the same type after an exponential random time Y with parameter λ (its lifetime). These two particles behave in the same way as the original one, i.e. after random times which are identically distributed as Y they split into 2 particles each, and so on. All lifetimes of the particles are assumed to be independent. Let $X(t)$ denote the number of particles in the container at time t.

Determine the absolute state probabilities $p_j(t) = P(X(t) = j)$; $j = 1, 2, ...$; of the stochastic process $\{X(t), t \geq 0\}$.

6.12) A pure death process has death rates $\mu_0 = 0$, $\mu_1 = 2$, and $\mu_2 = 1$. Given that $X(0) = 3$, determine $p_j(t) = P(X(t) = j)$ for $j = 0, 1, 2, 3$.

6.13) A linear death process $\{X(t), t \geq 0\}$ has death rates $\mu_j = j\,\mu; j = 0, 1, ...$
(1) Given that $X(0) = 2$, determine the distribution function of the time to entering state 0.
(2) Given that $X(0) = n,\ n > 1$, determine the expected value of the time at which the process enters state 0.

6.14) At time $t = 0$ there are an infinite number of molecules of type a and $2n$ molecules of type b in a two-component gas mixture. After an exponential random time with parameter μ any molecule of type b combines, independently of the others, with a molecule of type a to form a molecule ab.
(1) What is the probability that there are still j free molecules of type b in the container at time t?
(2) What is the expected time till there are only n free molecules of type b in the container?

6.15) At time $t = 0$ a cable consists of 5 identical, intact wires. The cable is subject to a constant load of 100 *kp* such that in the beginning each wire bears a load of 20 *kp*. Given a load of w *kp* per wire, the time to breakage of a wire (its lifetime) is exponential with expected value $1/\mu = 1000/w$ weeks. When one or more wires are broken, the load of 100 *kp* is uniformly distributed over the remaining intact ones. For any fixed number of wires, their lifetimes are assumed to be independent and identically distributed.
(1) What is the probability that all wires are broken at time $t = 50$ [*weeks*] ?
(2) What is the expected time until the cable breaks completely?

6.16)* Let $\{X(t), t \geq 0\}$ be a pure death process with $X(0) = n$ and positive death rates $\mu_1, \mu_2, ..., \mu_n$.
Prove that, if Y is an exponential random variable with parameter λ and independent of the death process, then

$$P(X(Y) = 0) = \prod_{i=1}^{n} \frac{\mu_i}{\mu_i + \lambda}$$

6.17) Let a birth- and death process have state space $\mathbf{Z} = \{0, 1, ..., n\}$ and transition rates $\lambda_j = (n-j)\,\lambda$ and $\mu_j = j\,\mu;\ j = 0, 1, ..., n.$
Determine its stationary state probabilities.

6.18) Check whether, or under what restrictions, a birth- and death process with transition rates

$$\lambda_j = \frac{j+1}{j+2}\lambda \text{ and } \mu_j = \mu;\ j = 0, 1, ...,$$

has a stationary state distribution.

6.19) A birth- and death process has transition rates

$$\lambda_j = (j+1)\lambda \text{ and } \mu_j = j^2\mu;\ j = 0, 1, ...;\ 0 < \lambda < \mu.$$

Confirm that this process has a stationary state distribution and determine it.

6.20) A computer is connected to three terminals (for example, measuring devices). It can simultaneously evaluate data records from only two terminals. When the computer is processing two data records and in the meantime another data record has been produced, then this new data record has to wait in a buffer when the buffer is empty. Otherwise the

new data record is lost. (The buffer can store only one data record.) The data records are processed according to the FCFS-queueing discipline. The terminals produce data records independently and in accordance with a homogeneous Poisson processes with intensity λ. The processing times of data records from all terminals are independent (even if the computer is busy with two data records at the same time) and have an exponential distribution with parameter μ. They are, moreover, independent of the input. Let $X(t)$ be the number of data records in the computer and buffer at time t.
(1) Verify that $\{X(t), t \geq 0\}$ is a birth- and death process, determine its transition rates and draw the transition graph.
(2) Determine the stationary loss probability, i.e. the probability that, in the steady state, a data record is lost.

6.21) Under otherwise the same assumptions as in exercise 6.20, it is assumed that a data record which has been waiting in the buffer for a random patience time, will be deleted as being no longer up to date. The *patience times* of all data records are assumed to be independent, exponential random variables with parameter ν. They are also independent of all arrival and processing times of the data records.
Determine the stationary loss probability.

6.22) Under otherwise the same assumptions as in exercise 6.21 it is assumed that a data record will be deleted when its total sojourn time in the buffer and computer exceeds a random time Z, where Z has an exponential distribution with parameter ν. Thus, the interruption of a current service of a data record is possible.
Determine the stationary loss probability.

6.23) A small filling station in a rural area provides diesel for agricultural machines. It has one diesel pump and waiting capacity for 5 machines. On average, 8 machines per hour arrive for diesel. An arriving machine immediately leaves the station without fuel when pump and all waiting places are occupied. The mean time a machine occupies the pump is 5 *minutes*. It is assumed that the station behaves like a $M/M/s/m$-queueing system.
(1) Determine the stationary loss probability.
(2) Determine the stationary probability that an arriving machine waits for diesel.

6.24) Consider a two-server loss system. Customers arrive according to a homogeneous Poisson process with intensity λ. A customer is always served by server 1 when this server is idle, i.e. an arriving customer goes only then to server 2, when server 1 is busy. The service times of both servers are idd exponential random variables with parameter μ. Let $X(t)$ be the number of customers in the system at time t.
Determine the stationary distribution of $\{X(t), t \geq 0\}$

6.25) A 2- server loss system is subject to a homogeneous Poisson input with intensity λ. The situation considered in the previous exercise is generalized as follows: If both servers are idle, a customer goes to server 1 with probability p and to server 2 with probability $1-p$. Otherwise a customer goes to the idle server (if there is any). The service times of the servers 1 and 2 are independent, exponential random variables with parameters μ_1 and μ_2, respectively. All arrival and service times are independent.
Describe the behaviour of the system by a suitable homogeneous Markov chain and draw the transition graph.

6.26) A single-server waiting system is subject to a homogeneous Poisson input with intensity $\lambda = 30$ $[hours^{-1}]$. If there are up to 3 customers in the system, the service times have an exponential distribution with mean $1/\mu = 2$ *minutes*. If there are more than 3 customers in the system, the service times are exponential with mean $1/\mu = 1$ *minute*. All arrival and service times are independent.
(1) Determine the stationary state probabilities.
(2) Determine the expected length of the waiting queue in the steady state.

6.27) Taxis and customers arrive at a taxi rank in accordance with two independent homogeneous Poisson processes with intensities $\lambda_1 = 4$ an hour and $\lambda_2 = 3$ an hour, respectively. Potential customers, who find 2 waiting customers, do not wait for service, but leave the rank immediately. (Groups of customers, who will use the same taxi, are considered to be one customer.) On the other hand, arriving taxis, who find already two taxis waiting, leave the rank as well.
What is the average number of customers waiting at the rank?

6.28) A transport company has 4 trucks of the same type. There are 2 maintenance teams for repairing the trucks after a failure. Each team can repair only one truck at a time and each failed truck is handled by only one team. The times between failures of the trucks (lifetimes) are exponential with parameter λ. The repair times are exponential with parameter μ. All life- and repair times are assumed to be independent. Let

$$\rho = \lambda/\mu = 0.2$$

What is the most efficient way of organizing the work: 1) to make both maintenance teams responsible for the maintenance of all 4 trucks so that any team which is free can repair any failed truck, or 2) to assign 2 definite trucks to each team?

6.29) Ferry boats and customers arrive at a ferry station in accordance with two independent homogeneous Poisson processes with intensities λ and μ, respectively. If there are k customers at the ferry station, when a boat arrives, then it departs with $min(k, n)$ passengers (n is the capacity of each boat). If $k > n$, then the remaining $k - n$ customers wait for the next boat. The sojourn times of the boats at the station are assumed to be negligibly small.
Model the situation by a suitable homogeneous Markov chain $\{X(t), t \geq 0\}$ and draw the transition graph.

6.30) Customers arrive at a waiting system of type $M/M/1/\infty$ with intensity λ. As long there are less than n customers in the system, the server remains idle. As soon as the n th customer arrives, the server resumes its work and stops working only then when all customers (including newcomers) have been served. After that the server again waits until the waiting queue has reached length n and so on. Let $1/\mu$ be the expected service time of a customer and $X(t)$ the number of customers in the system at time t.
(1) Give the transition graph of the Markov chain $\{X(t), t \geq 0\}$.
(2) Given that $n = 2$, compute the stationary state probabilities. (Make sure that they exist.)

6.31) At time $t = 0$ a computer system consists of n computers. As soon as a computer fails, it is with probability $1 - p$ separated from the system by an automatic switching device. If a failed computer is not separated from the system (this happens with probability p), then

the entire system fails. The lifetimes of the computers are independent and have an exponential distribution with parameter λ. Thus, this distribution does not depend on the system state. Provided the switching device has operated properly when required, the system is available as long as there is at least one computer available. Let $X(t)$ be the number of computers which are available at time t. By convention, if, due to the switching device, the entire system has failed in $[0, t)$, then $X(t) = 0$.
(1) Draw the transition graph of the Markov chain $\{X(t), t \geq 0\}$.
(2) Given $n = 2$, compute the expected lifetime $E(X_s)$ of the system.

6.32) A waiting-loss system of type $M/M/1/2$ is subject to two independent Poisson inputs 1 and 2 with intensities λ_1 and λ_2 (type 1- and type 2- customers), respectively. An arriving type 1-customer, who finds the server busy and the waiting places occupied, displaces a possible type 2-customer from its waiting place (such a type 2-customer is lost), but ongoing service of a type 2-customer is not interrupted. When a type 1-customer and a type 2-customer are waiting, then the type 1-customer will always be served first, regardless of the order of their arrivals. The service times of type 1- and type 2- customers are independent and have exponential distributions with parameters μ_1 and μ_2, respectively.
Describe the behaviour of the system by a homogeneous Markov chain, determine the transition rates, and draw the transition graph.

6.33) A queueing network consists of two servers **1** and **2** in series. Server **1** is subject to a homogeneous Poisson input with intensity $\lambda = 5$ an hour. A customer is lost if server **1** is busy. From server **1** a customer goes to server **2** for further service. If server **2** is busy, the customer is lost. The service times of servers 1 and 2 are exponential with expected values 6 *minutes* and 12 *minutes*. All arrival and service times are independent.
What percentage of customers (with respect to the total input at server 1) is served by both servers?

6.34) A queueing network consists of 3 nodes (queueing systems) **1**, **2** and **3**, each of type $M/M/1$. The external inputs into the nodes have respective intensities 4, 8 and 12 customers an hour. The respective expected service times at the nodes are 4, 2 and 1 *minutes*. After having been served by node **1**, a customer goes to nodes **2** and **3** with equal probabilities 0.4 or leaves the system with probability 0.2. From node **2**, a customer goes to node **3** with probability 0.9 or leaves the system with probability 0.1. From node **3**, a customer goes to node **1** with probability 0.2 or leaves the system with probability 0.8. The external inputs and the service times are independent.
(1) Check whether this queueing network is a *Jackson-network*.
(2) Determine the stationary state probabilities of the network.

6.35) A closed queueing network consists of 3 nodes. Each one has 2 servers. There are 2 customers in the network. After having been served at a node, a customer goes to one of the others with equal probability. All service times are independent and have an exponential distribution with parameter μ.
What is the stationary probability to find both customers at the same node?

6.36) Depending on the demand, a conveyor belt operates at 3 different speed levels 1, 2, and 3. A transition from level i to level j is made with probability p_{ij}, where

$$p_{12} = 0.8, \; p_{13} = 0.2, \; p_{21} = p_{23} = 0.5, \; p_{31} = 0.4, \text{ and } p_{32} = 0.6$$

The respective mean times the conveyor belt operates at levels 1, 2, or 3 between the transitions are $\mu_1 = 45$, $\mu_2 = 30$, and $\mu_3 = 12$ [*hours*].
Determine the stationary percentages of time in which the conveyor belt operates at levels 1, 2, and 3 by modeling the situation as a semi-Markov process.

6.37) The mean lifetime of a system is 620 hours. There are two failure types: Repairing the system after a type 1-failure requires 20 hours on average and after a type 2-failure 40 hours on average. 20% of all failures are type 2-failures. There is no dependence between the system lifetime and the subsequent failure type. Upon each repair the system is "as good as new". The repaired system immediately resumes its work. This process is continued indefinitely. All life- and repair times are independent.
Describe the situation by a semi-Markov process with 3 states, draw the transition graph, and determine the stationary state probabilities.

7 Wiener Processes

7.1 Definition and Properties

In 1828 the English botanist *R. Brown* published his observations on the movement of microscopically small organic and inorganic particles in liquids. (Originally he was only interested in the behaviour of pollen in liquids in order to investigate the fructification process of phanerogams.) He observed that, independent of their nature, the particles moved in a permanent, apparently irregular way. Hence Brown initially supposed that he had found an elementary form of life which is common to all particles. Although the strange movement of particles in liquids had already been detected before Brown, it is generally called *Brownian motion.*

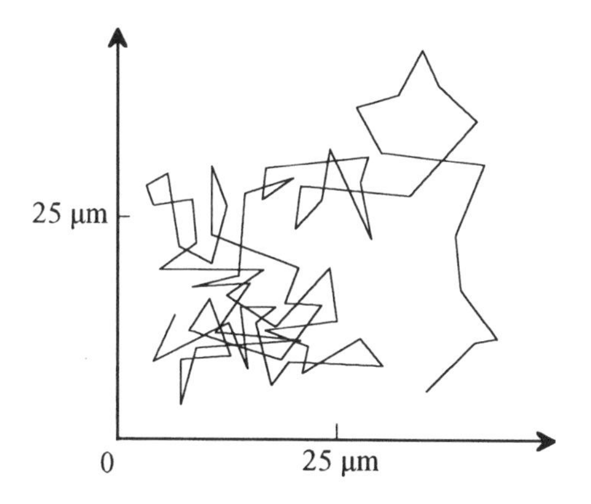

Figure 7.1 Sample path of a two-dimensional Brownian motion with measurements after 30 sec each (Perrin, 1916)

Analogous experiments were also carried out later (Figure 7.1). Figure 7.2 shows the typical graph of a one-dimensional Brownian motion. The first approaches to mathematically modeling the Brownian motion were made by *L. Bachelier* (1900) and *A. Einstein* (1905). Both found the normal distribution to be appropriate for describing the one-dimensional Brownian motion and gave a physical explanation of the observed phenomenon: The chaotic movement of sufficiently small particles in liquids and in gases is due to the huge number of impacts with the surrounding molecules, even in small time intervals. (Assuming average physical conditions, there are about 10^{21} collisions per second between a particle and the surrounding molecules in a liquid.) *N. Wiener* (1918) was the first to present a sufficiently

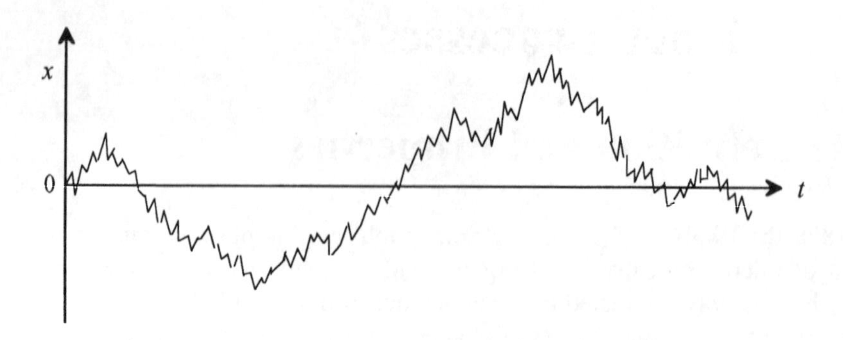

Figure 7.2 Sample path of the one-dimensional Brownian motion

general mathematical treatment of the Brownian motion. He defined and analysed a stochastic process which has served up till now as a mathematical model of Brownian motion. This process is known as the *Wiener process* or as the *Brownian motion process*. This chapter only deals with the one-dimensional Wiener process.

Definition 7.1 (*Wiener process*) A continuous-time stochastic process $\{X(t), t \geq 0$ with state space $\mathbf{Z} = (-\infty, +\infty)$ is called a *Wiener process* if it has the following properties:

1) $X(0) = 0$
2) $\{X(t), t \geq 0\}$ has stationary, independent increments.
3) $X(t)$ is normally distributed with

$$E(X(t)) = 0 \text{ and } Var(X(t)) = \sigma^2 t, \ t > 0. \qquad \bullet$$

In view of the stationarity of the increments, $X(t) - X(s)$ is normally distributed with expected value 0 and variance $\sigma^2 |t-s|$:

$$X(t) - X(s) = N(0, \sigma^2 |t-s|), \quad s, t \geq 0 \tag{7.1}$$

Since the Wiener process has independent increments, it is a Markov process. If $\sigma = 1$, then $\{X(t), t \geq 0\}$ is called a *standard Wiener process*.

Continuity and differentiability From

$$E\left(|X(t) - X(s)|^2\right) = Var(X(t) - X(s)) = \sigma^2 |t-s| \tag{7.2}$$

it follows that

$$\lim_{h \to 0} E\left(|X(t+h) - X(t)|^2\right) = \lim_{h \to 0} \sigma^2 |h| = 0$$

Corollary The Wiener process is mean-square continuous.

It is possible to prove that the random event "a sample path $x = x(t)$ of a Wiener process has points of discontinuity" has probability 0. An equivalent formulation of this fact is: Almost all sample paths of a Wiener process are continuous functions of t, $t \geq 0$. Thus, in practice one will never observe a sample path of a Wiener process with points of discontinuity. Hence it may surprise that a corresponding property of the sample paths of a Wiener process with respect to their differentiability does not hold. This can be illustrated by means of (7.2): For any sample path $x = x(t)$, the difference $x(t+h) - x(h)$ is approximately equal to $\sigma\sqrt{h}$. Therefore,

$$\frac{dx(t)}{dt} = \lim_{h\to 0} \frac{x(t+h) - x(t)}{h} \approx \lim_{h\to 0} \frac{\sigma\sqrt{h}}{h} = \infty$$

Hence it can be anticipated that the sample paths of a Wiener process are nowhere differentiable. More exactly,

Almost all sample paths of a Wiener process are nowhere differentiable.

The *variation* of a sample path (as well as of any real function) $x = x(t)$ in the interval $[0, \tau]$ is defined as the limit

$$\lim_{n\to\infty} \sum_{k=1}^{2^n} \left| x\left(\frac{k\tau}{2^n}\right) - x\left(\frac{(k-1)\tau}{2^n}\right) \right| \tag{7.3}$$

The non-differentiability of the sample paths implies that this limit cannot be finite. Hence, any sample path of a Wiener process is of *unbounded variation*. This property becomes practically interesting if one realizes that, because of the continuity of the sample paths, the limit (7.3) is the length of the graph of $x = x(t)$ in the interval $[0, \tau]$. Therefore, in any subinterval of $[0, \tau]$, even if its length is arbitrarily small, all sample paths of a Wiener process have an infinite length. Consequently, the sample paths of a Wiener process must be strongly dentate (in the sense of the structure of leaves), but this structure must continue to the infinitesimal. This result corresponds to the physical interpretation of the Wiener process: The numerous and rapid bombardements of particles in liquids or gases by the surrounding molecules cannot lead to a smooth sample path. Unfortunately, the unbounded variation of the sample paths implies that the particles have to move with an infinitely large velocity. This may arise doubts whether the Wiener process is an adequate model for describing Brownian motion. However, nowadays the enormous importance of the Wiener process is mainly due to the fact that it is one of the basic stochastic processes, whose role can be compared with that of the normal distribution in probability theory. It is also due to its fruitful application in various fields, for example, in time series analysis for modeling economical, technological, sociological and other processes, in finance (modeling fluctuations of share values), in reliability theory (wear modeling), and in communication theory (modeling signals, in particular noise).

Wiener-process and random walk With respect to the physical background of the one-dimensional Wiener process it is not surprising that it has a close relationship to the random walk of a particle along the real axis. Modifying the random walk described in example 5.1 it is now assumed that, after Δt time units, the particle jumps exactly Δx length units to the right or to the to the left, each with probability 1/2. If $X(t)$ is the position of the particle at time t, then, given $X(0) = 0$,

$$X(t) = (X_1 + X_2 + \cdots + X_{[t/\Delta t]})\Delta x, \tag{7.4}$$

where

$$X_i = \begin{cases} +1 & \text{if the } i\text{th jump goes to the right} \\ -1 & \text{if the } i\text{th jump goes to the left} \end{cases}$$

($[t/\Delta t]$ denotes the greatest integer less than or equal to $t/\Delta t$.) The random variables X_i are independent of each other with probability distribution

$$P(X_i = 1) = P(X_i = -1) = 1/2$$

Since $E(X_i) = 0$ and $Var(X_i) = 1$ and in view of formulas (7.4), (1.38) and (1.40),

$$E(X(t)) = 0; \quad Var(X(t)) = (\Delta x)^2\, [t/\Delta t] \tag{7.5}$$

The aim now consists in studying the behaviour of the process $\{X(t), t \geq 0\}$ as $\Delta x \to 0$ and $\Delta t \to 0$. In order to obtain the desired result, the variables Δx and Δt are assumed to satisfy $\Delta x = \sigma\sqrt{\Delta t}$, where σ is a positive constant. Then, from (7.5), taking the limit as $\Delta t \to 0$, the stochastic process $\{X(t), t \geq 0\}$ is seen to have properties

$$E(X(t)) = 0, \quad Var(X(t)) = \sigma^2 t$$

Due to its construction, the process $\{X(t), t \geq 0\}$ has independent and homogeneous increments. Moreover, by the central limit theorem, $X(t)$ is normally distributed for all $t > 0$. Hence the stochastic process of the "infinitesimal random walk" $\{X(t), t \geq 0\}$ is a Wiener process.

Multidimensional and conditional distributions Let $\{X(t), t \geq 0\}$ be a Wiener process and $f_t(x)$ the probability density of $X(t)$, $t>0$. From property 3 of definition 7.1,

$$f_t(x) = \frac{1}{\sqrt{2\pi t}\,\sigma}\, e^{-\frac{x^2}{2\sigma^2 t}}, \quad t > 0 \tag{7.6}$$

If $f_{s,t}(x_1, x_2)$ denotes the joint probability density of $(X(s), X(t))$ with $0 < s < t$ (see section 1.3.1), then

$$f_{s,t}(x_1, x_2)\, dx_1 dx_2 = P(X(s) = x_1,\, X(t) = x_2)\, dx_1 dx_2$$

The relationship

$$P(X(s)=x_1,\, X(t)=x_2)\, dx_1 dx_2 = P(X(s)=x_1,\, X(t)-X(s)=x_2-x_1)\, dx_1 dx_2,$$

the independence of the increments, and the fact that $X(t)-X(s)$ has probability density $f_{t-s}(x)$ imply that

$$f_{s,t}(x_1,x_2)\, dx_1 dx_2 = P(X(s)=x_1)\, P(X(t)-X(s)=x_2-x_1)\, dx_1 dx_2$$

$$= f_s(x_1) f_t(x_2-x_1)\, dx_1 dx_2$$

Hence,

$$f_{s,t}(x_1,x_2) = f_s(x_1) f_{t-s}(x_2-x_1) \tag{7.7}$$

Substituting (7.6) into (7.7) yields after simplification

$$f_{s,t}(x_1,x_2) = \frac{1}{2\pi\sigma^2\sqrt{s(t-s)}} \exp\left\{-\frac{1}{2\sigma^2 s(t-s)}\left(t x_1^2 - 2s x_1 x_2 + s x_2^2\right)\right\} \tag{7.8}$$

Comparing this density with the density of the bivariate normal distribution considered in example 1.4 (page 27) yields that $(X(s), X(t))$ is jointly normally distributed with correlation coefficient $\rho = +\sqrt{s/t}$, $0 < s < t$. Therefore, if $0 < s < t$, then

$$\rho(s,t) = +\sqrt{s/t}$$

is the correlation function and

$$C(s,t) = Cov(X(s), X(t)) = \sigma^2 s$$

the covariance function of the Wiener process. Since the roles of s and t can be exchanged,

$$C(s,t) = \sigma^2 \min(s,t) \tag{7.9}$$

However, the covariance function of the Wiener process can more easily be obtained directly: In view of the independence of the increments,

$$Cov(X(s),\, X(t)-X(s)) = 0, \quad 0 < s \le t$$

Hence,

$$Cov(X(s),\, X(t)) - Cov(X(s), X(s)) = 0$$

Thus,

$$Cov(X(s),\, X(t)) = Cov(X(s),\, X(s))$$

or, equivalently,

$$C(s,t) = Var(X(s)) = \sigma^2 s, \quad 0 < s \le t$$

Let $0 < s < t$ and $X(t) = b$. From (1.29), the conditional density of $X(s)$, given that $X(t) = b$, is

$$f_{X(s)}(x|X(t) = b) = \frac{f_{s,t}(x, b)}{f_t(b)} \tag{7.10}$$

Substituting (7.6) and (7.8) into (7.10) yields

$$f_{X(s)}(x|X(t) = b) = \frac{1}{\sqrt{2\pi\frac{s}{t}(t-s)}\,\sigma} \exp\left\{-\frac{1}{2\sigma^2\frac{s}{t}(t-s)}\left(x - \frac{s}{t}b\right)^2\right\} \tag{7.11}$$

But this is the density of a normally distributed random variable with parameters

$$E(X(s)|X(t) = b) = \frac{s}{t}b, \quad Var(X(s)|X(t) = b) = \sigma^2\frac{s}{t}(t-s) \tag{7.12}$$

It can easily be verified that the conditional variance assumes its maximum at $s = t/2$.

Consider now the n-dimensional probability density $f_{t_1,t_2,\dots,t_n}(x_1, x_2, \dots, x_n)$ of the random vector $(X(t_1), X(t_2), \dots, X(t_n))$, $0 < t_1 < t_2 < \dots < t_n < \infty$. By generalizing (7.7),

$$f_{t_1,t_2,\dots,t_n}(x_1, x_2, \dots, x_n) = f_{t_1}(x_1)\, f_{t_2-t_1}(x_2 - x_1) \dots f_{t_n-t_{n-1}}(x_n - x_{n-1})$$

Making use of the representation of $f_t(x)$ given by (7.6) yields

$$f_{t_1,t_2,\dots,t_n}(x_1, x_2, \dots, x_n) = \frac{\exp\left\{-\frac{1}{2\sigma^2}\left[\frac{x_1^2}{t_1} + \frac{(x_2-x_1)^2}{t_2-t_1} + \dots + \frac{(x_n-x_{n-1})^2}{t_n-t_{n-1}}\right]\right\}}{(2\pi)^{n/2}\sigma^n\sqrt{t_1(t_2-t_1)\dots(t_n-t_{n-1})}} \tag{7.13}$$

Transformation of this density analogously to the two-dimensional case shows that the random vector $(X(t_1), X(t_2), \dots, X(t_n))$ has an n-dimensional normal distribution. However, this result also follows from theorem 1.1, since each $X(t_i)$ can be represented as a sum of independent, normally distributed random variables (increments) in the following way:

$$X(t_i) = X(t_1) + (X(t_2) - X(t_1)) + \dots + (X(t_i) - X(t_{i-1})); \quad i = 2, 3, \dots, n$$

Therefore, all multidimensional distributions of Wiener processes are multidimensionally normal distributions. Thus, the Wiener process is a Gaussian process (definition 2.6). According to example 1.5, Gaussian processes are completely determined by their trend and covariance functions. Since the trend function of a Wiener process is identically zero, the following statement holds:

The Wiener process is uniquely characterized by its covariance function.

Brownian bridge The *Brownian bridge* $\{B(t), t \in [0,1]\}$ is a stochastic process which is derived from the Wiener process $\{X(t), t \geq 0\}$ by letting $B(t) = X(t)$, $0 \leq t \leq 1$, given that $X(1) = 0$.

From (7.11), the one-dimensional probability densities of the Brownian bridge are

$$f_{X(t)}(x) = \frac{1}{\sqrt{2\pi t(1-t)}\,\sigma} \exp\left\{-\frac{x^2}{2\sigma^2 t(1-t)}\right\}, \quad 0 < t < 1$$

In particular,

$$E(X(t)) = 0, \quad Var(X(t)) = \sigma^2 t(1-t), \quad 0 \leq t \leq 1$$

The two-dimensional probability densities of the Brownian bridge can be determined from

$$f_{t_1,t_2}(x_1, x_2) = \frac{f_{t_1,t_2,t_3}(x_1, x_2, 0)}{f_{t_3}(0)}$$

with $t_1 = s$, $t_2 = t$ and $t_3 = 1$. Taking into account (7.6) and (7.13) ,

$$f_{s,t}(x_1, x_2) = \frac{\exp\left\{-\frac{1}{2\sigma^2}\left[\frac{t}{s(t-s)}x_1^2 - \frac{2}{t-s}x_1 x_2 + \frac{1-s}{(t-s)(1-t)}x_2^2\right]\right\}}{2\pi\sigma^2\sqrt{s(t-s)(1-t)}}, \quad 0 < s < t < 1$$

A comparision with example 1.4 shows that the correlation and covariance function of the Brownian bridge are given by

$$\rho(s,t) = \sqrt{\frac{s(1-t)}{t(1-s)}}, \quad C(s,t) = \sigma^2 s(1-t), \quad 0 < s < t < 1$$

The Brownian bridge is a Gaussian process whose trend function is identically 0. Hence it is, like the Wiener process, uniquely determined by its covariance function.

7.2 First Passage Times

By definition, the Wiener process starts at $X(0) = 0$. Let $L(a)$ be the random time point at which the Wiener process reaches level a for the first time: $L(a)$ is called a *first passage time* of the process $\{X(t), t \geq 0\}$ with respect to level a. (A first passage time has already been introduced in section 4.7 for cumulative stochastic processes.) Hence, $L(a)$ is characterized by $X(L(a)) = a$. Since the sample paths of a Wiener process are continuous functions, the first passage time $L(a)$ is uniquely determined (Figure 7.3).

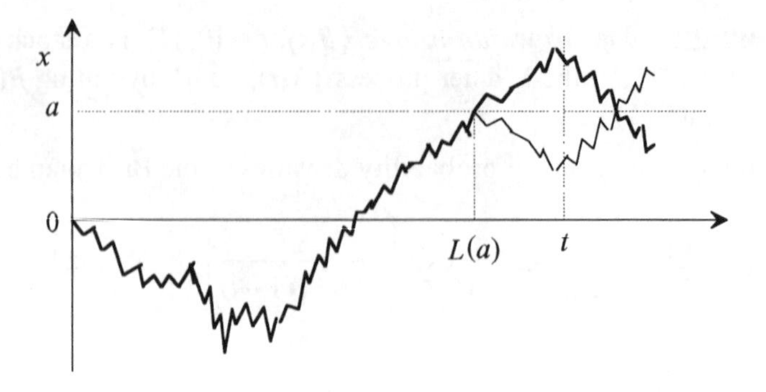

Figure 7.3 Illustration of the first passage time and the reflection principle

Next the aim is to determine the distribution function of $L(a)$. Applying the total probability rule yields, for all $a > 0$,

$$\begin{aligned} P(X(t) \geq a) = \; & P(X(t) \geq a | L(a) \leq t)\, P(L(a) \leq t) \\ & + P(X(t) \geq a | L(a) > t)\, P(L(a) > t) \end{aligned} \tag{7.14}$$

The second term on the right hand side of this formula vanishes, since, by definition of $L(a)$, the conditional probability

$$P(X(t) \geq a | L(a) > t)$$

is equal to 0 for all $t > 0$. Since $X(L(a)) = a$, and for symmetry reasons,

$$P(X(t) \geq a | L(a) \leq t) = \frac{1}{2} \tag{7.15}$$

This situation is illustrated in Figure 7.3: Two sample paths of the Wiener process which coincide up to reaching the level a and which after $L(a)$ are mirror symmetric with respect to the straight line $x(t) \equiv a$, have the same chance of occuring. (The probability of this event is, nevertheless, zero.) This heuristic argument is known as the *reflection principle.* Thus, from (7.14), (7.15) and (7.6),

$$\begin{aligned} F_{L(a)}(t) &= P(L(a) \leq t) \\ &= 2\, P(X(t) \geq a) \\ &= \frac{2}{\sqrt{2\pi t}\,\sigma} \int_a^{\infty} e^{-\frac{x^2}{2\sigma^2 t}}\, dx \end{aligned}$$

For symmetry reasons, the probability distributions of $L(a)$ and $L(-a)$ are identical for any a. Therefore,

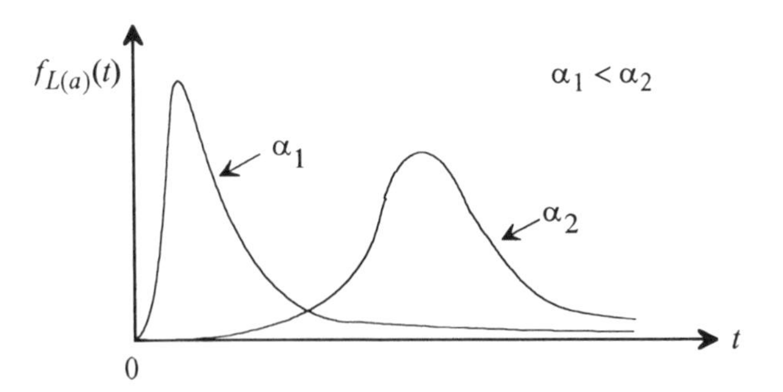

Figure 7.4 Probability density of the first passage time

$$F_{L(a)}(t) = \frac{2}{\sqrt{2\pi t}\,\sigma} \int_{|a|}^{\infty} e^{-\frac{x^2}{2\sigma^2 t}} dx, \quad t > 0$$

The probability distribution determined by this distribution function is a special case of the *inverse Gaussian distribution,* the general structure of which will be introduced in section 7.3.3. The relationship to the normal distribution (Gaussian distribution) becomes clear after substituting $u^2 = x^2/(\sigma^2 t)$:

$$F_{L(a)}(t) = \frac{2}{\sqrt{2\pi}} \int_{\frac{|a|}{\sigma\sqrt{t}}}^{\infty} e^{-u^2/2} du, \quad t > 0$$

Hence, the distribution function of the first passage time $L(a)$ can be written in the form

$$F_{L(a)}(t) = 2\left[1 - \Phi\left(\frac{|a|}{\sigma\sqrt{t}}\right)\right], \quad t > 0 \tag{7.16}$$

$\Phi(u)$ is as usual the distribution function of $N(0,1)$ (standard normal random variable.) Differentiation with respect to t yields the probability density of $L(a)$:

$$f_{L(a)}(t) = \frac{|a|}{\sqrt{2\pi}\,\sigma\, t^{3/2}} \exp\left\{-\frac{a^2}{2\sigma^2 t}\right\}, \quad t > 0 \tag{7.17}$$

Figure 7.4 qualitatively shows the graph of $f_{L(a)}(t)$ for two values of $\alpha = |a|/\sigma$. $E(L(a))$ and $Var(L(a))$ do not exist.

As a corollary, the probability distribution of

$$M(t) = \max\{X(s), 0 \le s \le t\} \tag{7.18}$$

is obtained as follows: For $t > 0$,

$$1 - F_{M(t)}(x) = P(M(t) \geq x)$$
$$= P(L(x) \leq t)$$
$$= 2\left[1 - \Phi\left(\frac{x}{\sigma\sqrt{t}}\right)\right], \quad x \geq 0$$

Hence, for $t > 0$, distribution function and probability density of $M(t)$ are

$$F_{M(t)}(x) = 2\,\Phi\left(\frac{x}{\sigma\sqrt{t}}\right) - 1, \quad x \geq 0 \tag{7.19}$$

$$f_{M(t)}(x) = \frac{2}{\sqrt{2\pi t}\,\sigma} e^{-x^2/(2\sigma^2 t)}, \quad x \geq 0 \tag{7.20}$$

Example 7.1 A sensor for measuring high temperatures gives an unbiased indication of the true temperature. At the start, the measurement is absolutely correct. In the course of time its accuracy deteriorates. Let $X(t)$ be the random deviation of the temperature indicated by the sensor at time t from the true temperature. From historical observations it is known that, for all $t > 0$, the deviation $X(t)$ is normally distributed with expectation 0 and variance $0.01\,t$. Specifically,

$$\sigma = \sqrt{Var(X(1))} = 0.1 \ \left[\text{in } {}^0C/\sqrt{24\,h}\,\right]$$

What is the probability that, within a year (365 days), $X(t)$ exceeds the critical level $-5\ {}^oC$, i.e. that the sensor under-reads by more than 5^0C?

Assuming $(X(t),\ t \geq 0)$ is a Wiener process, the desired probability $P(L(-5) < 365)$ is equal to

$$P(L(-5) < 365) = P(L(5) < 365) = F_{L(5)}(365)$$
$$= 2\left[1 - \Phi\left(\frac{5}{0.1\sqrt{365}}\right)\right]$$
$$= 2\,[1 - \Phi(2.617)]$$
$$= 0.009$$

If the accuracy of the sensor is allowed to exceed the critical value of -5^0C with probability 0.05, then the sensor has to be exchanged by a new one after a time $\tau_{0.05}$ given by

$$P(L(-5) \leq \tau_{0.05}) = 0.05$$

According to (7.16), $\tau_{0.05}$ satisfies equation

$$2\left[1-\Phi\left(\frac{5}{0.1\sqrt{\tau_{0.05}}}\right)\right]=0.05$$

or, equivalently,

$$\frac{5}{0.1\sqrt{\tau_{0.05}}}=\Phi^{-1}(0.975)=1.96$$

Thus, $\tau_{0.05}=651$ [days]. □

The proofs of the following formulas are straightforward consequences from the martingale stopping theorem (*Karlin/Taylor* (1981), *Durrett* (1999)).

Let $L(a,b)$ be the random time at which the Wiener process $\{X(t), t \geq 0\}$ reaches either value a or b, $b<0<a$, for the first time. Then

$$P(L(a)<L(b))=P(X(L(a,b))=a)$$

is the probability that the process assumes the value a before value b. This probability is given by

$$P(L(a)<L(b))=P(X(L(a,b))=a)=\frac{|b|}{a+|b|} \tag{7.21}$$

The expected value of $L(a,b)$ is related to $X^2(t)$ by

$$E(L(a,b))=\frac{1}{\sigma^2}E(X^2(L(a,b))$$

Hence, from (7.21),

$$E(L(a,b))=\frac{1}{\sigma^2}\left[a^2P(L(a)<L(b))+b^2P(L(a)>L(b))\right]$$

$$=\frac{1}{\sigma^2}\left[a^2\frac{|b|}{a+|b|}+b^2\frac{a}{a+|b|}\right]$$

Thus,

$$E(L(a,b))=\frac{1}{\sigma^2}a\,|b|$$

Therefore, contrary to the expected values of $L(a)$ and $L(b)$, the expected value of $L(a,b)$ exists.

For example, let $X(t)$ be the cumulative profit or loss, respectively, a gambler has achieved after t time units. If he interrupts the game after having achieved either a profit of a or a loss of b, then (7.21) is the probability that he finishes his game with a profit of a. Or with reference to example 7.1: The probability that the sensor reads 10^0C high before it reads 2^0C low, is equal to 2/(10+2) = 1/6. If, for instance, in example 7.1 the tolerance region for $X(t)$ is $[-5\,^0C,\ 5\,^0C]$, then $X(t)$ on average leaves this region after $E(Y(-5,5))=25/0.01=2500$ days.

Example 7.2 Let $p_{(1,t]}$ be the probability that the Wiener process $\{X(t), t \geq 0$ crosses the x-axis at least once in the interval $(1, t]$, $1 < t$. To determine $p_{(1,t]}$, note that for symmetry reasons and in view of (7.19), for any $a > 0$,

$$\begin{aligned}
&P(X(s) = 0 \quad \text{for an } s \text{ with } 1 < s \leq t \,|\, X(1) = a) \\
&= P(X(s) = 0 \quad \text{for an } s \text{ with } 1 < s \leq t \,|\, X(1) = -a) \\
&= P(X(s) \leq -a \quad \text{for an } s \text{ with } 0 < s \leq t-1) \\
&= P(X(s) \geq a \quad \text{for an } s \text{ with } 0 < s \leq t-1) \\
&= P(M(t-1) \geq a) \\
&= \frac{2}{\sqrt{2\pi\,(t-1)}\,\sigma} \int_a^\infty e^{-\frac{x^2}{2\sigma^2(t-1)}}\, dx\,,
\end{aligned}$$

where $M(t-1)$ is the maximum of the Wiener process in the interval $[0, t-1]$. Taking the mathematical expectation of this probability with respect a (the negative values of a are taken into account by the factor 2) gives $p_{[1,t]}$:

$$p_{[1,t]} = 2 \int_0^\infty P(X(s) = 0 \text{ for an } s \text{ with } 1 < s \leq t \,|\, X(1) = a) f_{X(1)}(a)\, da$$

$$= \frac{2}{\pi\,\sqrt{t-1}\,\sigma^2} \int_0^\infty \int_a^\infty e^{-\frac{x^2}{2\sigma^2(t-1)}}\, dx\; e^{-\frac{a^2}{2\sigma^2}}\, da$$

Substituting

$$x = u\sigma\sqrt{t-1} \quad \text{and} \quad v = \frac{a}{\sigma}$$

in the inner and outer intergrals, repectively, yields

$$p_{[1,t]} = \frac{2}{\pi} \int_0^\infty \int_{\frac{v}{\sqrt{t-1}}}^\infty e^{-\frac{u^2+v^2}{2}}\, du\, dv$$

The integration can be simplified by a transition to polar coordinates (r, ϕ). Then the (u, v)-domain of integration has to be transformed as follows:

$$\left\{0 < v < \infty,\; \frac{v}{\sqrt{t-1}} < u < \infty\right\} \rightarrow \left\{0 < r < \infty,\; \arctan\frac{1}{\sqrt{t-1}} < \phi < \frac{\pi}{2}\right\}$$

Since

$$\int_0^\infty r\, e^{-r^2/2}\, dr = 1,$$

the desired probability becomes

$$p_{[1,t]} = \frac{2}{\pi} \int_0^{\infty} \int_{\arctan \frac{1}{\sqrt{t-1}}}^{\pi/2} e^{-r^2/2}\, r\, d\phi\, dr$$

$$= \frac{2}{\pi}\left[\frac{\pi}{2} - \arctan \frac{1}{\sqrt{t-1}}\right] \int_0^{\infty} r e^{-r^2/2}\, dr$$

$$= 1 - \frac{2}{\pi} \arctan \frac{1}{\sqrt{t-1}}$$

$$= \frac{2}{\pi} \arccos \frac{1}{\sqrt{t}}$$

When replacing the time unit 1 by τ, $0 < \tau < t$, this formula immediately yields the probability $p_{[\tau, t]}$ that the Wiener process crosses the x-axis at least once in the interval $[\tau, t]$:

$$p_{[\tau, t]} = \frac{2}{\pi} \arccos \sqrt{\frac{\tau}{t}}$$ □

7.3 Transformations of the Wiener Process

7.3.1 Elementary Transformations

Transforming the Wiener process leads to stochastic processes which are significant in their own right. They are of considerable theoretical importance and, moreover, useful stochastic models for describing many practical phenomena. Some transformations again yield the Wiener process. Theorem 7.1 compiles three transformations of this type.

Theorem 7.1 If $\{X(t), t \geq 0\}$ is a standard Wiener process, then the following processes are also standard Wiener processes:

(1) $\{U(t), t \geq 0\}$ with $U(t) = c\, X(t/c^2), \quad c > 0$

(2) $\{V(t), t \geq 0\}$ with $V(t) = X(t+h) - X(h), \quad h > 0$

(3) $\{W(t), t \geq 0\}$ with $W(t) = \begin{cases} t X(1/t) & \text{for } t > 0 \\ 0 & \text{for } t = 0 \end{cases}$

Proof As in *Kannon* (1979), the theorem is proved by verifying properties 1) to 3) in definition 7.1.

Obviously,

$$U(0) = V(0) = W(0)$$

Since the Wiener process has independent, normally distributed increments, the processes (1) to (3) have the same property. The trend functions of the processes (1) to (3) are identically zero. Therefore, in order to establish the stationarity of the increments, it is sufficient to show that thevariances of the increments of the processes (1) to (3) in any interval $[s, t]$, $s < t$, are equal to $t - s$.

(1) $$\begin{aligned} Var(U(t) - U(s)) &= E([U(t) - U(s)]^2) \\ &= E(U^2(t)) - 2Cov(U(s)U(t)) + E(U^2(s)) \\ &= c^2\left[E(X^2(t/c^2)) - 2Cov(X(s/c^2), X^2(t/c^2)) + E(X^2(s/c^2))\right] \\ &= c^2\left[\frac{t}{c^2} - 2\frac{s}{c^2} + \frac{s}{c^2}\right] \\ &= t - s \end{aligned}$$

(2) $$\begin{aligned} Var(V(t) - V(s)) &= E([X(t+h) - X(s+h)]^2) \\ &= (t+h) - 2(s+h) + (s+h) \\ &= t - s \end{aligned}$$

(3) $$\begin{aligned} Var(W(t) - W(s)) &= E([tX(1/t) - sX(1/s)]^2) \\ &= t^2 \cdot \frac{1}{t} - 2st \cdot \frac{1}{t} + s^2 \cdot \frac{1}{s} \\ &= t - s \end{aligned}$$ ■

If $\{X(t), t \geq 0\}$ is a Wiener process, then, with probability 1,

$$\lim_{t \to \infty} \frac{1}{t} X(t) = 0 \tag{7.22}$$

(For a proof, see, e.g., *Lawler* (1999). If t is replaced by $1/t$, then, taking the limit as $t \to \infty$, is equivalent to taking the limit as $t \to 0$. Hence, with probability 1,

$$\lim_{t \to \infty} tX\left(\frac{1}{t}\right) = 0 \tag{7.23}$$

A consequence of formula (7.22) is that any Wiener process $\{X(t), t \geq 0\}$ crosses the x-axis at least once with probability 1 in the interval $[\tau, \infty)$, $\tau > 0$, and hence even countably frequently. Since $\{tX(1/t), t \geq 0\}$ is also a Wiener process, it must have the same property. Therefore, for any $\tau > 0$, a Wiener process $\{X(t), t \geq 0\}$ crosses the x-axis in $(0, \tau]$ countably frequently with probability 1.

7.3.2 Ornstein-Uhlenbeck Process

As already mentioned in section 7.1, the sample paths of a Wiener process are nowhere differentiable. Hence, particles whose motion in liquids or gases is modeled by a Wiener process theoretically have an infinite velocity. To overcome this unrealistic situation, *Ornstein* and *Uhlenbeck* developed a stochastic process for describing the velocity of particles in liquids and gases.

Definition 7.3 Let $\{X(t),\ t \ge 0\}$ be a Wiener process with parameter σ. Then the stochastic process $\{V(t),\ -\infty < t < \infty\}$ defined by

$$V(t) = e^{-\alpha t} X(e^{2\alpha t})$$

is said to be an *Ornstein-Uhlenbeck process* with parameters α and σ; $\alpha > 0$. ●

Because of (7.6) the density of $V(t)$ is easily obtained:

$$f_{V(t)}(x) = \frac{1}{\sqrt{2\pi}\,\sigma} e^{-x^2/(2\sigma^2)}, \quad -\infty < x < \infty$$

Thus, $V(t)$ is normally distributed with parameters

$$E(V(t)) = 0, \quad Var(V(t)) = \sigma^2 \tag{7.24}$$

In particular, $V(t)$ is standard normal if $\{X(t),\ t \ge 0\}$ is the standard Wiener process. Since $\{X(t),\ t \ge 0\}$ is Gaussian, the Ornstein-Uhlenbeck process has the same property. (This is a corollary from theorem 1.1.) Hence, the multidimensional distributions of the Ornstein-Uhlenbeck process are multidimensional normal distributions. Moreover, there is a unique correspondence between the sample paths of the Wiener process and the corresponding Ornstein-Uhlenbeck-process. Thus, the Ornstein-Uhlenbeck process, like the Wiener process, is Markovian. From (7.24), the trend function of the Ornstein-Uhlenbeck process is seen to be identically zero. Its covariance function is

$$C(s,t) = \sigma^2 e^{-\alpha(t-s)}, \quad s \le t \tag{7.25}$$

This is proved as follows: For $s \le t$,

$$\begin{aligned} C(s,t) &= Cov(V(s), V(t)) = E(V(s)V(t)) \\ &= e^{-\alpha(s+t)} E(X(e^{2\alpha s})\, X(e^{2\alpha t})) \\ &= e^{-\alpha(s+t)} Cov(X(e^{2\alpha s}),\, X(e^{2\alpha t})) \\ &= e^{-\alpha(s+t)} \sigma^2 e^{2\alpha s} \quad \text{(from (7.9)} \\ &= \sigma^2 e^{-\alpha(t-s)} \end{aligned}$$

In particular, letting $s = t$ yields $Var(V(t)) \equiv \sigma^2$.

Corollary The Ornstein-Uhlenbeck process is wide-sense stationary. Since it is a Gaussian process, it is also strictly stationary.

Thus, the stationary Ornstein-Uhlenbeck process arises from the nonstationary Wiener process by a time transformation and standardization. In contrast to the Wiener process, the Ornstein-Uhlenbeck process has the following properties:

1) The increments of the Ornstein-Uhlenbeck process are not independent.
2) The sample paths of the Ornstein-Uhlenbeck process are everywhere mean-square differentiable.

7.3.3 Wiener-Process with Drift

Definition 7.4 A stochastic process $\{W(t), t \geq 0\}$ is called a *Wiener-process with drift* if it has the following properties:

1) $W(0) = 0$,
2) $\{W(t), t \geq 0\}$ has stationary, independent increments,
3) Every increment $W(t) - W(s)$ has a normal distribution with expected value $\mu(t-s)$ and variance $\sigma^2 |t-s|$. ●

Equivalently, $\{W(t), t \geq 0\}$ is a Wiener process with drift if

$$W(t) = \mu t + X(t),$$

where $\{X(t), t \geq 0\}$ is a Wiener process with $\sigma^2 = Var(X(1))$. The constant μ is called *drift parameter*. Thus, a Wiener process with drift arises by superimposing the Wiener process on a deterministic function. This deterministic function is a straight line and coincides with its trend function: $m(t) = \mu t$.

If properties 2) and 3) are fulfilled but $W(0) = w \neq 0$, then $\{W(t), t \geq 0\}$ is said to be a *Wiener process with drift starting at w*.

The one-dimensional density functions of the Wiener process with drift are

$$f_{W(t)}(x) = \frac{1}{\sqrt{2\pi t}\,\sigma} e^{-\frac{(x-\mu t)^2}{2\sigma^2 t}}; \quad -\infty < x < \infty, \quad t > 0 \tag{7.26}$$

Wiener processes with drift are, amongst other applications, used for modeling wear parameters, maintenance costs of technical systems, productivity criteria and capital increments over given time periods, as well as for modeling physical noise processes. Generally speaking, Wiener processes with drift can be successfully applied to modeling practical situations in which causally linear processes are permanently disturbed by random influences.

In view of these applications it is not surprising that first passage times of Wiener processes with drift play an important role both with respect to theory and practice. If a denotes a critical level which the Wiener process with drift should (not) reach or cross, let $L(a)$ be its corresponding first passage time. Since the Wiener process with drift has independent increments and is Gaussian, the following relationship between the probability densities of $W(t)$ and $L(a)$ holds:

$$f_{L(a)}(t) = \frac{a}{t} f_{W(t)}(a), \quad a > 0, \ \mu > 0$$

(For more general assumptions guaranteeing the validity of this formula, see *Franz* (1977)). Hence, the probability density of $L(a)$ is

$$f_{L(a)}(t) = \frac{a}{\sqrt{2\pi}\,\sigma\, t^{3/2}} \exp\left\{-\frac{(a-\mu t)^2}{2\sigma^2 t}\right\}, \quad t > 0 \tag{7.27}$$

For symmetry reasons, the probability density of the first passage time $L(a)$ of a Wiener process starting at w, $a > w$, can be obtained from (7.27) by replacing a with $a - w$. (A direct proof of (7.27) was given by *Scheike* (1992).)

The probability density $f_{L(a)}(t)$ is that of the *inverse Gaussian distribution* with parameters μ, σ^2 and a. For $\mu = 0$, this density simplifies to the first passage time density of Wiener processes (7.17). If $a < 0$ and $\mu < 0$, formula (7.27) yields the density of the corresponding first passage time $L(a)$ by substituting $|a|$ and $|\mu|$ for a and μ, respectively.

Expected value and variance of $L(a)$ are

$$E(L(a)) = \frac{a}{\mu}, \qquad Var(L(a)) = \frac{a\sigma^2}{\mu^3} \tag{7.28}$$

Contrary to the Wiener process ($\mu = 0$), now expected value and variance of $L(a)$ exist. Let $F_{L(a)}(t)$ be the distribution function of $L(a)$ and

$$\bar{F}_{L(a)}(t) = 1 - F_{L(a)}(t)$$

Assuming $a > 0$ and $\mu > 0$, integration of (7.27) yields

$$\bar{F}_{L(a)}(t) = \Phi\left(\frac{a-\mu t}{\sqrt{t}\,\sigma}\right) - e^{-2a\mu}\,\Phi\left(-\frac{a+\mu t}{\sqrt{t}\,\sigma}\right), \quad t > 0 \tag{7.29}$$

If the second term on the right-hand side of (7.29) is sufficiently small, then one obtains an interesting result: The Birnbaum-Saunders distribution as a limit distribution of first passage times of cumulative stochastic processes (theorem 4.10) approximately coincides with the inverse Gaussian distribution. After some tedious algebra, the Laplace transform of $f_{L(a)}(t)$ is seen to be

$$E(e^{-sL(a)}) = \int_0^\infty e^{-st} f_{L(a)}(t)\,dt = e^{-\frac{a}{\sigma^2}\left(\sqrt{2\sigma^2 s+\mu^2}-\mu\right)} \tag{7.30}$$

Now let $b < 0 < a$ and $\mu \neq 0$. Then the Wiener process with drift reaches level a before level b with probability

$$P(L(a) < L(b)) = \frac{1 - e^{-2\mu b/\sigma^2}}{e^{-2\mu a/\sigma^2} - e^{-2\mu b/\sigma^2}} \tag{7.31}$$

(*Karlin, Taylor* (1981)). More generally, for a Wiener process with drift starting at $W(0) = w$, $b < w < a$, and $\mu \neq 0$,

$$P(L(a) < L(b) \mid W(0) = w) = \frac{e^{-2\mu w/\sigma^2} - e^{-2\mu b/\sigma^2}}{e^{-2\mu a/\sigma^2} - e^{-2\mu b/\sigma^2}}$$

Let

$$M = \max_{t \in (0,\infty)} W(t)$$

As b tends to $-\infty$, then (7.31) converges towards the probability that the Wiener process with drift will ever reach level a, i.e. towards the probability of the random event "$M > a$":

$$\lim_{b \to -\infty} P(L(a) < L(b)) = P(M > a)$$

It follows that $P(M > a) = 1$ for $\mu > 0$ and

$$P(M > a) = e^{-\frac{2|\mu|}{\sigma^2} a}, \quad a > 0, \tag{7.32}$$

for $\mu < 0$. Thus, if $\mu < 0$, then M has an exponential distribution with parameter $\lambda = 2|\mu|/\sigma^2$.

Example 7.3 The price of a share at time t is

$$Z(t) = z_0 + W(t), \tag{7.33}$$

where $\{W(t), t \geq 0\}$ is a Wiener process with drift with negative drift parameter μ. Thus, z_0 is the initial price of the share: $z_0 = Z(0)$. At time $t = 0$ a speculator acquires the right to buy the share at price z_1, $z_1 \geq z_0$, at any time point in the future, independently of the current market value. Using the terminology in finance, the speculator owns an *American call option* with *strike price* z_1 on the share. It is assumed that the option has no expiry date. Although the value of the share is decreasing on average, the investor hopes to profit from random fluctuations in the share price. The difference $z_1 - z_0$ may be interpreted as the cost of acquiring the option.

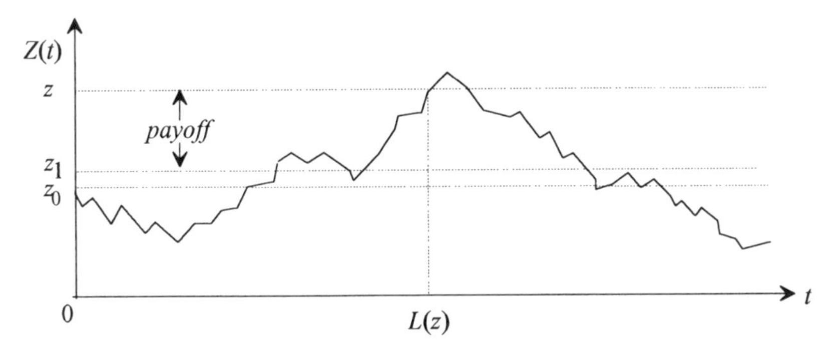

Figure 7.5 Payoff from random share price fluctuations

If the speculator buys the share (he "exercises" the option) at that time point when its price reaches the value z, $z > z_1$, for the first time, then his payoff (profit) immediately on exercising is $z - z_1$. By following this policy, the speculator's expected payoff amounts to

$$G(z) = (z - z_1)p(z) + 0 \cdot (1 - p(z)) = (z - z_1)p(z),$$

where $p(z)$ is the probability that the share price will ever reach level z. Because the option has no expiry date, this probability is given by (7.32) with $a = z - z_0$. Thus, letting

$$\lambda = 2|\mu|/\sigma^2,$$

the expected payoff amounts to

$$G(z) = (z - z_1)\, e^{-\lambda(z - z_0)} \tag{7.34}$$

The condition $dG(z)/dz = 0$ yields the optimum level $z = z^*$:

$$z^* = z_1 + 1/\lambda \tag{7.35}$$

The corresponding maximum expected profit is

$$G(z^*) = \frac{1}{\lambda e^{\lambda(z_1 - z_0) + 1}} \tag{7.36}$$

The greater the variance parameter σ^2 and the smaller the average price decrease per unit time μ, the greater is the expected profit. If, in particular, $z_1 = z_0$, then

$$G(z^*) = \frac{1}{\lambda e}$$

Note that the formulas given also hold if $z_1 < z_0$. It can be easily verified that in this case $G(z^*) > z_0 - z_1$. Otherwise, the speculator would exercise the option immediately after having acquired it.

Discounted payoff Let α, the risk free discount rate, be positive. The payoff from exercising the option at time t given the share has undiscounted price z, $z > z_1$, is $e^{-\alpha t}(z - z_1)$. Since under the policy considered the speculator buys the share at time $L(z - z_0)$ (= first passage time of the Wiener process with drift $\{W(t), t \geq 0$ with respect to level $z - z_0$) his random discounted payoff amounts to

$$e^{-\alpha L(z-z_0)}(z - z_1)$$

Hence, his expected discounted payoff is

$$G_\alpha(z) = (z - z_1)\int_0^\infty e^{-\alpha t} f_{L(z-z_0)}(t)\, dt\ , \tag{7.37}$$

where $f_{L(z-z_0)}(t)$ is given by (7.27) with $a = z - z_0$. The integral in (7.37) is equal to the Laplace transform of $f_{L(z-z_0)}(t)$ at $s = \alpha$. Thus, formula (7.30) immediately yields the expected discounted payoff

$$G_\alpha(z) = (z - z_1)\ \exp\left\{-\frac{z - z_0}{\sigma^2}\left(\sqrt{2\sigma^2\alpha + \mu^2} - \mu\right)\right\} \tag{7.38}$$

Functionally, the expected undiscounted payoff and the expected discounted payoff given by (7.34) and (7.38), respectively, are identical. Hence the optimal parameters with respect to $G_\alpha(z)$ are again given by (7.35) and (7.36) with λ replaced by

$$\gamma = \frac{1}{\sigma^2}\left(\sqrt{2\sigma^2\alpha + \mu^2} - \mu\right) \tag{7.39}$$

Note that minimizing $G_\alpha(z)$ makes also sense for a positive drift parameter μ. □

Example 7.4 Since, for a negative drift parameter, the sample paths of a stochastic process $\{Z(t), t \geq 0\}$ of structure (7.33) eventually become negative with probability 1, (7.33) has only limited application to modeling share prices. Hence it seems to be more realistic to assume that the share price at time t is given by

$$Z(t) = z_0\, e^{W(t)}$$

The sample paths of the stochastic process $\{Z(t), t \geq 0\}$ are obviously positive. The other assumptions as well as the formulation of the problem and the notation introduced in example 7.3 remain valid. In particular, the price of the share at time $t = 0$ is again equal to z_0.

The random event "$Z(t) \geq z$" with $z > z_0$ is equivalent to

$$W(t) \geq \ln \frac{z}{z_0}$$

The probability that the share price will ever reach the level z is therefore given by

$$p(z) = e^{-\lambda \ln \frac{z}{z_0}} = \left(\frac{z_0}{z}\right)^{\lambda}$$

Thus, if the speculator exercises the option as soon as the share price is z, his expected payoff is

$$G(z) = (z - z_1)\left(\frac{z_0}{z}\right)^{\lambda} \tag{7.40}$$

The optimum level $z = z^*$ is

$$z^* = \frac{\lambda}{\lambda - 1} z_1 \tag{7.41}$$

To ensure that $z^* > 0$, the additional assumption $\lambda = 2|\mu|/\sigma^2 > 1$ has to be made. The corresponding maximum expected payoff is

$$G(z^*) = \left(\frac{\lambda - 1}{z_1}\right)^{\lambda - 1} \left(\frac{z_0}{\lambda}\right)^{\lambda} \tag{7.42}$$

Discounted profit The undiscounted payoff $z - z_1$ is made when $W(t) = \ln \frac{z}{z_0}$ for the first time. Thus, the expected discounted payoff is

$$G_\alpha(z) = (z - z_1)\left(\frac{z_0}{z}\right)^{\gamma} \tag{7.43}$$

with γ given by (7.39). The functional forms of the expected undiscounted payoff (7.40) and the expected discounted payoff (7.43) are identical. Hence, the optimum values z^* and $G_\alpha(z^*)$ can be immediately obtained from (7.41) and (7.42), respectively, with λ replaced by γ. Note that the condition $\gamma > 1$ is equivalent to $2(\alpha - \mu) > \sigma^2$. As in the previous example, a positive drift parameter μ needs not be excluded. □

Comment Under different model assumptions, examples 7.3 and 7.4 pursue the same aim, namely maximizing the profit of the investor. However, their corresponding optimal characteristics z^*, $G(z^*)$ and $G_\alpha(z^*)$ may widely differ. The appropriate specification of the mathematical model to fit the practical situation is, therefore, of great importance.

Geometric Wiener process with drift The stochastic process $\{Z(t), t \geq 0\}$ given by

$$Z(t) = e^{W(t)}$$

is referred to as the *geometric Wiener process with drift.* The expected value of $(Z(t))^s$ coincides with the moment generating function of $W(t)$. Since $W(t)$ has a normal distribution, its moment generating function is

$$E\left(e^{s\,W(t)}\right) = \exp\left\{s\,t\left(\mu + \tfrac{1}{2}\sigma^2 s\right)\right\}$$

Substituting $s=1$ and $s=2$ yields the expected value and the second moment of $Z(t)$, respectively:

$$E(Z(t)) = e^{t(\mu+\sigma^2/2)} \tag{7.44}$$

$$E(Z^2(t)) = e^{2t(\mu+\sigma^2)}$$

By applying (1.10),

$$Var(Z(t)) = e^{t(2\mu+\sigma^2)}(e^{t\sigma^2}-1)$$

In the following example, the geometric Wiener process with positive drift is used to model the cumulative maintenance cost arising over a time period. It is a formal disadvantage of this model assumption that the corresponding cumulative maintenance cost does not have nondecreasing sample paths. However, the problem to be analyzed is not based on sample paths generated by the process, but on its trend function and expected first passage time. Both have "reasonable" properties with respect to the application considered.

Example 7.5 A system starts working at time $t=0$. Let $A(t)$ be the cumulative maintenance cost of the system over the time interval $[0, t]$. It is assumed that the sample paths of the stochastic process $\{A(t),\ t>0\}$ are continuous. Two maintenance policies are considered.

***Policy* 1** The system is replaced by a new one as soon as the cumulative maintenance cost $A(t)$ reaches level a.

Note that $A(t)$ includes not only repair costs, but also costs due to monitoring, servicing, stockkeeping as well as personnel costs. Hence, from the modeling point of view, policy 1 is superior to the "repair cost limit replacement policy" (policy 5 of section 3.2). Assuming an infinite planning horizon, negligibly small replacement times and independent replacement cycles, the long-run total maintenance cost rate is given by

$$K(a) = \frac{a+c}{E(L(a))}, \tag{7.45}$$

where c denotes the cost of a replacement and $L(a)$ the first passage time of the cost process $\{A(t),\ t\geq 0\}$ with respect to level a. (Note that formula (7.45) is a special case of (4.46)). $A(t)$ is assumed to have structure

$$A(t) = a_0\left[e^{W(t)}-1\right], \tag{7.46}$$

where $\{W(t),\ t\geq 0\}$ is a Wiener process with drift with positive parameters μ and σ^2. From (7.28), since

$$A(t) = a \ \text{ if and only if } \ W(t) = \ln(a+a_0) - \ln a_0,$$

the expected value of $L(a)$ is

$$E(L_A(a)) = \frac{\ln(a+a_0) - \ln a_0}{\mu}$$

Therefore,

$$K(a) = \frac{a + c}{\ln(a+a_0) - \ln a_0}\,\mu$$

A cost-optimum limit a satisfies equation

$$\ln\frac{a+a_0}{a_0} = \frac{a + c}{a+a_0}$$

Obviously, a unique solution $a = a^*$ exists. The corresponding minimum long-run total maintenance cost rate is

$$K(a^*) = (a^* + a_0)\,\mu$$

It is interesting to compare policy 1 with the common economic lifetime approach.

***Policy* 2 (*economic lifetime*)** If the system is replaced by a new one after a constant time τ, then the corresponding long-run total maintenance cost rate is

$$K(\tau) = \frac{m(\tau)+c}{\tau}, \tag{7.47}$$

where $m(\tau) = E(A(\tau))$. A replacement interval $\tau = \tau^*$ which is optimal with respect to $K(\tau)$ is called the *economic lifetime*. It satisfies the necessary condition

$$\tau\, m'(\tau) - m(\tau) = c$$

With $A(t)$ given by (7.46), formula (7.44) yields

$$K(\tau) = \frac{a_0\left[e^{(\mu+\sigma^2/2)\tau} - 1\right] + c}{\tau}$$

The corresponding economic lifetime τ^* is uniquely determined.

***Comparision of policies* 1 *and* 2** Using the notation $K(\tau,\sigma)$, $m(\tau,\sigma)$ and $\tau^*(\sigma)$ instead of $K(\tau)$, $m(\tau)$ and τ^*, $K(\tau,0)$ is seen to be

$$K(\tau,0) = \frac{a_0[e^{\mu\tau} - 1] + c}{\tau}$$

Letting

$$a = a_0(e^{\mu\tau} - 1),$$

it is readily verified that minimizing $K(a)$ with respect to a and minimizing $K(\tau,0)$ with respect to τ are equivalent problems. Hence,

$$K(\tau^*,\sigma) \ge K(\tau^*(0),0) = K(a^*)$$

Note also that $m(\tau,\sigma) \ge m(\tau,0)$ and, therefore, $K(\tau,\sigma) \ge K(\tau,0)$. Hence, under the assumptions stated, the following corollary holds:

Corollary Applying the economic lifetime on condition that the cumulative maintenance cost evolves deterministically according to $m(t, 0)$ is equivalent to applying the optimum maintenance cost limit a^*. Thus, policy 1 equalizes the cost-increasing influence of the random fluctuations of individual repair costs which are ignored under policy 2. □

The following example is to illustrate that optimizing replacement intervals on the basis of cumulative maintenance cost limits does not need full information on the underlying stochastic process $\{A(t), t \geq 0\}$ of the cumulative maintenance cost development.

Example 7.6 The same problem with the same notation as in the previous example is considered. The one-dimensional probability distribution of the cost process $\{A(t), t \geq 0\}$ is assumed to be known:

$$F_t(x) = P(A(t) \leq x) \quad \text{for all } t \geq 0$$

Since the sample paths of the process $\{A(t), t \geq 0\}$ are nondecreasing,

$$P(A(t) \leq a) = P(L(a) \geq t)$$

Hence, the trend function and expected first passage time with respect to level a of $\{A(t), t \geq 0\}$ are respective given by

$$m(t) = \int_0^\infty (1 - F_t(x))\, dx$$

$$E(L(a)) = \int_0^\infty F_t(a)\, dt$$

Specifically, let $A(t)$ have a Rayleigh-distribution with probability density

$$f_t(x) = \frac{2x}{\lambda^2 t^{2y}} \exp\left\{-\left(\frac{x}{\lambda t^y}\right)^2\right\}; \quad x \geq 0, \; y > 1, \; \lambda > 0$$

***Policy* 1** The expected value of $L(a)$ is

$$E(L(a)) = \int_0^\infty \int_0^a f_t(x)\, dx\, dt = \int_0^a \int_0^\infty f_t(x)\, dt\, dx$$

Integration yields

$$E(L(a)) = \left(\frac{1}{\lambda}\right)^{1/y} \Gamma\left(1 - \frac{1}{2y}\right) a^{1/y} = k_1 a^{1/y}$$

Minimizing the corresponding long-run total maintenance cost rate (7.45) yields the optimum limit a^* and the minimum cost rate $K(a^*)$:

$$a^* = \frac{c}{y-1}, \quad K(a^*) = \frac{y}{k_1}\left(\frac{c}{y-1}\right)^{(y-1)/y}$$

***Policy* 2** The expected value of $A(t)$ is

$$m(t) = \frac{\sqrt{\pi}}{2}\lambda t^y = k_2 t^y$$

Minimizing the corresponding long-run total maintenance cost rate (7.47) yields

$$\tau^* = \left(\frac{c}{k_2(y-1)}\right)^{1/y}, \quad K(\tau^*) = y k_2^{1/y}\left(\frac{c}{y-1}\right)^{(y-1)/y}$$

***Comparision of policies* 1 *and* 2** For all $y > 1$, the inequality $K(a^*) < K(\tau^*)$ is seen to be equivalent to

$$\frac{2}{\sqrt{\pi}} < U(x), \quad 0.5 \le x < 1 \tag{7.48}$$

with

$$U(x) = [\Gamma(x)]^{\frac{1}{2(1-x)}}$$

Note that $U(x)$ is a decreasing function in $[0.5 \le x < 1]$ with

$$U(0.5) = \sqrt{\pi} > 2/\sqrt{\pi} \text{ and } \lim_{x\to 1} U(x) = e^{E/2} > 2/\sqrt{\pi},$$

where $E \approx 0.5772$ is the Euler number. Hence, inequality (7.48) holds for all $y \ge 1$ so that, as in the previous example, policy 1 is superior to policy 2. In particular, if $1.1 \le y \le 5$, then average cost savings between 25 and 9% are achieved by applying the optimum cost limit a^* instead of the economic lifetime τ^*. □

Point estimation The parameters of a probability distribution are generally estimated from samples taken from this distribution. However, when an inverse Gaussian random variable is the first passage time of a Wiener process with drift, which allows observation of its development, then its parameters can be also estimated by scanning sample paths of the process. Using this approach, in what follows maximum-likelihood estimators $\hat{\mu}$ and $\hat{\sigma}^2$ of the parameters μ and σ^2 of a Wiener process with drift will be constructed. This is equivalent to estimating the parameters μ and σ^2 of the corresponding inverse Gaussian distribution.

Let $\{W(t), t \ge 0\}$ be a Wiener process with drift which starts at $W(0) = w$ and let

$$w_i = w_i(t); \ i = 1, 2, \ldots, n$$

be n of its sample paths which have been observed in n independent random experiments. The sample path $w_i = w_i(t)$ is scanned at the time points $t_{i1}, t_{i2}, \ldots, t_{im_i}$ with $0 < t_{i1} < t_{i2} < \ldots < t_{im_i}$; $m_i \ge 2$, $i = 1, 2, \ldots, n$. The outcomes are

$$w_{ij} = w(t_{ij}); \quad j = 1, 2, \ldots, m_i; \quad i = 1, 2, \ldots, n$$

Let

$$m = \Sigma_{i=1}^{n} m_i$$

be the total number of measurements and

$$\Delta w_{ij} = w_{ij} - w_{ij-1}, \quad \Delta t_{ij} = t_{ij} - t_{ij-1}$$

with $j = 2, 3, \dots, m_i$; $i = 1, 2, \dots, n$. If w is a constant, then $w = w_i(0)$; $i = 1, 2, \dots, n$. In this case, the maximum-likelihood estimators of μ and σ^2 are given by

$$\hat{\mu} = \frac{\sum_{i=1}^{n} w_{i m_i} - n w}{\sum_{i=1}^{n} t_{i m_i}}$$

$$\hat{\sigma}^2 = \frac{1}{m} \left\{ \sum_{i=1}^{n} \frac{(w_{i1} - \hat{\mu} t_{i1} - w)^2}{t_{i1}} + \sum_{i=1}^{n} \sum_{j=2}^{m_i} \frac{(\Delta w_{ij} - \hat{\mu} \Delta t_{ij})^2}{\Delta t_{ij}} \right\}$$

Note that these estimators are biased. The structure of the estimator $\hat{\mu}$ confirms the intuitively obvious fact that for estimating μ only the initial value w and the last tuples $(t_{i m_i}, w_{i m_i})$ of each sample path are relevant.

If w is a random variable, then the maximum-likelihood estimator of its expected value is

$$\hat{w} = \frac{\sum_{i=1}^{n} w_{i1} t_{i1}^{-1} - n \sum_{i=1}^{n} w_{i m_i} \left(\sum_{i=1}^{n} t_{i m_i} \right)^{-1}}{\sum_{i=1}^{n} t_{i1}^{-1} - n^2 \left(\sum_{i=1}^{n} t_{i m_i} \right)^{-1}} \tag{7.49}$$

The following maximum-likelihood estimators are obtained on condition that w is random.

Special case n=1 In this case only one sample path is available for estimating. Let $t_1, t_2, \dots, t_m$ and $w_1, w_2, \dots, w_m$ denote the time points at which the sample path is scanned and the corresponding outcomes, respectively. Letting

$$\Delta t_j = t_j - t_{j-1} \text{ and } \Delta w_j = w_j - w_{j-1},$$

the bias-corrected maximum-likelihood estimators of μ and σ^2 are

$$\hat{\mu} = \frac{w_m - \hat{w}}{t_m}$$

$$\hat{\sigma}^2 = \frac{1}{m-2} \left\{ \frac{(w_1 - \hat{\mu} t_1 - \hat{w})^2}{t_1} + \sum_{j=2}^{m} \frac{(\Delta w_j - \hat{\mu} \Delta t_j)^2}{\Delta t_j} \right\}$$

Special case $m_i = 1;\ i = 1, 2, ..., n$ In this case the estimation is based on n sample paths, but each sample path is only scanned at one time point. Hence, $m = n$. (Thus, the assumption $m_i \geq 2$ stated above is dropped.) The bias-corrected maximum-likelihood estimators of μ and σ^2 are

$$\hat{\mu} = \frac{\sum_{i=1}^{m} w_i - m\,\hat{w}}{\sum_{i=1}^{m} t_i} \tag{7.50}$$

$$\hat{\sigma}^2 = \frac{1}{m-2} \sum_{i=1}^{m} \frac{(w_i - \hat{\mu}\,t_i - \hat{w})^2}{t_i} \tag{7.51}$$

Example 7.7 *Pieper* (1988) measured the mechanical wear of 35 identical items (cylinder running bushes) used in diesel engines of ships over a time span of 11355 hours, each at one time point. The estimates of w, μ and σ^2 obtained from (7.49) to (7.51) are

$$\hat{w} = 36.145\ \mu m, \quad \hat{\mu} = 0.0029\ \mu m/h, \quad \hat{\sigma}^2 = 0.137\ \mu m^2/h$$

The wear $W(t)$ is assumed to evolve according to a Wiener process with drift starting at $W(0) = 36.145$:

$$W(t) = 36.145 + 0.0029t + X(t), \tag{7.52}$$

where $\{X(t),\ t \geq 0\}$ is a Wiener process with variance parameter $\sigma^2 = 0.137$.

Hint If the model (7.52) is correct, then the test function

$$T(t) = \frac{W(t) - 0.0029\,t - 36.145}{\sqrt{0.137\,t}}$$

has a standard normal distribution for all t (according to property 3 of definition 7.1). In particular, this must hold for all scanned points t_i. Hence, model (7.52) can be supported or rejected by a chi-square goodness of fit test.

Let $a = 1000\ \mu m$ be an upper critical level whose exceeding leads to a drift failure. Then the first passage time $L = L(1000)$ is the random time to drift failures of the wear parts. According to (7.28), estimates of its expected value, variance and standard deviation are

$$E(L) \approx \frac{1000 - 36.145}{0.0029} = 332364\ [h]$$

$$Var(L) \approx \frac{(1000 - 36.145) \cdot 0.137}{(0.0029)^3} = 5.41425 \cdot 10^9\ \left[h^2\right]$$

$$\sqrt{Var(L)} \approx 73581\ [h]$$

Let $t = \hat{\tau}_\varepsilon$ be the time point at which the wear parts must be preventively replaced in order to avoid drift failures with a given probability ε. With the survival function $\bar{F}(t)$ in (7.29), this time span satisfies

$$\bar{F}(\hat{\tau}_\varepsilon) = \varepsilon$$

Since

$$e^{-2(a-\hat{w})\hat{\mu}} \approx e^{-5.6},$$

the second term in (7.29) may be neglected. Therefore, $t = \hat{\tau}_\varepsilon$ is the solution of equation

$$\Phi\left(\frac{\hat{a} - \hat{\mu}\, t}{\sqrt{t}\,\hat{\sigma}}\right) = \varepsilon,$$

where $\hat{a} = a - \hat{w}$. Denoting the ε-percentile of the standard normal distribution by u_ε, then this equation is equivalent to

$$\frac{\hat{a} - \hat{\mu}\, t}{\sqrt{t}\,\hat{\sigma}} = u_\varepsilon$$

Thus,

$$\hat{\tau}_\varepsilon = \frac{\hat{a}}{\hat{\mu}} + \frac{1}{2}\left(\frac{u_\varepsilon \hat{\sigma}}{\hat{\mu}}\right)^2 - \frac{u_\varepsilon \hat{\sigma}}{\hat{\mu}}\sqrt{\frac{\hat{a}}{\hat{\mu}} + \left(\frac{u_\varepsilon \hat{\sigma}}{2\hat{\mu}}\right)^2}$$

If, in particular, $\varepsilon = 0.95$, then $u_{0.95} = 1.65$ and

$$\hat{\tau}_{0.95} = 231121\ [h]$$

Thus, with probability 0.95, the wear will remain below the critical level of 1000 μm within an operating time of 231121 hours . □

Wiener processes with drift were first considered by *Schrödinger* (1915) and *Smoluchowski* (1915). Both found the first passage time distribution of these processes. However, the notation *inverse Gaussian distribution* was proposed by *Tweedie* only in 1956. In 1978, *Folks* and *Chhikara* published tables of the percentiles of the inverse Gaussian distribution. As a distribution of first passage times, the inverse Gaussian distribution naturally plays a significant role as a statistical model for lifetimes of systems which are subject to drift failures (*Pieper* and *Tiedge* (1983)). *Folks* and *Chhikara* (1989) give a survey of the theory and of further applications: distribution of the water level of dams, duration of strikes, length of employment times of people in a company, wind velocity, and cost caused by system breakdowns. *Seshadri* (1999) presents an up to date and comprehensive treatment of the inverse Gaussian distribution.

7.3.4 Integral Transformations

Integrated Wiener Process If $\{X(t),\, t \ge 0\}$ is a Wiener process, then its sample paths $x = x(t)$ are mean-square continuous. Hence, the integrals

$$u(t) = \int_0^t x(y)\, dy$$

exist. They are realizations of the *random integral*

$$U(t) = \int_0^t X(y)\, dy$$

The stochastic process $\{U(t),\, t \ge 0\}$ is called the *integrated Wiener process*. This process can be a suitable model for practical situations where the observed sample paths seem to be "smoother" than those of the Wiener process. Analogously to the definition of the Riemann integral, for any n-dimensional vector $(t_1, t_2, ..., t_n)$ with $0 = t_0 < t_1 < ... < t_n = t$ and $\Delta t_i = t_i - t_{i-1};\ i = 1, 2, \dots, n$, the random integral $U(t)$ is defined as the limit

$$U(t) = \lim_{\substack{n\to\infty \\ \Delta t_i \to 0}} \left\{ \sum_{i=1}^{n} X(t_{i-1})\, \Delta t_i \right\} \tag{7.53}$$

(Note that passing to the limit refers here and in what follows to mean-square convergence.) Thus, the random variable $U(t)$, being the limit of a sum of normally distributed random variables, is itself normally distributed. More generally, by theorem 1.1, the integrated Wiener process is a Gaussian process. Therefore, the integrated Wiener process is uniquely characterized by its trend and covariance function. Its trend function is identically zero:

$$m(t) = E\left(\int_0^t X(y)\, dy\right) = \int_0^t E(X(y))\, dy \equiv 0$$

Its covariance function $C(s, t) = Cov(U(s), U(t)),\ s \le t$, is obtained as follows:

$$C(s, t) = E\left\{\int_0^s X(z)\, dz \int_0^t X(y)\, dy\right\}$$

$$= \int_0^s \int_0^t E(X(z)\, X(y))\, dy\, dz$$

Since $E(X(y)) = 0,\ y \ge 0$, and $Cov(X(y), X(z)) = \sigma^2 \min(y, z)$,

$$C(s, t) = \int_0^s \int_0^t Cov(X(y), X(z))\, dy\, dz$$

$$= \sigma^2 \int_0^s \int_0^t \min(y, z)\, dy\, dz$$

$$= \sigma^2 \int_0^s \left[\int_0^z y\, dy + \int_z^t z\, dy\right] dz$$

Thus,

$$C(s,t) = \frac{\sigma^2}{6}(3t - s)s^2, \quad s \le t$$

Letting $s = t$ yields

$$Var(U(t)) = \frac{\sigma^2}{3}t^3$$

The structure of the covariance function implies that the integrated Wiener process is nonstationary. But it can be shown that the stochastic process $\{V(t), t \ge 0\}$ defined by $V(t) = U(t+\tau) - U(t)$ for any $\tau > 0$, is stationary. (Note that for a Gaussian process strong and wide-sense stationarity are equivalent.)

White noise As already mentioned in section 7.1, almost all sample paths of a Wiener process are nowhere differentiable. Hence, a stochastic process of the form $\{Z(t), t \ge 0\}$ with

$$Z(t) = \frac{dX(t)}{dt} = X'(t) \quad \text{or} \quad dX(t) = Z(t)\,dt$$

cannot be introduced by taking the limit in a difference quotient. However, a definition via an integral is possible. To establish an heuristic approach to this definition, let $g(t)$ be any function with a continuous derivative $g'(t)$ in $[a, b]$. Further, let $t_0, t_1, \ldots, t_n$ define a partition of this interval:

$$a = t_0 < t_1 < \ldots < t_n = b \quad \text{and} \quad \Delta t_i = t_i - t_{i-1}; \;\; i = 1, 2, \ldots, n$$

Then the *stochastic integral* $\int_a^b g(t)\,dX(t)$ is defined as the limit

$$\int_a^b g(t)\,dX(t) = \lim_{\substack{n\to\infty \\ \max\limits_{i=1,2,\ldots,n} \Delta t_i \to 0}} \left\{ \sum_{i=1}^{n} g(t_{i-1})\,[X(t_{i-1} + \Delta t_i) - X(t_{i-1})] \right\} \tag{7.54}$$

The sum in (7.54) is written as follows:

$$\sum_{i=1}^{n} g(t_{i-1})(X(t_{i-1} + \Delta t_i) - X(t_{i-1})) = g(b)X(b) - g(a)X(a) - \sum_{i=1}^{n} X(t_i)[g(t_i) - g(t_{i-1})]$$

Taking the limit on both sides as in (7.54) yields

$$\int_a^b g(t)\,dX(t) = g(b)\,X(b) - g(a)\,X(a) - \int_a^b X(t)g'(t)dt \tag{7.55}$$

Formula (7.55) is reminiscent of the well-known formula of partial integration. This explanation of the stochastic integral is usually preferred to (7.54). As a limit of a sum of normal random variables, the stochastic integral also has a normal distribution. Its expected value results from (7.55):

$$E\left(\int_a^b g(t)\,dX(t)\right) = 0 \tag{7.56}$$

Since $Var(X(t)-X(s)) = \sigma^2|t-s|$,

$$Var\left(\sum_{i=1}^{n} g(t_{i-1})\,[X(t_{i-1}+\Delta t_i)-X(t_{i-1})]\right) = \sum_{i=1}^{n} g^2(t_{i-1})\,Var(X(t_{i-1}+\Delta t_i)-X(t_{i-1})$$

$$= \sigma^2 \sum_{i=1}^{n} g^2(t_{i-1})\,\Delta t_i$$

Passing to the limit as in (7.54) yields the variance of the stochastic integral:

$$Var\left(\int_a^b g(t)\,dX(t)\right) = \sigma^2 \int_a^b g^2(t)\,dt \tag{7.57}$$

Formula (7.55) motivates the following definition.

Definition 7.3 (*White noise*) Let $\{X(t),\, t \geq 0\}$ be the Wiener process. A stochastic process $\{Z(t),\, t \geq 0\}$ is called *white noise* if it satisfies

$$\int_a^b g(t)\,Z(t)\,dt = g(b)\,X(b) - g(a)\,X(a) - \int_a^b X(t)g'(t)dt \tag{7.58}$$

for any function $g(t)$ which has a continuous derivative $g'(t)$ in $[a, b]$. ●

(The term "white noise" is motivated in the following chapter.) If the first derivative of $X(t)$ did exist, then $Z(t) = dX(t)/dt$ would satisfy (7.58) anyhow. Thus, the $Z(t)$ introduced in definition 7.3 can be interpreted as a "generalized derivative" of $X(t)$, because it also exists if the differential quotient does not exist. But even this interpretation is not sufficient to facilitate the intuitive understanding of the white noise. To get an idea of the nature of white noise, its covariance function is determined. Before doing this, two special functions are introduced (Appendix 2).

The *(Dirac) delta function* $\delta(t)$ is defined by

$$\delta(t) = \lim_{h\to 0} \begin{cases} 1/h & \text{for } -h/2 \leq t \leq +h/2 \\ 0 & \text{elsewhere} \end{cases} \tag{7.59}$$

Symbolically,

$$\delta(t) = \begin{cases} \infty & \text{for } t = 0 \\ 0 & \text{elsewhere} \end{cases}$$

The delta function has a characteristic property which is sometimes used to its definition: For any function $f(t)$,

$$\int_{-\infty}^{+\infty} f(t)\,\delta(t-t_0)\,dt = f(t_0) \tag{7.60}$$

This can be proved as follows:

$$\int_{-\infty}^{+\infty} f(t)\,\delta(t-t_0)\,dt = \int_{-\infty}^{+\infty} f(t+t_0)\,\delta(t)\,dt$$

$$= \lim_{h\to 0} \int_{-h/2}^{+h/2} f(t+t_0)\frac{1}{h}\,dt$$

$$= \frac{1}{2}\left\{ \lim_{h\to 0} \frac{F(t_0+h/2)-F(t_0)}{h/2} + \lim_{h\to 0} \frac{F(t_0)-F(t_0-h/2)}{h/2} \right\}$$

$$= \frac{1}{2}\{f(t_0)+f(t_0)\}$$

$$= f(t_0),$$

where $F(t)$ is an antiderivative of $f(t)$. The *Heavyside function* $H(t)$ is defined by

$$H(t) = \begin{cases} 1 & \text{for } t \ge 0 \\ 0 & \text{for } t < 0 \end{cases} \tag{7.61}$$

The delta function can be formally considered the first derivative of the Heavyside function if an infinite slope is assigned to jump point $t = 0$ of $H(t)$:

$$\delta(t) = \frac{dH(t)}{dt} \tag{7.62}$$

Provided that the order of differentiation and integration can be also exchanged in case of generalized derivatives, the covariance function of the white noise is obtained as follows:

$$C(s,t) = Cov(Z(s), Z(t))$$

$$= Cov\left(\frac{\partial X(s)}{\partial s}, \frac{\partial X(t)}{\partial t}\right)$$

$$= \frac{\partial}{\partial s}\frac{\partial}{\partial t} Cov(X(s), X(t))$$

$$= \frac{\partial}{\partial s}\frac{\partial}{\partial t} \min(s,t)$$

$$= \frac{\partial}{\partial s} H(s-t)$$

Hence,

$$C(s,t) = \delta(s-t)$$

Therefore, since $C(s, t)$ is a measure of the statistical dependence between $Z(s)$ and $Z(t)$, there is no dependence between $Z(s)$ and $Z(t)$, $s \ne t$, even if $|s-t|$ is arbitrarily small. Thus, one cannot expect that the white noise process exists in the real world (see also section 8.3).

Example 7.8 Let $\{N(t), t \geq 0\}$ be a homogeneous Poisson process with intensity λ and $\{S(t), t \geq 0\}$ be the shot noise introduced in example 3.4:

$$S(t) = \sum_{i=1}^{N(t)} h(t - T_i),$$

where $h(t)$ denotes the response of the system to the Poisson events arriving at the time points T_i. (In examples 2.9 and 3.4 the corresponding underlying point process is referred to as a *pulse process*.)

The shot noise in vacuum tubes of example 2.9 is now considered in more detail: In the vacuum tube a current impulse is initiated as soon as the cathode emits an electron. If e denotes the charge on an electron and if an emitted electron arrives at the anode after z time units, then the current impulse induced by an electron is known to be

$$h(t) = \begin{cases} \frac{\alpha e}{z^2} t & \text{for} \quad 0 \leq t \leq z \\ 0 & \text{elsewhere} \end{cases},$$

where α is a tube-specific constant. $S(t)$ is, therefore, the total current flowing in the tube at time t. Now the covariance function of $S(t)$ can immediately be obtained from (3.16):

$$C(s, t) = \begin{cases} \frac{\lambda(\alpha e)^2}{3z}\left[1 - \frac{3|s-t|}{2z} + \frac{|s-t|^3}{2z^3}\right] & \text{for} \quad |s-t| \leq z \\ 0 & \text{elsewhere} \end{cases}$$

Since

$$\lim_{z \to 0} C(s, t) = \delta(s - t),$$

this shot noise approximately behaves as white noise if the transition time z is sufficiently small. □

Exercises

Notation In all exercises, $\{X(t), t \geq 0\}$ denotes the Wiener process with $Var(X(1)) = \sigma^2$.

7.1) Verify that the probability density $f_t(x)$ of $X(t)$,

$$f_t(x) = \frac{1}{\sqrt{2\pi t}\,\sigma} e^{-x^2/(2\sigma^2 t)}, \quad t > 0,$$

satisfies the *thermal conduction equation*

$$\frac{\partial f_t(x)}{\partial t} = c \frac{\partial^2 f_t(x)}{\partial x^2} \quad (c \text{ is a constant}).$$

7.2) Verify that the conditional probability density of $X(t)$ given $X(s) = y$ is

$$f_t(x|X(s) = y) = \frac{1}{\sqrt{2\pi\,(t-s)}\,\sigma} \exp\left(-\frac{1}{2(t-s)\sigma^2}(x-y)^2\right), \quad 0 \le s < t$$

7.3) Prove that the stochastic process $\{B(t),\ 0 \le t \le 1\}$ with $B(t) = X(t) - t X(1)$ is the Brownian bridge.

7.4) Let $\{B(t),\ 0 \le t \le 1\}$ be the Brownian bridge. Prove that $\{U(t),\ t \ge 0\}$ given by

$$U(t) = (t+1) B\left(\tfrac{t}{t+1}\right)$$

is the standard Wiener process.

7.5) Verify that $X(s) + X(t)$ has probability density

$$f_{X(s)+X(t)}(x) = \frac{1}{\sqrt{2\pi(t+3s)}\,\sigma} \exp\left\{-\frac{1}{2}\frac{x^2}{(t+3s)\sigma^2}\right\}, \quad -\infty < x < +\infty$$

7.6) Show that for any positive integer n the sum $S(n) = X(1) + X(2) + \ldots + X(n)$ has expected value

$$E(S(n)) = 0 \text{ and variance } Var(S(n)) = \frac{n\,(n+1)\,(2n+1)}{6}\sigma^2$$

Hint For any sequence of random variables $X_1, X_2, \ldots, X_n$ with $Var(X_i) < \infty$, $i = 1, 2, \ldots, n$,

$$Var(X_1 + X_2 + \ldots + X_n) = \sum_{i=1}^{n} Var(X_i) + 2 \sum_{\substack{i,j=1 \\ i<j}}^{n} Cov(X_i, X_j)$$

7.7) Prove that for any positve τ the Gaussian stochastic process $\{V(t),\ t \ge 0\}$ defined by $V(t) = X(t+\tau) - X(t)$ is stationary.

7.8) Prove that the Ornstein-Uhlenbeck process does not have independent increments.

7.9) Starting from $x = 0$, a particle makes independent jumps of length $\Delta x = \sigma\sqrt{\Delta t}$ to the right or to the left every Δt time units. The respective probabilities of jumps to the right and to the left are

$$p = \frac{1}{2}\left(1 + \frac{\mu}{\sigma}\sqrt{\Delta t}\right) \text{ and } 1-p, \text{ where } \sqrt{\Delta t} \le \left|\frac{\sigma}{\mu}\right|, \quad \sigma > 0$$

Show that as $\Delta t \to 0$ the position of the particle at time t is governed by a Wiener process with drift with parameters μ and σ.

7.10) Let $\{W(t),\ t \ge 0\}$ be a Wiener process with drift with parameters μ and $\sigma = 1$. Verify that

$$E\left(\int_0^t [W(s)]^2\, ds\right) = \frac{t}{2}(t + 2\mu)$$

Hint Note that $E\left(\int_0^t [W(s)]^2\, ds\right) = \int_0^t E[W(s)]^2\, ds$.

7.11) Show that for $c > 0$ and $d > 0$

$$P(X(t) \le c\,t + d \text{ for all } t \ge 0) = 1 - e^{-2cd/\sigma^2}.$$

Hint Make use of formula (7.32).

7.12) The price of a share at time t is

$$Z(t) = z_0 e^{W(t)+|W(t)|}, \quad t \ge 0,$$

where $\{W(t), t \ge 0\}$ is a Wiener process with drift with negative drift parameter μ. ($|W(t)|$ denotes the absolute value of $W(t)$.) At time $t = 0$ a speculator acquires an American call option with strike price z_1 on this share. The option has no expiry date (for terminology see example 7.3.) Check when/whether the speculator should exercise his option to make

(1) maximum expected profit without discounting and

(2) maximum expected discounted profit (risk-free positive discount rate α).

7.13) The value of a share at time t is

$$Z(t) = z_0 + W(t),$$

where $\{W(t), t \ge 0\}$ is a Wiener process with drift with positive drift parameter μ, $z_0 > 0$. At time $t = 0$ a speculator acquires an American call option on this share with finite expiry date T.
When should he exercise the option to make maximum expected undiscounted profit?

7.14) Let $A(t)$ be the cumulative maintenance cost of a system over the interval $(0, t]$ (excluding replacement costs) and $Z(t) = A(t)/t$ the corresponding maintenance cost rate. Assume that

$$Z(t) = z_0 (W(t))^r, \quad r > 0,$$

where $\{W(t), t \ge 0\}$ is a Wiener process with drift with positive drift parameter μ. The system is replaced by an equivalent new one as soon as $Z(t)$ reaches level z.

(1) Given a constant replacement cost c, determine a level $z = z^*$ which is optimum with respect to the long-run total maintenance cost per unit time $K(z)$. ($K(z)$ includes replacement costs.) Check under which conditions an optimum level z^* exists.

(2) Compare $K(z^*)$ with the minimum long-run total maintenance cost per unit time $K(\tau^*)$ which arises by applying the corresponding economic lifetime τ^*.

(*Hint*: Verify that $K(z) = z + c/E(L(z))$, where $L(z)$ is the first passage time of $\{Z(t), t \ge 0\}$ with respect to level z.)

7.15) With the notation and assumptions of the previous exercise, let $A(t)$ be given by

$$A(t) = a_0 (W(t))^r, \quad r > 0.$$

The system is replaced by an equivalent new one as soon as $A(t)$ reaches level a.

(1) Determine a level $a = a^*$ which is optimum with respect to the long-run total maintenance cost per unit time $K(a)$. Check under which conditions an optimum a^* exists.

(2) Compare $K(a^*)$ with the minimum long-run total maintenance cost per unit time $K(\tau^*)$ which arises when applying the corresponding economic lifetime τ^*.

Hint See example 7.5

7.16) Let $\sigma = 1$ and $U(t) = \int_0^t X(s)\, ds$.

(1) Determine the covariance between $X(t)$ and $U(t)$.
(2) Verify

$$E(U(t)|X(t) = x) = \frac{tx}{2} \quad \text{and} \quad Var(U(t)|X(t) = x) = \frac{t^3}{12}$$

Hint Make use of the fact that the random vector $(X(t), U(t))$ has a two-dimensional normal distribution.

7.17) Let α be any constant. Show that for $\sigma = 1$

$$E(e^{\alpha U(t)}) = e^{\alpha^2 t^3/6}$$

with $U(t)$ defined as in exercise 7.16.

Hint Make use of the moment generating function of the normal distribution.

8 SPECTRAL ANALYSIS OF STATIONARY PROCESSES

8.1 Foundations

Covariance functions of wide-sense stationary stochastic processes can be represented by their *spectral densities*. These *spectral representations* of covariance functions have proved a useful analytic tool in many technical and physical applications.

The mathematical treatment of spectral representations and the application of the results, particularly in electrotechnics and electronics, is facilitated by introducing the concept of a complex stochastic process. $\{X(t),\ t \in \mathbf{R}\}$ is a *complex stochastic process* if $X(t)$ is given by

$$X(t) = Y(t) + iZ(t), \quad \mathbf{R} = (-\infty, +\infty),$$

where $\{Y(t),\ t \in \mathbf{R}\}$ and $\{Z(t),\ t \in \mathbf{R}\}$ are two real-valued stochastic processes and $i = \sqrt{-1}$. Thus, the probability distribution of $X(t)$ is given by the joint probability distribution of the random vector $(Y(t), Z(t))$, $\mathbf{R} = (-\infty, +\infty)$. The trend- and covariance function of $\{X(t),\ t \geq 0\}$ are defined by

$$m(t) = E(X(t)) = E(Y(t)) + i\,E(Z(t)) \tag{8.1}$$

$$C(s,t) = Cov\,(X(s), X(t)) = E\Big([X(s) - E(X(s))]\big[\overline{X(t) - E(X(t))}\big]\Big) \tag{8.2}$$

If $X(t)$ is real, then (8.1) and (8.2) coincide with (2.4) and (2.6), respectively.

Notation If $z = a + ib$, then z and $\bar{z} = a - ib$ are *conjugate complex numbers*. The *modulus* of z, denoted by $|z|$, is defined as $|z| = \sqrt{z\bar{z}} = \sqrt{a^2 + b^2}$.

A complex stochastic process $\{X(t),\ t \geq 0\}$ is a *second-order process* if

$$E(|X(t)|^2) < \infty \text{ for all } t \in \mathbf{R}$$

Analogously to definition 2.2, a second-order complex process $\{X(t),\ t \in \mathbf{R}\}$ is said to be *wide-sense stationary* if it has the following properties:

1) $m(t) \equiv m$

2) $C(s,t) = C(0, t-s)$

In this case $C(s,t)$ simplifies to a function of one variable $C(\tau) = C(0, t-s)$, where $\tau = t - s$. m is a complex constant.

Ergodicity If the process $\{X(t), t \in \mathbf{R}\}$ is strictly stationary, then one anticipates that, for any of its sample paths $x(t) = y(t) + iz(t)$, its constant trend function can be obtained from

$$m = \lim_{T \to \infty} \frac{1}{2T} \int_{-T}^{+T} x(t)\, dt \tag{8.3}$$

This representation of the trend as an improper integral uses the full information which is contained in one sample path of the process. On the other hand, if N sample paths of the process $x_1(t), x_2(t), \ldots, x_N(t)$ are each only scanned at time point t_0 and if these values are obtained independently of each other, then $m = E(X(t_0))$ can be obtained from

$$m = \lim_{N \to \infty} \frac{1}{N} \sum_{k=1}^{N} x_k(t_0) \tag{8.4}$$

The equivalence of formulas (8.3) and (8.4) allows a simple physical interpretation: the mean of a stochastic process at a given time point is equal to its mean over the whole observation period. With respect to their practical application, this is the most important property of the *ergodic stochastic processes*. Besides the representation (8.2), for any sample path $x = x(t)$, the covariance function of an ergodic process can be obtained from

$$C(\tau) = \lim_{T \to \infty} \frac{1}{2T} \int_{-T}^{+T} [x(t) - m][\overline{x(t+\tau) - m}]\, dt \tag{8.5}$$

The exact definition of ergodic stochastic processes cannot be given here. In the technical literature, the ergodicity of stationary processes is frequently simply defined by properties (8.3) and (8.5). The application of formula (8.5) is useful if the sample path of an ongoing stochastic process is being recorded continuously. The estimated value of $C(\tau)$ becomes the better the larger the time span of observation $[-T, +T]$.

Assumption This chapter deals only with wide-sense stationary processes. Hence, the attribute "wide-sense" is generally omitted. Moreover, without of loss of generality, the trend function of all processes considered is identically zero.

In view of this assumption, representation (8.2) of the covariance function simplifies to

$$C(\tau) = C(t, t+\tau) = E(X(t)\overline{X(t+\tau)}) \tag{8.6}$$

In what follows, the *Euler's formulae* is needed:

$$e^{\pm ix} = \cos x \pm i \sin x \tag{8.7}$$

Solving for $\sin x$ and $\cos x$ yields

$$\sin x = \frac{1}{2i}\left(e^{ix} - e^{-ix}\right), \qquad \cos x = \frac{1}{2}\left(e^{ix} + e^{-ix}\right) \tag{8.8}$$

8.2 Processes with Discrete Spectrum

In this section the general structure of stationary stochastic processes with discrete spectra is developed. Next the simple stochastic process $\{X(t),\ t \geq 0\}$ with

$$X(t) = a(t)\,X, \tag{8.9}$$

is considered, where X is a complex random variable and $a(t)$ a complex function, $a(t) \not\equiv$ constant. For $\{X(t), t \geq 0\}$ to be stationary, the two conditions

$$E(X) = 0 \quad \text{and} \quad E(|X|^2) < \infty$$

are necessary. Moreover, because of (8.5), the function

$$E(X(t)\,\overline{X(t+\tau)}\,) = a(t)\,\overline{a(t+\tau)}\,E(|X|^2) \tag{8.10}$$

is not allowed to depend on t. Letting $\tau = 0$, this implies

$$a(t)\,\overline{a(t)} = |a(t)|^2 = |a|^2 = \text{constant}$$

Therefore, $a(t)$ has structure

$$a(t) = |a|\,e^{i\,\omega(t)}, \tag{8.11}$$

where $\omega(t)$ is a <u>real</u> function. Substituting (8.11) into (8.10) shows that the difference $\omega(t + \tau) - \omega(t)$ must not depend on t. Thus, if $\omega(t)$ is assumed to be differentiable, then $\omega(t)$ satisfies equation $d\,[\omega(t+\tau) - \omega(t)]/dt = 0$, or, equivalently,

$$\frac{d}{dt}\,\omega(t) = \text{constant}$$

Hence, $\omega(t) = \omega\,t + \phi$, where ω and ϕ are constants. (Note that for proving this result it is only necessary to assume the continuity of $\omega(t)$.) Thus,

$$a(t) = |a|\,e^{i\,(\omega t+\phi)}$$

If in (8.9) the random variable X is multiplied by $|a|e^{i\phi}$ and $|a|e^{i\phi}X$ is again denoted as X, then the desired result assumes the following form:

A stochastic process $\{X(t),\ t \geq 0\}$ defined by (8.9) is stationary if and only if

$$X(t) = X\,e^{i\,w\,t}, \tag{8.12}$$

where $E(X) = 0$ and $E(|X|^2) < \infty$.

Letting $s = E(|X|^2)$, the corresponding covariance function is

$$C(\tau) = s\,e^{-i\,\omega\tau}$$

Remark Apart from a constant factor, the paramater s is physically equal to the mean energy of the oscillation per unit time (mean power).

The real part $\{Y(t), t \geq 0\}$ of the stochastic process $\{X(t), t \in \mathbf{R}\}$ given by (8.12) describes a cosine oscillation with random amplitude and phase. Its sample paths are, therefore, given by

$$y(t) = a\cos(\omega t + \phi),$$

where a and ϕ are realizations of possibly dependent random variables A and Φ. The parameter ω is the *circular frequency* of the oscillation.

Generalizing the situation dealt with so far, a linear combination of two stationary processes of structure (8.12) is considered:

$$X(t) = X_1 e^{i\omega_1 t} + X_2 e^{i\omega_2 t} \tag{8.13}$$

X_1 and X_2 are two complex random variables with expected values 0 whereas ω_1 and ω_2 are two constant real numbers with $\omega_1 \neq \omega_2$. The covariance function of the stochastic process $\{X(t), t \in \mathbf{R}\}$ defined by (8.13) is

$$\begin{aligned} C(t, t+\tau) &= E(X(t)\overline{X(t+\tau)}) \\ &= E\left(\left[X_1 e^{i\omega_1 t} + X_2 e^{i\omega_2 t}\right]\left[\bar{X}_1 e^{-i\omega_1(t+\tau)} + \bar{X}_2 e^{-i\omega_2(t+\tau)}\right]\right) \\ &= E\left(\left[X_1\bar{X}_1 e^{-i\omega_1\tau} + X_1\bar{X}_2 e^{i(\omega_1-\omega_2)t - i\omega_2\tau)}\right]\right) \\ &+ E\left(\left[X_2\bar{X}_2 e^{-i\omega_2\tau} + X_2\bar{X}_1 e^{i(\omega_2-\omega_1)t - i\omega_1\tau)}\right]\right) \end{aligned}$$

Hence, the process $\{X(t), t \in \mathbf{R}\}$ is stationary if and only if X_1 and X_2 are uncorrelated. (Two complex random variables X and Y are said to be *uncorrelated* if they satisfy the condition $E(X\bar{Y}) = 0$ or, equivalently, $E(Y\bar{X}) = 0$.) In this case, the covariance function of $\{X(t), t \in \mathbf{R}\}$ is given by

$$C(\tau) = s_1 e^{-i\omega_1\tau} + s_2 e^{-i\omega_2\tau}, \tag{8.14}$$

where

$$s_1 = E(|X_1|^2), \quad s_2 = E(|X_2|^2)$$

Generalizing equation (8.13) leads to

$$X(t) = \sum_{k=1}^{n} X_k e^{i\omega_k t} \tag{8.15}$$

with real numbers ω_k satisfying $\omega_j \neq \omega_k$ for $j \neq k$; $i, j = 1, 2, \ldots, n$. If the X_k are uncorrelated and have expected value 0, then it can easily be shown by induction that the process $\{X(t), t \in \mathbf{R}\}$ is stationary. Its covariance function is

$$C(\tau) = \sum_{k=1}^{n} s_k e^{-i\omega_k\tau}, \tag{8.16}$$

where

$$s_k = E(|X_k|^2); \quad k = 1, 2, \dots, n$$

In particular,

$$C(0) = E(|X(t)|^2) = \sum_{k=1}^{n} s_k \tag{8.17}$$

The oscillation $X(t)$ given by (8.15) is an additive superposition of n harmonic oscillations. According to (8.17), its mean power is equal to the sum of the mean powers of these n harmonic oscillations.

Now let $X_1, X_2, \dots$ be a countably infinite sequence of uncorrelated complex random variables with $E(X_k) = 0$; $k = 1, 2, \dots$; and

$$\sum_{k=1}^{\infty} E\left(|X_k|^2\right) = \sum_{k=1}^{\infty} s_k < \infty \tag{8.18}$$

Under these assumptions, the equation

$$X(t) = \sum_{k=1}^{\infty} X_k e^{i\omega_k t}, \quad \omega_j \neq \omega_k \text{ for } j \neq k, \tag{8.19}$$

defines a stationary process $\{X(t), t \in \mathbf{R}\}$ with covariance function

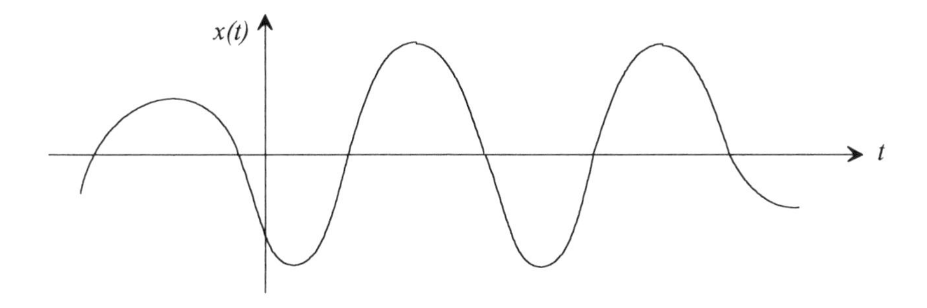

Figure 8.1 Sample path of a real narrow-band process

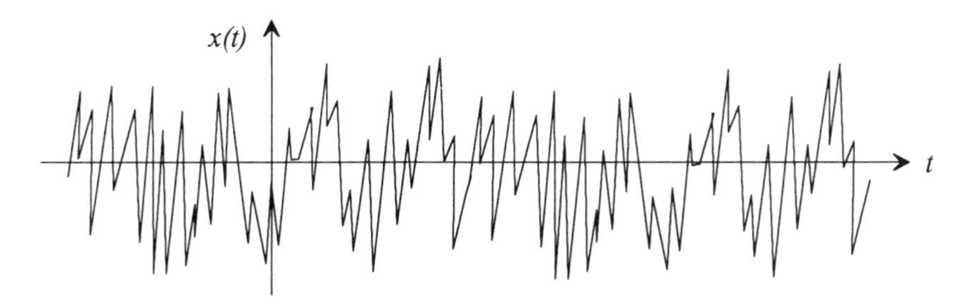

Figure 8.2 Sample path of a real wide-band process for large n

$$C(\tau) = \sum_{k=1}^{\infty} s_k e^{-i\omega_k \tau} \tag{8.20}$$

The sets $\{\omega_1, \omega_2, ..., \omega_n\}$ and $\{\omega_1, \omega_2, ...\}$ are said to be the *spectra* of the stochastic processes $\{X(t), t \in \mathbf{R}\}$ defined by (8.15) and (8.19), respectively. If all ω_k are sufficiently close to a single value ω, then $\{X(t), t \in \mathbf{R}\}$ is called a *narrow-band process* (Figure 8.1), otherwise it is called a *wide-band process* (Figure 8.2). The covariance function of stationary processes with discrete spectrum does not tend to zero as $|\tau| \to \infty$. However, it can be shown that, with respect to convergence in mean-square, any stationary process $\{X(t), t \geq 0\}$ can be sufficiently closely approximated by a stationary process of structure (8.15) in any finite time interval $[-T \leq t \leq +T]$.

Using property (7.60) of the delta function $\delta(t)$, the covariance function (8.20) can be written in the form

$$C(\tau) = \sum_{k=1}^{\infty} s_k \int_{-\infty}^{+\infty} e^{i\omega\tau} \delta(\omega - \omega_k)\, d\omega$$

Hence, symbolically,

$$C(\tau) = \int_{-\infty}^{+\infty} e^{i\omega\tau} s(\omega)\, d\omega\,, \tag{8.21}$$

where

$$s(\omega) = \Sigma_{k=1}^{\infty} s_k\, \delta(\omega - \omega_k) \tag{8.22}$$

The (generalized) function $s(\omega)$ is called the *spectral density* of the stationary process. $C(\tau)$ is obviously the Fourier transform of $s(\omega)$.

Real stationary processes In contrast to a stochastic process with structure (8.12), a stationary process $\{X(t), t \in \mathbf{R}\}$ with structure (8.13), i.e.

$$X(t) = X_1 e^{i\omega_1 t} + X_2 e^{i\omega_2 t}\,,$$

can be real. To see this, let

$$X_1 = \tfrac{1}{2}(A + iB) \quad \text{and} \quad X_2 = \bar{X}_1 = \tfrac{1}{2}(A - iB)$$

with $\omega_1 = -\omega_2 = \omega$, where A and B are two real random variables with expected values 0. Substituting these X_1 and X_2 into (8.13) yields

$$X(t) = A\cos\omega t - B\sin\omega t$$

If A and B are uncorrelated, then, letting $s = E(|X_1|^2) = E(|X_2|^2)$, the covariance function is seen to be

$$C(\tau) = 2\,s\cos\omega\tau$$

More generally, it can be shown that equation (8.15) with n terms defines a real stationary process if n is even and pairs of the X_k are complex conjugates.

8.3 Processes with Continuous Spectrum

8.3.1 Spectral Representation of the Covariance Function

Let $\{X(t), t \in \mathbf{R}\}$ be a complex stationary process with covariance function $C(\tau)$. Then there exists a real, nondecreasing, and bounded function $S(\omega)$ such that $C(\tau)$ has the following form:

$$C(\tau) = \int_{-\infty}^{+\infty} e^{i\omega\tau}\, dS(\omega) \tag{8.23}$$

(This fundamental relationship is associated with the names of *Bochner, Khinchin* and *Wiener*.) $S(\omega)$ is called the *spectral function* of the process. The definition of the covariance function implies that for all t

$$C(0) = S(\infty) - S(-\infty) = E(|X(t)|^2) < \infty$$

Given $C(\tau)$, the spectral function is uniquely determined apart from a constant c. Usually c is selected in such a way that $S(-\infty) = 0$. If the first derivative $s(\omega) = dS(\omega)/d\omega$ of the spectral function exists, then

$$C(\tau) = \int_{-\infty}^{+\infty} e^{i\omega\tau}\, s(\omega)\, d\omega \tag{8.24}$$

$s(\omega)$ is called the *spectral density* of the process. Since $S(\omega)$ is nondecreasing and bounded, the spectral density has properties

$$s(\omega) \ge 0, \quad C(0) = \int_{-\infty}^{+\infty} s(\omega)\, d\omega < \infty \tag{8.25}$$

Conversely, it can be shown that every function $s(\omega)$ with properties (8.25) is the spectral density of a stationary process.

Remark Frequently the function $f(\omega) = s(\omega)/2\pi$ is referred to as the spectral density. An advantage of this representation is that $\int_{-\infty}^{+\infty} f(\omega)\, d\omega$ is the mean power of the oscillation.

The set $\{\omega, s(\omega) > 0\}$ with its marginal points $\inf_{\omega \in \mathbf{S}} \omega$ and $\sup_{\omega \in \mathbf{S}} \omega$ is said to be the (*continuous*) *spectrum* of the process. Its *bandwidth* is defined as

$$w = \sup_{\omega \in \mathbf{S}} \omega - \inf_{\omega \in \mathbf{S}} \omega$$

Determining the covariance function is generally much simpler than determining the spectral density. (Note: In numerical evaluations, integrate in (8.24) only over the spectrum.) Hence the inversion of the relationship (8.24) is of importance. It is known from the theory of the Fourier integral that this inversion is always possible if the following inequality holds:

$$\int_{-\infty}^{+\infty} |C(t)|\, dt < \infty \tag{8.26}$$

In this case,

$$s(\omega) = \frac{1}{2\pi} \int_{-\infty}^{+\infty} e^{-i\omega t} C(t) dt \qquad (8.27$$

The intuitive interpretation of assumption (8.26) is that $C(\tau)$ must sufficiently fast converge to 0 as $|\tau| \to \infty$. The stationary processes occurring in electrotechnics and communication generally satisfy this condition. Integration of $s(\omega)$ over the interval $[\omega_1, \omega_2]$, $\omega_1 < \omega_2$, yields

$$S(\omega_2) - S(\omega_1) = \frac{i}{2\pi} \int_{-\infty}^{+\infty} \frac{e^{-i\omega_2 t} - e^{-i\omega_1 t}}{t} C(t)\, dt \qquad (8.28)$$

Comment This formula is also valid if the spectral density does not exist. But in this case the additional assumption has to be made that at each point of discontinuity ω_0 of $S(\omega)$ the spectral function is assigned the value

$$S(\omega_0) = \frac{1}{2}[S(w_0 + 0) - S(\omega_0 - 0)]$$

Note that the delta function $\delta(t)$ satisfies condition (8.26). If $\delta(t)$ is substituted for $C(t)$ in (8.27), then (7.60) yields

$$s(\omega) = \frac{1}{2\pi} \int_{-\infty}^{+\infty} e^{-i\omega t} \delta(t)\, dt \equiv \frac{1}{2\pi} \qquad (8.29)$$

The formal inversion of this relationship according to (8.24) provides a complex representation of the delta function:

$$\delta(t) = \frac{1}{2\pi} \int_{-\infty}^{+\infty} e^{i\omega t}\, d\omega \qquad (8.30)$$

The time-discrete analogues to formulas (8.26) and (8.27) are

$$\sum_{t=-\infty}^{+\infty} |C(t)| < \infty \qquad (8.31)$$

$$s(\omega) = \frac{1}{2\pi} \sum_{t=-\infty}^{+\infty} e^{-it\omega} C(t) \qquad (8.32)$$

Real stationary processes Since for real stationary processes $C(\tau) = C(-\tau)$, the covariance function can be written in the form

$$C(\tau) = [C(\tau) + C(-\tau)]/2$$

Substituting (8.24) for $C(\tau)$ into this equation and using (8.8) yields

$$C(\tau) = \int_{-\infty}^{+\infty} \cos \omega\tau\; s(\omega)\, d\omega$$

Because $\cos \omega\tau = \cos(-\omega\tau)$, this formula can be written as

$$C(\tau) = 2 \int_0^{+\infty} \cos \omega\tau\; s(\omega)\, d\omega \qquad (8.33)$$

Analogously, (8.27) yields the spectral density in the form

$$s(\omega) = \frac{1}{2\pi} \int_{-\infty}^{+\infty} \cos\omega t \, C(t)\, dt$$

Since $s(\omega) = s(-\omega)$,

$$s(\omega) = \frac{1}{\pi} \int_{-\infty}^{+\infty} \cos\omega t \, C(t)\, dt \tag{8.34}$$

However, even in case of real processes it is sometimes more convenient to use the formulas (8.24) and (8.27) instead of (8.33) and (8.34), respectively.

In many applications, the *correlation time* τ_0 is of interest. It is defined by

$$\tau_0 = \frac{1}{C(0)} \int_0^{\infty} C(t)\, dt \tag{8.35}$$

If $|\tau| \le \tau_0$, then there is a significant correlation (dependence) between $X(t)$ and $X(t+\tau)$. If $|\tau| > \tau_0$, then the correlation between $X(t)$ and $X(t+\tau)$ decreases quickly as $|\tau|$ tends to infinity.

Example 8.1 Let $\{\ldots, X_{-1}, X_0, X_1, \ldots\}$ be the purely random sequence (discrete white noise) defined in example 2.10. Its covariance function is given by

$$C(0) = \sigma^2, \quad C(\tau) = 0 \ \text{ for } \tau = \pm 1, \pm 2, \ldots \tag{8.36}$$

Hence, from (8.32),

$$s(\omega) = \sigma^2/2\pi$$

Thus, the discrete white noise has a constant spectral density. This result is in accordance with (8.29), since the the covariance function of the discrete white noise as given by (6.36) is equivalent to $C(t) = \sigma^2\,\delta(t)$. □

Example 8.2 The first-order autoregressive sequence considered in example 2.14 has covariance function

$$C(\tau) = c\, a^{|\tau|}; \quad \tau = 0, \pm 1, \ldots,$$

where a and c are real constants with $|a| < 1$. The corresponding spectral density is obtained from (8.32) as follows:

$$s(\omega) = \frac{1}{2\pi} \sum_{\tau=-\infty}^{\infty} C(\tau)\, e^{-i\tau\omega}$$

$$= \frac{c}{2\pi}\left[\sum_{\tau=-\infty}^{-1} a^{-\tau} e^{-i\tau\omega} + \sum_{\tau=0}^{\infty} a^{\tau} e^{-i\tau\omega}\right]$$

$$= \frac{c}{2\pi}\left[\sum_{\tau=1}^{\infty} a^{\tau} e^{i\tau\omega} + \sum_{\tau=0}^{\infty} a^{\tau} e^{-i\tau\omega}\right]$$

Hence,

$$s(\omega) = \frac{c}{2\pi}\left[\frac{a\,e^{i\omega}}{1-a\,e^{i\omega}} + \frac{1}{1-a\,e^{-i\omega}}\right]$$ □

Example 8.3 The random telegraph signal considered in example 3.2 has covariance function

$$C(\tau) = ae^{-b|\tau|}, \quad a>0,\ b>0 \tag{8.37}$$

Since condition (8.26) is satisfied, the corresponding spectral density $s(\omega)$ can be obtained from (8.27):

$$s(\omega) = \frac{1}{2\pi}\int_{-\infty}^{+\infty} e^{-i\omega t}\,ae^{-b|t|}\,dt$$

$$= \frac{a}{2\pi}\left\{\int_{-\infty}^{0} e^{(b-i\omega)t}\,dt + \int_{0}^{\infty} e^{-(b+i\omega)t}\,dt\right\}$$

$$= \frac{a}{2\pi}\left\{\frac{1}{b-i\omega} + \frac{1}{b+i\omega}\right\}$$

Hence,

$$s(\omega) = \frac{a\,b}{\pi\,(\omega^2+b^2)}$$

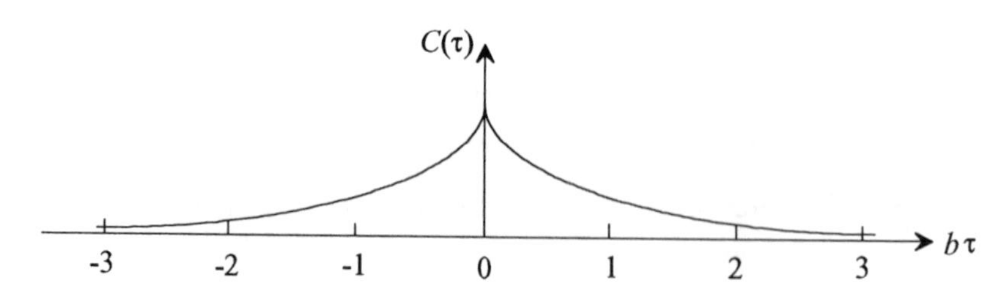

Figure 8.3 Covariance function in example 8.3

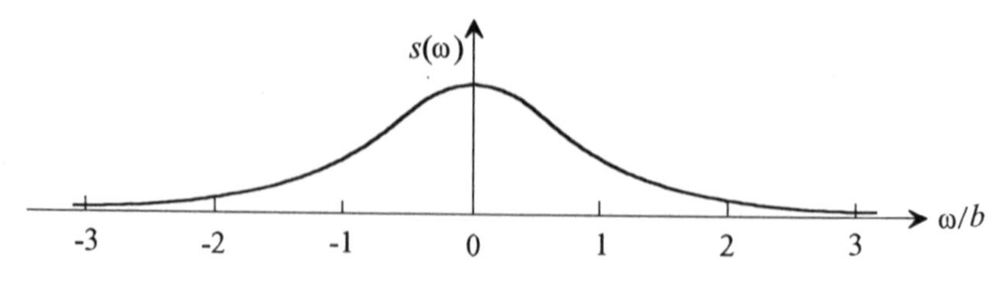

Figure 8.4 Spectral density in example 8.3

The corresponding correlation time is

$$\tau_0 = 1/b$$

This result is in accordance with Figure 8.3. Because of its simple structure, the covariance function (8.37) is sometimes even then applied if it only approximately coincides with the actual covariance function. □

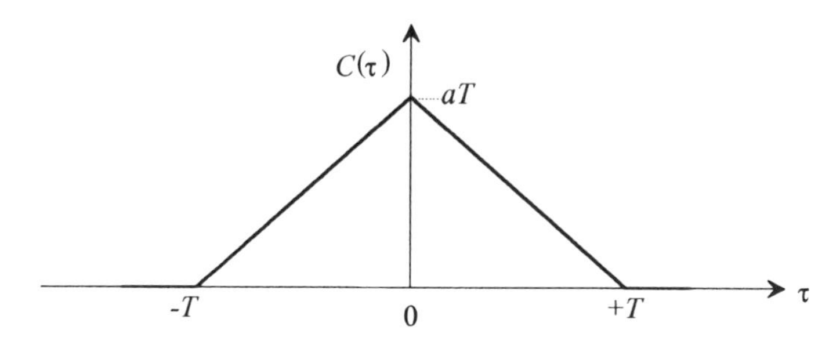

Figure 8.5 Covariance function in example 8.4

Example 8.4 Let

$$C(\tau) = \begin{cases} a(T-|\tau|) & \text{for } |\tau| \le T \\ 0 & \text{for } |\tau| > T \end{cases}, \quad a > 0, \; T > 0 \tag{8.38}$$

For example, the covariance function of the randomly delayed pulse code modulation considered in example 2.8 has this structure (see Figures 2.8 and 8.5). The corresponding spectral density follows from (8.27):

$$s(\omega) = \frac{a}{2\pi} \int_{-T}^{+T} e^{-i\omega t}(T-|t|)\,dt$$

$$= \frac{a}{2\pi}\left\{ T \int_{-T}^{+T} e^{-i\omega t}\,dt - \int_{0}^{+T} t e^{+i\omega t}\,dt - \int_{0}^{+T} t e^{-i\omega t}\,dt \right\}$$

$$= \frac{a}{2\pi}\left\{ \frac{2T}{\omega}\sin\omega T - 2\int_{0}^{T} t\cos\omega t\,dt \right\}$$

Hence,

$$s(\omega) = \frac{a}{\pi}\,\frac{1-\cos\omega T}{\omega^2}$$

Figure 8.6 shows the graph of $s(\omega)$. □

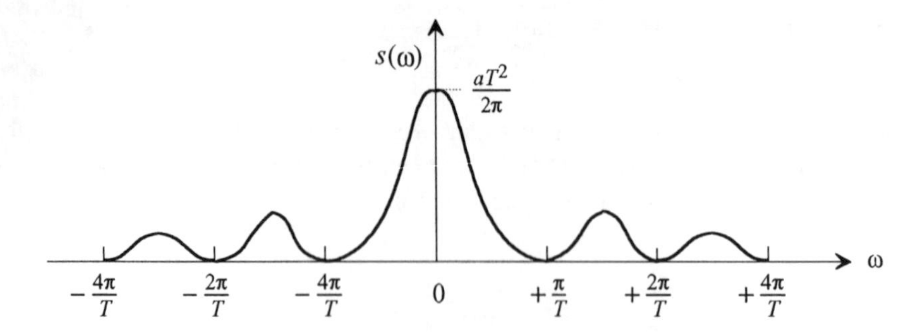

Figure 8.6 Spectral density in example 8.4

The previous examples should not give rise to the conjecture that for every function $f(\tau)$ which tends to zero as $|\tau| \to \infty$, a stationary stochastic process can be found with $f(\tau)$ being its covariance function. A slight modification of (8.38) yields a counter example:

$$f(\tau) = \begin{cases} a\left(T - \tau^2\right) & \text{for} \quad |\tau| \le T \\ 0 & \text{for} \quad |\tau| > T \end{cases}, \quad a > 0, \ T > 0$$

If this function is substituted for $C(\tau)$ in (8.27), then the resulting function $s(\omega)$ does not have properties (8.25). Therefore, $f(\tau)$ cannot be the spectral density of a stationary process.

Example 8.5 The stochastic processes considered in the examples 2.5 and 2.6 have covariance functions of the form

$$C(\tau) = a \cos \omega_0 \tau$$

Using (8.8), the corresponding spectral density is obtained as follows:

$$s(\omega) = \frac{a}{2\pi} \int_{-\infty}^{+\infty} e^{-i\omega t} \cos \omega_0 t \, dt$$

$$= \frac{a}{4\pi} \int_{-\infty}^{+\infty} e^{-i\omega t} \left(e^{i\omega_0 t} - e^{-i\omega_0 t}\right) dt$$

$$= \frac{a}{4\pi} \left\{ \int_{-\infty}^{+\infty} e^{i(\omega_0 - \omega)t} \, dt + \int_{-\infty}^{+\infty} e^{-i(\omega_0 + \omega)t} \, dt \right\}$$

Applying (8.30) yields a symbolic representation of $s(\omega)$ (Figure 8.7):

$$s(\omega) = \frac{a}{2}\{\delta(\omega_0 - \omega) + \delta(\omega_0 + \omega)\} \tag{8.39}$$

Making use of (7.60), the corresponding spectral function is seen to be

$$S(\omega) = \begin{cases} 0 & \text{for } \omega \le -\omega_0 \\ a/2 & \text{for } -\omega_0 < \omega \le \omega_0 \\ a & \text{for } \omega > \omega_0 \end{cases}$$

Thus, the spectral function is piecewise constant (Figure 8.8). □

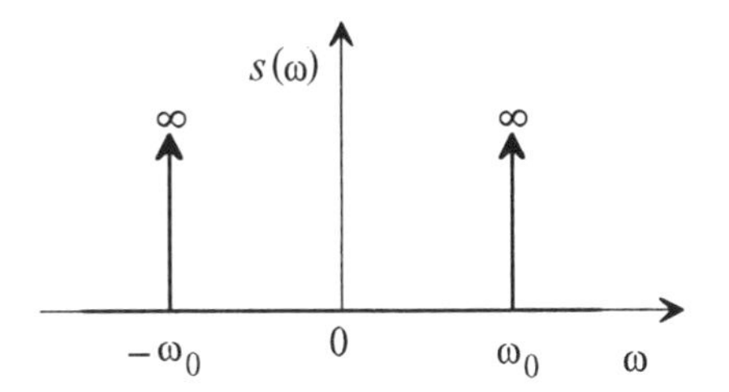

Figure 8.7 "Spectral density" in example 8.5

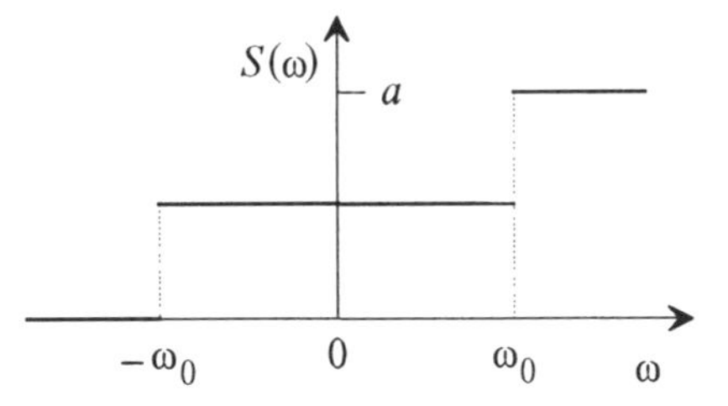

Figure 8.8 Spectralfunction in example 8.5

Comment Since in example 8.5 the covariance function does not tend to zero as $|\tau| \to \infty$, the condition (8.26), which is necessary for applying (8.27), is not satisfied. This fact motivates the occurrence of the delta function in (8.39). Hence, (8.39) as well as (8.22) are symbolic representations of the spectral density. The usefulness of such symbolic representations based on the delta function has been already illustrated in section 7.3.

If $C(\tau) = C_1(\tau)\, C_2(\tau)$, where $C_1(\tau)$ and $C_2(\tau)$ are the covariance functions of two stationary processes, then it can be shown that there exists a stationary process with covariance function $C(\tau)$. The following example considers a stationary process whose covariance function $C(\tau)$ is the product of the covariance functions of the stationary processes discussed in examples 8.3 and 8.5.

Example 8.6 Let $C(\tau)$ be given by the exponentially damped oscillation

$$C(\tau) = a\, e^{-b\,|\tau|} \cos\omega_0\tau \tag{8.40}$$

where $a > 0$, $b > 0$, and $\omega_0 > 0$. Thus, $C(\tau)$ satisfies condition (8.26) so that the corresponding spectral density can be obtained from (8.27):

$$s(\omega) = \frac{a}{\pi} \int_0^{\infty} e^{-bt} \cos(\omega t)\, \cos(\omega_0 t)\, dt$$

$$= \frac{a}{2\pi} \int_0^{+\infty} e^{-bt} [\cos(\omega - \omega_0)\, t + \cos(\omega + \omega_0)\, t]\, dt$$

Therefore,

$$s(\omega) = \frac{a\, b}{2\pi} \left\{ \frac{1}{b^2 + (\omega - \omega_0)^2} + \frac{1}{b^2 + (\omega + \omega_0)^2} \right\}$$

Functions of type (8.40) are frequently used to model covariance functions of stationary processes (possibly approximately) whose observed covariances periodically change their sign as τ increases. A practical example for such a stationary process is the fading of radio signals which are recorded by radar. □

White Noise In section 7.3, the (continuous) white noise process $\{Z(t), t \geq 0\}$ is defined as a generalized first derivative of the Wiener process. The introduction of a more general concept of differentiation proved necessary, since the sample paths of a Wiener process are nowhere differentiable with probability 1. There it has been also shown that the covariance function of the white noise, apart from a constant factor, coincides with the delta function. Thus, if the constant factor is denoted as $2\pi s_0$, the covariance of the white noise is

$$C(\tau) = 2\pi s_0 \, \delta(\tau) \tag{8.41}$$

The corresponding spectral density is now easily obtained from (8.27) and (7.60):

$$s(\omega) = \frac{1}{2\pi} \int_{-\infty}^{+\infty} e^{-i\omega t} \, 2\pi s_0 \, \delta(t) \, dt \equiv s_0$$

This result allows another characterization of the white noise:

The (continuous) white noise is a real, stationary, continuous-time stochastic process with constant spectral density.

The spectral density of white noise only satisfies the first of the conditions (8.25). Therefore, white noise cannot exist in practice, since its mean power would be infinite:

$$\int_{-\infty}^{+\infty} s(\omega) \, d\omega =$$

Nevertheless, the white noise is of great importance for approximately statistically modeling various phenomena in electronics, electrical engineering, communication, time series, econometrics and other applications. Its role can be compared with the one of a point mass in mechanics, which also only exists in the theory.

Because of (8.41), there is no correlation between $Z(t)$ and $Z(t+\tau)$, even for arbitrarily small $|\tau|, |\tau| > 0$. Thus, the white noise can be interpreted as the "most random" stochastic process, and this property explains its favourite role as a process for modeling random noise which is superimposed on a useful signal.

The (continuous) white noise is obviously the continuous-time analogue to the discrete white noise or the purely random sequence introduced in example 2.10. As shown in example 8.1, the discrete white noise also has a constant spectral density.

Hint The term "white noise" is due to a not fully justified comparision with the spectrum of the white light. This spectrum actually also has a wide-band structure, but its frequencies are not uniformly distributed over the entire bandwidth.

A stationary process $\{Z(t), t \geq 0\}$ can be approximately considered a white noise process if the covariance between $Z(t)$ and $Z(t+\tau)$ tends to 0 extremely quickly with increasing $|\tau|$. For example, if $Z(t)$ denotes the fluctuations of the absolute value of the force which affects particles in a liquid at time t and which causes their Brownian motion, then this force arises from the about 10^{21} collisions per second between the particles and the surrounding molecules of the liquid (assuming average temperature and average particle size). Hence, $Z(t)$ and $Z(t+\tau)$ are practically independent if

$$|\tau| \geq 10^{-18}$$

Thus, if a covariance function of type

$$C(\tau) = e^{-b|\tau|}, \; b > 1$$

(example 8.3) is assumed, then the parameter b must satisfy

$$b \geq 10^{19} \text{ sec}^{-1}$$

A similar fast drop of the covariance function can be observed if $\{Z(t), t \geq 0\}$ describes the fluctuations of the electromotive force in a conductor, which is caused by the thermal movement of electrons. The case of amperage fluctuations in vacuum tubes caused by the random emission times of electrons by the cathode has been already discussed in example 8.7.

White noise can be thought of as a sequence of extremely sharp pulses, which occur after extremely short time intervals, and which have independent, identically distributed amplitudes. The times in which the pulses rise and fall are so short that they cannot be registered by measuring instruments. Moreover, the response times of measurements are so large that during any response time a huge number of pulses occur which cannot be registered (Figure 8.9).

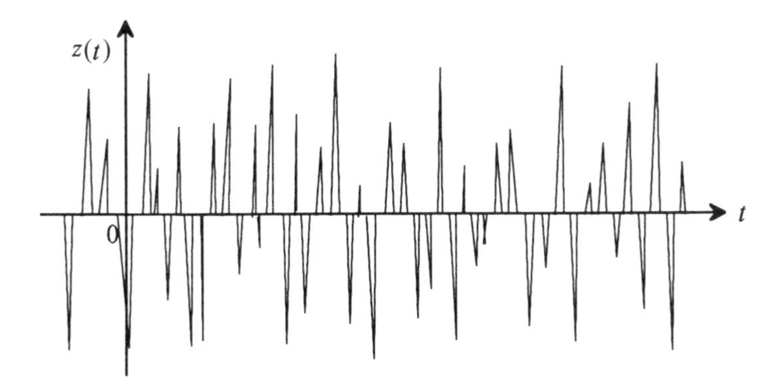

Figure 8.9 Illustration of the white noise (time axis extremely stretched)

Example 8.7 (*band-limited white noise*) As already pointed out, a stationary process with constant spectral density $s(\omega) = s_0$, $\omega \in (-\infty, +\infty)$, cannot exist. However, a stationary process with spectral density

$$s(\omega) = \begin{cases} s_0 & \text{for } \; -w/2 \le \omega \le +w/2 \\ 0 & \text{otherwise} \end{cases}$$

can (Figure 8.10 (*a*)). Using (8.8), its covariance function is (Figure 8.10 (*b*))

$$C(\tau) = \int_{-w/2}^{+w/2} e^{i\omega\tau} s_0 \, d\omega = 2 s_0 \frac{\sin w\tau/2}{\tau}$$

The average power of such a process is proportional to $C(0) = s_0 w$. The parameter w is the bandwidth of its spectrum. The white noise is obtained from band-limited white noise by passing to the limit as $w \to \infty$. □

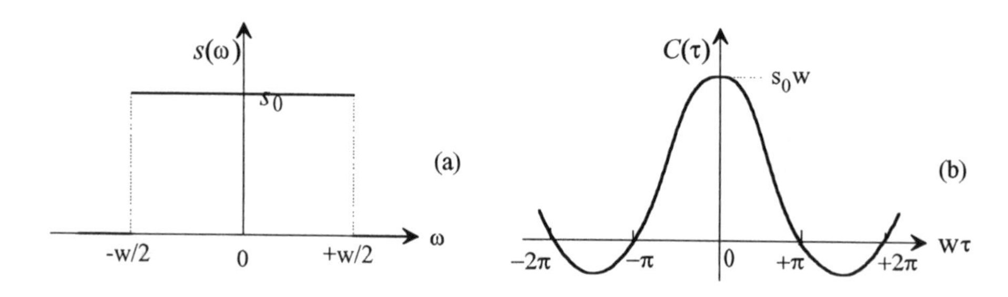

Figure 8.10 Spectral density (a) and covariance function (b) of the band-limited white noise

8.3.2 Spectral Representation of Stationary Processes*

Finally, some important results on the spectral analysis of wide-sense stationary processes are presented. This requires the introduction of a concept not needed up till now: a stochastic process $\{Y(t), t \in \mathbf{R}\}$ with $\mathbf{R} = \{-\infty, +\infty\}$ is said to have *orthogonal increments* if, for all disjoint intervals $[t_1, t_2)$ and $[t_3, t_4)$,

$$E\Big([Y(t_2) - Y(t_1)]\big[\overline{Y(t_4) - Y(t_3)}\big]\Big) = 0$$

In particular, a real-valued stochastic process with independent increments, whose trend function is identically zero, has orthogonal increments.

Let $\{X(t), t \in \mathbf{R}\}$ with $\mathbf{R} = \{-\infty, +\infty\}$ be a wide-sense stationary complex process with $E(X(t)) \equiv 0$. Then, as found by *A. N. Kolmogorov,* a second-order stochastic process $\{U(\omega), \omega \in \mathbf{R}\}$ with orthogonal increments exists such that

$$X(t) = \int_{-\infty}^{+\infty} e^{i\omega t}\, dU(\omega) \tag{8.42}$$

$\{U(\omega), \omega \in \mathbf{R}\}$ is called the *spectral process* of $\{X(t), t \in \mathbf{R}\}$. Apart from an additive constant, the spectral process is uniquely determined. If the constant is chosen such that $P(U(-\infty) = 0) = 1$, then the following relationships between $\{U(\omega), \omega \in \mathbf{R}\}$ and the spectral function $S(\omega)$ of $\{X(t), t \in \mathbf{R}\}$ hold:

$$E(U(\omega)) \equiv 0$$

$$E\left(|(U(\omega)|^2\right) = S(\omega)$$

$$E\left(|(dU(\omega)|^2\right) = dS(\omega)$$

Analogously to (7.54), the *stochastic Fourier-Stieltjes integral* in (8.42) is defined as follows: A finite interval $[a, b]$ is partitioned by the sequence

$$a = \omega_0 < \omega_1 < \ldots < \omega_n = b$$

into n disjoint intervals of lengths

$$\Delta\omega_k = \omega_k - \omega_{k-1}\,; \quad k = 1, 2, \ldots, n$$

Then, with respect to mean-square convergence,

$$\int_{-\infty}^{+\infty} e^{i\omega t}\, dU(\omega) = \lim_{\substack{a \to -\infty \\ b \to +\infty}} \; \lim_{\substack{n \to +\infty \\ \Delta\omega_k \to 0}} \sum_{k=1}^{n} e^{i\omega_{k-1} t}\,[U(\omega_k) - U(\omega_{k-1})]$$

Thus, a stationary process is the result of an additive superposition of harmonic oscillations in which the spectral process $\{U(\omega), \omega \in \mathbf{R}\}$ determines the random weights assigned to the occillations in the different frequency ranges. (The frequency ω_{k-1}, which represents the frequencies in the range $[\omega_{k-1}, \omega_k)$, is assigned the random weight $U(\omega_k) - U(\omega_{k-1})$.)

The spectral process is obtained by inversion of (8.42):

$$U(\omega_2) - U(\omega_1) = \frac{i}{2\pi} \int_{-\infty}^{+\infty} \frac{e^{-i\omega_2 t} - e^{-i\omega_1 t}}{t}\, X(t)\, dt$$

The spectral representation of a real-valued wide-sense stationary stochastic process $\{X(t), t \geq 0\}$ can be written in the following form:

$$X(t) = \int_0^{\infty} \cos\omega t\, dU(\omega) + \int_0^{\infty} \sin\omega t\, dV(\omega)\,,$$

where $\{U(t), t \geq 0\}$ and $\{V(t), t \geq 0\}$ are independent second-order processes with orthogonal increments and

$$E(U(t)) \equiv E(V(t)) \equiv 0$$

With respect to mean-square con vergence,

$$U(\omega) = \frac{1}{2\pi} \int_{-\infty}^{+\infty} \frac{\sin \omega t}{t} X(t)\, dt$$

$$V(\omega) = \frac{1}{2\pi} \int_{-\infty}^{+\infty} \frac{1 - \cos \omega t}{t} X(t)\, dt$$

The spectral representation of stationary processes is of practical importance if, for example, $\{X(t), t \in \mathbf{R}\}$ is the input signal of a linear filter, as well as for tackling prediction problems (*Hellstrom* (1984), *Ochi* (1990)).

Exercises

8.1) Define the stochastic process $\{X(t), t \in \mathbf{R}\}$ by $X(t) = A\cos(\omega t + \Phi)$, where Φ is uniformly distributed over $[0, 2\pi]$ and $E(A) = 0$. A and Φ are assumed to be independent random variables.
Check whether the covariance function of the wide-sense stationary process $\{X(t), t \in \mathbf{R}\}$ can be otained from (8.5).
Hint The covariance function of this process has been determined in example 2.5.

8.2) A wide-sense stationary continuous-time process has covariance function

$$C(\tau) = \sigma^2 e^{-\alpha|\tau|}\left[\cos\beta\,\tau - \frac{\alpha}{\beta}\sin\beta|\tau|\right]$$

Prove that its spectral density is given by

$$s(\omega) = \frac{2\sigma^2\alpha\,\omega^2}{\pi\left[\omega^2 + \alpha^2 + \beta^2 - 4\beta^2\omega^2\right]}$$

8.3) A wide-sense stationary continuous-time process has covariance function

$$C(\tau) = \sigma^2 e^{-\alpha|\tau|}\left[\cos\beta\,\tau + \frac{\alpha}{\beta}\sin\beta|\tau|\right]$$

Prove that its spectral density is given by

$$s(\omega) = \frac{2\sigma^2\alpha\,(\alpha^2 + \beta^2)}{\pi\left[\omega^2 + \alpha^2 - \beta^2 + 4\alpha^2\beta^2\right]}$$

8.4) A wide-sense stationary continuous-time process has covariance function

$$C(\tau) = a e^{-b\tau^2}; \quad a > 0,\ b > 0$$

Prove that its spectral density is given by

$$s(\omega) = \frac{a}{2\sqrt{\pi b}}\, e^{-\omega^2/4b}$$

8.5) Define the wide-sense stationary stochastic process $\{V(t),\ t \geq 0\}$ by

$$V(t) = X(t+1) - X(t),$$

where $\{X(t),\ t \geq 0\}$ is the standard Wiener process.
Prove that its spectral density is proportional to $(1 - \cos\omega)/\omega^2$.

8.6)* Define the stochastic process $\{V(t),\ t \geq 0\}$ by

$$V(t) = \int_{-\infty}^{t} e^{-\alpha(t-u)}\, dX(u), \quad \alpha > 0,$$

where $\{X(t),\ t \geq 0\}$ is the standard Wiener process.
Prove that $\{V(t),\ t \geq 0\}$ is a wide-sense stationary process and provide an integral representation of its spectral density.

8.7)* Let $\{X(t),\ t \geq 0\}$ be a Gaussian process which has the Markovian property.
Prove that the covariance function of such processes has structure $C(\tau) = a\, e^{-b|\tau|}$.
(The corresponding spectral density has been derived in example 8.3.)

8.8) A wide-sense stationary, continuous-time stochastic process has spectral density

$$s(\omega) = \sum_{k=1}^{n} \frac{\alpha_k}{\omega^2 + \beta_k^2}, \quad \alpha_k > 0$$

Prove that its covariance function is given by

$$C(\tau) = \pi \sum_{k=1}^{n} \frac{\alpha_k}{\beta_k} e^{-\beta_k |\tau|}$$

8.9) A wide-sense stationary, continuous-time stochastic process has spectral density

$$s(\omega) = \begin{cases} 0 & \text{for } |\omega| < \omega_0 \text{ or } |\omega| > 2\omega_0 \\ a^2 & \text{for } \omega_0 \leq |\omega| < 2\omega_0 \end{cases}, \quad \omega_0 > 0$$

Prove that its covariance function is given by

$$C(\tau) = 2a^2 \sin\omega_0\tau\, \frac{2\cos\omega_0\tau - 1}{\tau}$$

Appendix 1 Landau Order Symbol

In this book, the Landau order symbol $\mathrm{o}(x)$ represents any function $g(x)$ which has property

$$\lim_{x\to 0} \frac{g(x)}{x} = 0$$

In this case $g(x)$ is said to be $\mathrm{o}(x)$ as $x \to 0$. Therefore, the defining property of the symbol $\mathrm{o}(x)$ is

$$\lim_{x\to 0} \frac{\mathrm{o}(x)}{x} = 0$$

Thus, a function $g(x)$ which is $\mathrm{o}(x)$ as $x \to 0$, converges to 0 "faster" than $y = x$ as $x \to 0$, or, $\mathrm{o}(x)$ is negligibly small compared to x for x sufficiently small.

1) The function $g(x) = x^2$ is $\mathrm{o}(x)$ as $x \to 0$, since

$$\lim_{x\to 0} \frac{g(x)}{x} = \lim_{x\to 0} \frac{x^2}{x} = \lim_{x\to 0} x = 0$$

2) The function $g(x) = \sqrt{x}$ is not $\mathrm{o}(x)$ as $x \to 0$, since

$$\lim_{x\to 0} \frac{g(x)}{x} = \lim_{x\to 0} \frac{\sqrt{x}}{x} = \lim_{x\to 0} \frac{1}{\sqrt{x}} = \infty$$

3) The function $g(x) = \sin x$ is not $\mathrm{o}(x)$ as $x \to 0$, since

$$\lim_{x\to 0} \frac{g(x)}{x} = \lim_{x\to 0} \frac{\sin x}{x} = 1$$

Further properties are:

1) If $g_1(x)$ and $g_2(x)$ are $\mathrm{o}(x)$ as $x \to a$, then $g_1(x) + g_2(x)$ is $\mathrm{o}(x)$ as $x \to a$, since

$$\lim_{x\to a} \frac{g_1(x) + g_2(x)}{x} = \lim_{x\to a} \frac{g_1(x)}{x} + \lim_{x\to a} \frac{g_2(x)}{x} = 0 + 0 = 0$$

2) If $g(x)$ is $\mathrm{o}(x)$ as $x \to a$, then, for any constant c, $cg(x)$ is $\mathrm{o}(x)$ as $x \to a$, since

$$\lim_{x\to a} \frac{cg(x)}{x} = c \lim_{x\to a} \frac{g(x)}{x} = 0$$

Appendix 2 Dirac Delta Function

The *(Dirac) delta function* $\delta(t)$ is defined by

$$\delta(t) = \lim_{h \to 0} \begin{cases} 1/h & \text{for} \quad -h/2 \le t \le +h/2 \\ 0 & \text{otherwise} \end{cases}, \quad h > 0$$

Symbolically,

$$\delta(t) = \begin{cases} \infty & \text{for} \quad t = 0 \\ 0 & \text{otherwise} \end{cases}$$

A complex representation of the delta function is

$$\delta(t) = \frac{1}{2\pi} \int_{-\infty}^{+\infty} e^{i\omega t} \, d\omega$$

For any continuous function $f(t)$,

$$\int_{-\infty}^{+\infty} f(t)\, \delta(t - t_0)\, dt = f(t_0)$$

The *Heavyside function* is defined by

$$H(t) = \begin{cases} 1 & \text{for} \quad t \ge 0 \\ 0 & \text{for} \quad t < 0 \end{cases}$$

Symbolically, the delta function is the first derivative of the Heavyside function if an infinite slope is assigned to the jump point $t = 0$ of $H(t)$:

$$\delta(t) = \frac{dH(t)}{dt}$$

If $F(t)$ is a piecewise constant function with jump points of height 1 at $t_1, t_2, \ldots$, $t_1 < t_2 < \cdots$, then $F(t)$ and its "first derivative" $F'(t)$ are (Figure)

$$F(t) = \Sigma_{k=1}^{\infty} H(t - t_k) \quad \text{and} \quad \frac{dF(t)}{dt} = \Sigma_{k=1}^{\infty} \delta(t - t_k)$$

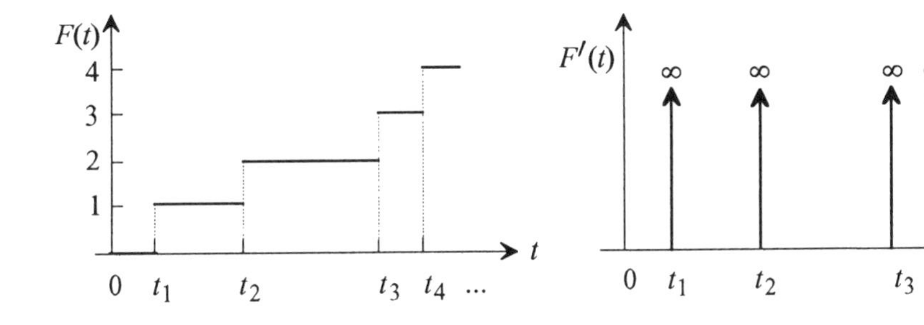

Answers to Selected Exercises

Chapter 1: Probability Theory

1.3) 0.6, 0.3, 0.4

1.4) (1) 0.925, 0.89, 0.85. 0.965, 0.15 (2) no

1.5) 0.68, 0.52, 0.852, 0.881, 0.1795, 0.8205 **1.7)** 0.0902

1.8) 13 **1.9)** (1) and (2): don't check (3) check

1.10) (1) 0.0278 (2) 0.6475 (3) 0.9979

1.11) (1) $2p^2(1+p+p^3)-5p^4$

1.12) (1) 0.7800 (2) 0.9744 (3) 0.1818

1.13) 0.4493 **1.14)** 0.54

1.15) (1) 0.4 (2) 0.4 **1.16)** (1) 0.1 (2) 0.7355

1.17) (1) 2.5 (2) 0.0579 **1.18)** 0.7165

1.19) (1) 0.0485 (2) 0.297 **1.20)** (1) 0.434 (2) 0.282

1.21) (1) $E(Z) = 200\,000$, $\sqrt{Var(Z)} = 12\,296$ (2) 0.792

1.23) (2) $f_X(x|y) = \frac{x+y}{y+1/2}$, $0 \le x, y \le 1$; $E(X|Y=0.5) = 7/12$, $E(X|Y=1) = 5/9$
(3) no

1.24) X and Y are independent **1.25)** 0.857

1.26) (1) 0.0322, 0.406 (2) 726.4kg **1.27)** 0.293

1.28) (1) $M(z) = \frac{1-z^{n+1}}{1-z} - 1$, $E(X) = \frac{n+1}{2}$ (2) no

1.29) $M(z) = \frac{pz}{1-(1-p)z}$, $E(X) = \frac{1-p}{p}$

1.30) (1) $M(z) = 1-p+pz$ (2) $M_n(z) = (1-p+pz)^n$

1.31) (1) $M(z) = (1-p+pz)^n$, $E(X) = np$, $Var(X) = np(1-p)$
(2) $p_i = p$; $i = 1, 2, ..., m$

1.32) (1) $\hat{f}(s) = \frac{1-e^{-sT}}{sT}$, $E(X) = \frac{T}{2}$ (2) no

Chapter 2: Stochastic Processes

2.1) (1) $m(t) = \frac{\sqrt{\pi}}{2}t$ (2) no **2.2)** (1) $m(t) = \mu t$

2.3) $C(\tau) = \frac{1}{2}E(A^2)\cos\omega\tau$, $\rho(\tau) = \cos\omega\tau$

2.5) $C(\tau) = \frac{1}{2}\sum_{i=1}^{\infty} a_i^2 \cos\omega\tau$, $\rho(\tau) = \cos\omega\tau$

2.6) (2) $C(s,t) = \begin{cases} 1, & nT \le s, t \le (n+\frac{1}{2})T, \; n = 0, 1, ... \\ 0, & \text{elsewhere} \end{cases}$

2.7) The trend function of the second order process $\{Z(t), t \ge 0\}$ is identically 0 and its covariance function is $C_Z(\tau) = C(\tau)\cos\omega\tau$

2.9) $C_U(s,t) = C_V(s,t) = C_X(s,t) + C_Y(s,t)$

2.10) $C(\tau) = \frac{25}{9}(0.8)^{|\tau|}, \quad \rho(\tau) = (0.8)^{|\tau|}$

2.11) (2) $C(\tau) = 4.829\,(0.9)^{|\tau|+1} - 0.921\,(-0.1)^{|\tau|+1}$

$\rho(\tau) = 1.088\,(0.9)^{|\tau|+1} - 0.209\,(-0.1)^{|\tau|+1}$

Chapter 3: Poisson Processes

3.1) (1) 0.4422 (2) 0.4422 **3.3)** 0.2739

3.4) (1) 0.9084 (2) $E(Y) = 1/4\, min, \; Var(Y) = (1/4)^2$

3.5) 0.1341 **3.8)** $C(\tau) = \begin{cases} \frac{\lambda}{2}(2\pi - |\tau|)\cos|\tau|, & 0 \le |\tau| \le 2\pi \\ 0, & \text{elsewhere} \end{cases}$

3.9) λ_2/λ_1 **3.10)** 0.639

3.11) (1) $\tau^* = \theta\left[\frac{c_p}{(\beta-1)c_m}\right]^{1/\beta}$ (2) not sensitive

Chapter 4: Renewal Theory

4.1) (1) $n^* = 85$ (2) $n^* = 87$ **(4.2)** (1) $H(t) = \frac{1}{4}\left(2\lambda t + e^{-2\lambda t} - 1\right)$

4.3) (1) $\mu = \frac{p}{\lambda_1} + \frac{1-p}{\lambda_2}, \quad \mu_2 = \frac{2p}{\lambda_1^2} + \frac{2(1-p)}{\lambda_2^2}$ **4.5)** 0.2642

4.9) $\mu = \sqrt{\pi}/2, \; \mu_2 = 1, \; \sigma^2 = 1 - \pi/4$

(1) $\lim_{t\to\infty} \int_0^t e^{-(t-x)^2} dH(x) = 1$ (2) $\lim_{t\to\infty}(H(t) - t/\mu) = 2/\pi - 1$

4.10) $\frac{1}{\mu}\int_0^t \bar{F}(x)\,dx$

4.11) (1) $P(V(t) > y \,|\, R(t) = x) = \frac{\bar{F}(x+y)}{\bar{F}(x)}$

(2) $P(V(t) > y \,|\, R(t + y/2) = x) = \frac{\bar{F}(x+y/2)}{\bar{F}(x)}$

4.13) $\frac{1}{3}(\lambda x + 2)e^{-\lambda x}$ **4.14)** (1) \$561.4 (2) 0.0096

4.15) (1) 0.9841 (2) 0.9971

4.17) (1) $K(\tau) = \frac{c_e F(\tau) + c_p \bar{F}(\tau)}{\int_0^\tau \bar{F}(t)\,dt}$

(2) $\lambda(\tau)\int_0^\tau \bar{F}(t)\,dt - F(\tau) = c/(1-c)$, where $c = c_p/c_e$

(3) $\tau^* = \frac{T}{1-c}\left[\sqrt{c(2-c)} - c\right]$

4.18) (1) $K(\tau) = \frac{c_p + c_e H(\tau)}{\tau}$ (2) $(1 + 3\lambda\tau)\,e^{-3\lambda\tau} = 1 - 9c/2$

Chapter 5: Discrete Markov Chains

5.1) (1) 0.5, 0.2 (2) 0.25, 0.25 (3) 0.5, 0.072

5.2) (1) $\mathbf{P}^{(2)} = \begin{pmatrix} 0.58 & 0.12 & 0.3 \\ 0.32 & 0.28 & 0.4 \\ 0.36 & 0.18 & 0.46 \end{pmatrix}$ (2) 0.42, 0

5.3) (1) 0.2864 (3) $\pi_0 = 0.4,\quad \pi_1 = \pi_2 = 0.3$ **5.4)** yes

5.5) $\pi_i = 0.25;\quad i = 0, 1, 2, 3$ (**P** is a doubly stochastic matrix)

5.8) (1) $\{1, 2, 3\}$, $\{3, 4\}$ (2) no inessential states

5.9) $\{0, 1, 2\}$ essential, $\{3\}$ inessential

5.10) (3) $\pi_0 = 50/150,\quad \pi_1 = 10/150,\quad \pi_2 = 40/150$

$\pi_3 = 13/150,\quad \pi_4 = 37/150$

5.11) (1) $\{0, 1\}$ essential, $\{2, 3, 4\}$ inessential

(2) $\{0, 1\}$ recurrent, $\{2, 3, 4\}$ transient

5.12) $\pi_i = p(1-p)^i,\quad i = 0, 1, ...$

5.15) (1) positive recurrent (2) transient
Hint: see example 5.14)

Chapter 6: Continuous Markov Chains

6.1) no **6.3)** $\pi_0 = \frac{2\lambda}{2\lambda+3\mu},\quad \pi_1 = \frac{2\mu}{2\lambda+3\mu},\quad \pi_2 = \frac{\mu}{2\lambda+3\mu}$

6.4) (1) 96% (2) 81%

6.5) state (i, j): i, j respective states of unit 1 and 2; 0 down, 1 operating

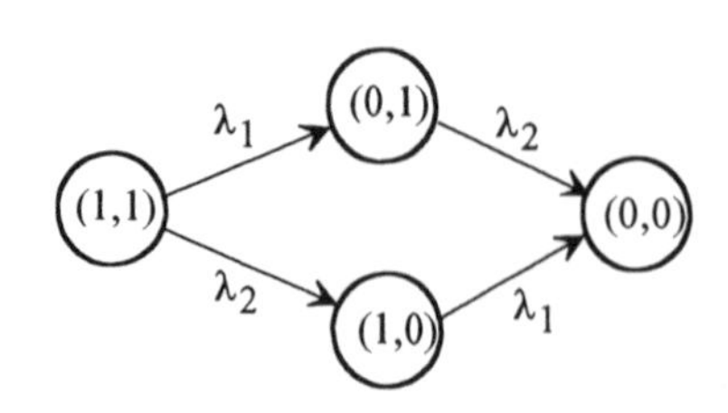

6.7) states: 0 system operating, 1 dangerous state, 2 system blocked
3 system blocked after dangerous failure

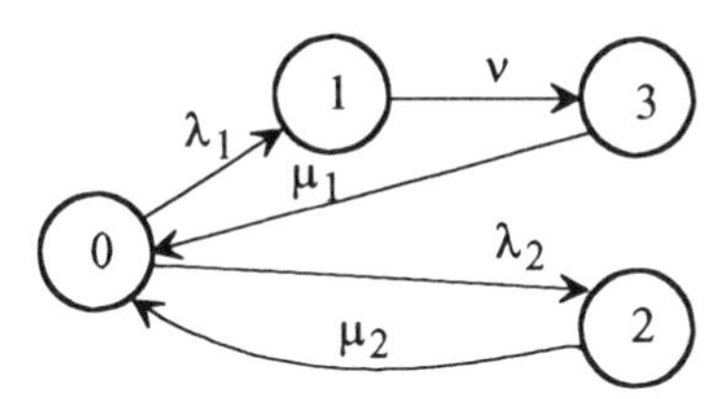

$$\pi_1 = \frac{\lambda_1}{\nu}\pi_0, \quad \pi_2 = \frac{\lambda_2}{\mu_2}\pi_0, \quad \pi_3 = \frac{\lambda_1}{\mu_1}\pi_0, \quad \pi_0 = \frac{1}{1+\frac{\lambda_1}{\nu}+\frac{\lambda_2}{\mu_2}+\frac{\lambda_1}{\mu_1}}$$

$P(\text{system blocked}) = \pi_2 + \pi_3$

6.9) $p_0(t) = e^{-2t}, \quad p_1(t) = 2\left(e^{-2t} - e^{-3t}\right), \quad p_2(t) = 3e^{-t}(1-e^{-t})^2$

6.10) (1) $(1-e^{-\lambda t})^2$ (2) $\frac{1}{\lambda}\left(1+\frac{1}{2}+\cdots+\frac{1}{n-1}\right)$

6.11) $p_j(t) = e^{-\lambda t}(1-e^{-\lambda t})^{j-1}; \quad j = 1, 2, \ldots$

6.14) (1) $\binom{2n}{j} e^{-j\mu t}(1-e^{-\mu t})^{2n-j}$ (2) $\frac{1}{\mu}\left(\frac{1}{2n}+\frac{1}{2n-1}+\cdots+\frac{1}{n+1}\right), \ n \geq 1$

6.15) (1) 0.56 (2) 50 weeks (*Hint*: $p_0(t) = P(\text{cable completely broken at time } t)$ is given by an Erlang distribution with parameters $n = 5$ and $\lambda = 0.1$.

6.17) see example 6.14, $r = n$ **6.18)** $\lambda < \mu$

6.20)

(1) States 0, 1, 2, 3; transitions 0→1, 1→2, 2→3 with rate 3λ; 1→0 with rate μ, 2→1 with rate 2μ, 3→2 with rate 2μ.

(2) $\pi_{loss} = \pi_3 = \dfrac{6.75\rho^3}{1+3\rho+4.5\rho^2+6.75\rho^3}, \quad \rho = \lambda/\mu$

6.21) $$\pi_{loss} = \pi_3 = \frac{13.5\rho^3\frac{1}{2+\nu/\mu}}{1+3\rho+4.5\rho^2+13.5\rho^3\frac{1}{2+\nu/\mu}}$$

6.23) $\pi_{loss} = 0.0311, \quad \pi_{wait} = 0.6149$

6.24) state (i,j): i, j customers at server 1, 2; $i,j = 0, 1$

$$\pi_{(1,0)} = \rho\,\pi_{(0,0)}, \quad \pi_{(1,1)} = \frac{\rho^2}{2}\pi_{(0,0)}, \quad \pi_{(0,1)} = \frac{\rho^2}{2(\rho+1)}\pi_{(0,0)}$$

6.26) (1) $\pi_0 = \pi_1 = \pi_2 = \pi_3 = 8/33$ (2) 1.2273 **6.28)** see example 6.14

6.31) (1)

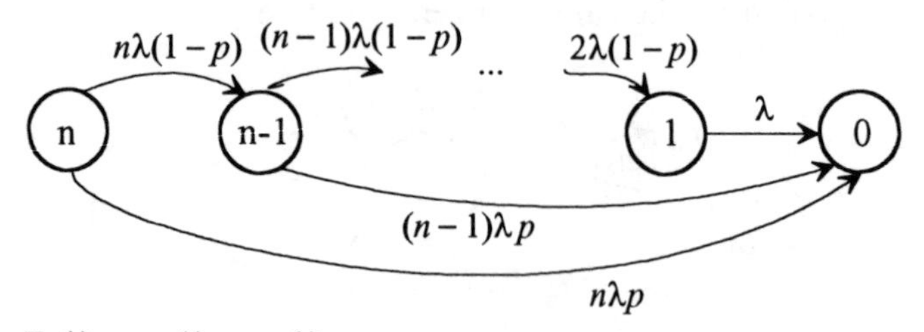

(2) $1 - F_s(t) = p_1(t) + p_2(t)$

$E(X_s) = 1/2\lambda + 2(1-p)[1/\lambda - 1/2\lambda]$

6.33) 2/3 **6.36)** 0.3153, 0.4144, 0.2703

6.37) states: 1 working, 2 repair after type 2 failure, 3 repair after type 1 failure
$P(X=1) = 360/372$, $P(X=2) = 4/372$, $P(X=3) = 8/372$

Chapter 7: Wiener Processes

Exercises given with solutions

Chapter 8: Stationary Processes

Exercises given with solutions

References

Anděl, J. (1984): Statistische Analyse von Zeitreihen. Akademie-Verlag, Berlin.

Bachelier, L. (1900): Theorie de la speculation.
Ann. Sci. Ec. Norm. Super.,17, 3, 21-26.

Beichelt, F. (1976): A general preventive maintenance policy.
Mathem. Operationsforschung und Statistik, 7, 927-932.

Beichelt, F. (1982): Replacement policy based on the maintenance cost rate.
IEEE Trans. Reliab., R-31, 391-393.

Beichelt, F. (1993*a*): Zuverlässigkeits- und Instandhaltungstheorie.
B. G. Teubner, Stuttgart.

Beichelt, F. (1993*b*): A unifying approach to a class of replacement policies with minimal repair. Naval Research Logistics, 40 (1993) 1, 58-75.

Beichelt, F. (1995): Stochastik für Ingenieure - Eine Einführung in die Wahrscheinlichkeitstheorie und Mathematische Statistik. B. G. Teubner, Stuttgart.

Beichelt, F. (1997: Stochastische Prozesse für Ingenieure. B. G. Teubner, Stuttgart.

Beichelt, F.; Franken, P. (1983): Zuverlässigkeit und Instandhaltung - Mathematische Methoden. Verlag Technik, Berlin; Carl Hanser Verlag, München-Wien.

Beveridge, W. H. (1921): Weather and harvest cycles. Econ. J., 31, 429-452.

Brown, R. (1828): A brief account of microscopial observations made in the months of June, July, and August, 1827, on particles contained in the pollen of plants; and on the general existence of active molecules in organic and inorganic bodies. Phil. Mag., Series 2, No. 4, 161-173.

Chhikara, R. S.; Folks, J. L. (1988): The Inverse Gaussian Distribution.
Marcel Dekker, Inc., New York, Basel.

Chung, K. L. (1960): Markov Chains with Stationary Transition Probabilities.
Springer-Verlag, Berlin.

Cramér, H.; Leadbetter, M. R. (1967): Stationary and Related Stochastic Processes. Wiley, New York.

Družinin, G. V. (1977): Reliability of Automatic Systems, Energija, Moscov
(in Russian).

Einstein, A. (1905): Über die von der molekularkinetischen Theorie der Wärme geforderte Bewegung von in ruhenden Flüssigkeiten suspendierten Teilchen.
Ann. Phys., 17, 549-560.

Feller, W. (1968): An Introduction to Probability Theory and its Applications. Vol. I (3rd ed.)., Wiley, New York.

Feller, W. (1971): An Introduction to Probability Theory and its Applications. Vol. II (2nd ed.)., Wiley, New York.

Fischer, K. (1984): Zuverlässigkeits- und Instandhaltungstheorie. VEB Transpress-Verlag für Verkehrswesen, Berlin.

Folks, J. L.; *Chhikara, R. S.* (1978): The inverse Gaussian distribution and its statistical application-a review. J. Royal Stat. Soc., B 40, 263-289.

Folks, J. L.; *Chhikara, R. S.* (1989):

Franken, P. (1963): Utočnenije pridelnoj teoremy dlja superpozicii nezavisimych processov vosstanovlenija. Teor. Verojatn. i Primen., 8, 341-349.

Franz, J. (1977): Niveaudurchgangszeiten zur Charakterisierung sequentieller Schätzverfahren. Mathem. Operationsforsch. u. Statistik, Ser. Statistics, 8, 499-510.

Gaede, K. W. (1977): Zuverlässigkeit. Mathematische Modelle. Carl Hanser Verlag, München-Wien.

Gardiner, C. W. (1997): Handbook of Stochastic Methods. Springer, New York, Berlin.

Gardner, W. A. (1989): Introduction to Random Processes with Applications to Signals and Systems. Mc Graw-Hill Publishing Company, New York.

Gelenbe, E.; *Pujolle, G.* (1987): Introduction to Queueing Networks. Wiley, New York.

Gnedenko, B. W.; König, D. (1983, 1984): Handbuch der Bedienungstheorie I, II. Akademie-Verlag, Berlin.

Gut, A. (1990): Cumulative shock models. Adv. Appl. Prob., 22, 504-506.

Hellstrom, C. W. (1984): Probability and Stochastic Processes for Engineers. Macmillan Publishing Company, New York; Collier Macmillan Publishers, London.

Jaglom, A. M. (1962): An Introduction to the Theory of Stationary Random Functions. Prentice-Hall, Englewood Cliffs.

Kannan, D. (1979): An Introduction to Stochastic Processes. North Holland; New York, Oxford.

Karlin, S.; *Taylor, H. M.* (1981): A Second Course to Stochastic Processes. Academic Press, New York.

Karlin, S.; *Taylor, H. M.* (1994): An Introduction to Stochastic Modeling. Academic Press, New York.

Kulkarni, V. G. (1995): Modeling and Analysis of Stochastic Systems. Chapman & Hall, London, New York.

Kijma, M. (1996): Markov Processes for Stochastic Modeling. Chapman and Hall, London, New York.

Lawler, G. F. (1995): Introduction to Stochastic Processes. Chapman & Hall, London, New York.

Mac Donald, D. K. C. (1962): Noise and Fluctuations. Wiley, New York.

Matthes, K. (1962): Ergodizitätseigenschaften rekurrenter Ereignisse I. Mathem. Nachr., 24, 109-119.

Montgomery, D. C.; Runger, G. C. (1994): Applied Statistics and Probability for Engineers. Wiley, New York.

Ochi, M. K. (1990): Applied Probability and Stochastic Processes in Engineering and Physical Sciences. Wiley, New York.

Partzsch, L. (1984): Vorlesungen zum eindimensionalen Wienerschen Prozeß. B.G. Teubner, Leipzig.

Perrin, J. (1916): Atoms. Van Nostrand, Princeton.

Pieper, V. (1988): Zuverlässigkeitsuntersuchungen auf der Grundlage von Niveauüberschreitungsuntersuchungen bei stochastischen Prozessen und der Modellierung von Abnutzungsvorgängen. Dissertation (B). TU Magdeburg.

Pieper, V.; Tiedge, J. (1983): Zuverlässigkeitsmodelle auf der Grundlage stochastischer Modelle von Verschleißprozessen. Mathem. Operationsf. u. Statistik, Ser. Statistics, 14, 485-502.

Ross, S. M. (1979): Applied Probability Models with Optimization Applications. Holden-Day, San Francisco.

Ross, S. M. (1989): Introduction to Probability Models (4th ed.) Academic Press, New York.

Ross, S. M. (1996): Stochastic Processes (2nd ed.), Wiley, New York.

Saur, C. H.; Chandi, K. M. (1981): Computer Systems Perfomance Modeling. Prentice Hall, Englewood Cliffs.

Scheike, T. H. (1992): A boundary-crossing result for Brownian motion. J. Appl. Prob., 29, 448-453.

Schrödinger, E. (1915): Zur Theorie der Fall- und Steigversuche an Teilchen mit Brownscher Bewegung. Physikal. Zeitschr.,16, 289-295.

Seshadri, V. (1999): The Inverse Gaussian Distribution. Statistical Theory and Applications. Springer, New York, Berlin.

Smoluchowski, M. (1915): Notiz über die Berechnung der Brownschen Molekularbewegung bei der Ehrenhaft-Millikanschen Versuchsanordnung. Physikalische Zeitschr., 16, 318-321.

Taylor, H. (1967/68): Evaluating a call-option and optimal timing strategy in the stock market. Management Science, 12, 111-120.

Tijms, H. C. (1994): Stochastic Models - An Algorithmic Approach. Wiley, New York.

Tweedie, M. C. K. (1956): Some statistical properties of inverse Gaussian distributions. Virginia J. Sci., 7, 160-165.

van Dijk, N. (1993): Queueing Networks and Product Forms. Wiley, New York.

Walrand, J. (1988): An Introduction to Queueing Networks. Prentice Hall, Englewood Cliffs.

Wiener, N. (1923): Differential space ... , J. Math. and Phys., 2, 131-174.

Index

absolute distribution **171**
absorbing state **146 ,173**
age renewal policy **133**, 244
aging **10**, 11
alternating renewal process **120**, 132, 244
American call option **274**
arithmetic distribution **115**
arithmetic random variable **115**
ARMA-models **66**
arrival intensity (rate) **211**
autocorrelation function **49**
autocovariance function **49**
autoregressive sequence **63, 64**, 302
availability **122**
- average **122**
- interval **113, 122**
- long-run **123**, 184, 245, 249
- point **122**, 177
- stationary **123**, 184, 245, 249
band-limited white noise **308**
bandwidth **299**, 308
Bayes' theorem **4**
Beta distribution **7**
binary random variable **5**
binomial moment **107**
Binomial distribution **6**
- negative **6**
Birnbaum-Saunders-distribution **130**, 273
birth- and death processes **163, 191**
- linear **197**
- pure birth process **192**
- pure death process **195**
wide-band process **298**
Brownian motion **257**
Brownian motion process **258**
Brownian bridge **263**
Campbell's theorem **82**, 97
central limit theorem **30**
Chapman-Kolmogorov equations **139, 170**
classes of states (Markov chain) **147**, 181
- closed **145**
- essential **148**
- inessential **148**
- minimal **146**
closed set of states **145**
- minimal **146**
compound Poisson process **91**
conditional distribution function **16**
conditional probability **3**
conditional probability density **16**
convolution **25**
correlation coefficient **18**
correlation function **48**
covariance **17**
covariance function **48**, 293
cumulative stochastic process **125**
degree of server utilisation **211**
de Morgan, rules of **2**
density **6**
Dirac delta function **287**, 304, 306, **313**
Dirichlet's formula **35**
discrete Markov chain **137**, 242
- homogeneous **138**
- irreducible **146**
- reducible **146**
discrete random variable **5**
discrete white noise **60, 301**, 306
drift failure **128**, 283
drift parameter **272**
economic lifetime **279**, 281

elementary event **1**
emergency renewal **95**, **133**, 244
Engset's loss system **215**
Engset's waiting system **219**
equilibrium state probabilities **141**
equivalence classes **147**
ergodicity **294**
Erlang distribution 7, **26**, 37, 73,102, 105
Erlang's loss formula **213**
Erlang's phase method **181**
essential state **148**
Euler's formulae **294**
events, random **1**
- mutually exclusive **2**
- independent **3**
- exhaustive **3**
exponential distribution 7, **11**, 105
fading **46**
failure probability **8**
failure rate **11**, 90
- integrated **11**
first passage probabilities **152**
first passage time **128**, 153, **263**, 273
flow (of demands) **209**
gambler's ruin problem **164**
geometric Wiener process with drift **277**
hazard function **11**, 86
Heavyside function 288, **313**
imbedded Markov chain **208**, 242
impatient customers **221**
inessential state **148**
initial distribution **139**, **171**
- stationary **140**, **171**
input (into a queueing system) **209**
input (into a queueing network) **229**
- external **229**
- internal **229**
insensitivity **214**
interval reliability **113**, **122**
- stationary **123**
inverse Gaussian distribution 265, **273**
irreducible Markov chain **146**, 151
Jackson queueing network **230**
joint probability distribution **13**, **15**
Kendall's notation **210**
Kolmogorov's differential equations **175**
Laplace-transformation **34**
Laplace-transforms **34**, 104, 106, 123
Landau order symbol 70, **312**
linear birth- and death process **197**
linear birth process **193**
linear death process **196**
loss probability **213**, 220, 222
loss system 209, **212**
marginal distribution **13**, **16**
Markovian property **52**
Markov chain **137**, **169**
- continuous-time **169**
- discrete-time **137**
- embedded **208**, 242
- sojourn times of a **207**
- stationary **159**, 171, **183**
Markovian system **187**
mean-square continuous **52**
memoryless property **10**, 113
minimal repair **89**, **90**
moving averages **61**, 62, 145
narrow-band process **298**
noise
- industrial **46**
- metereological **46**
- white **60**, **286**, **306**
normal distribution 7, **27**, 101, 105, 117
- bivariate **18**
- *n*-dimensional **21**
null-recurrent **153**

Ornstein-Uhlenbeck-process **271**
orthogonal increments **308**
output (of a queueing system) **209**
paradox of the renewal theory **116**
patient time **221**
period (of a Markov chain) **150**, 173
point process **69**
- homogeneous **70**
- simple **70**, 73
- stationary **70**
Poisson events **72**
Poisson process **69**
- compound **91**
- homogeneous **70**, 172
- inhomogeneous **85**
- simple **70**
positive recurrent **153**
preventive renewal **89**, 94, 244
priority systems 211, **223**
probability distribution **4**, 5, **47**
- of a random variable **4**, 5
- of a stochastic process **47**
pulse code modulation **56**
- randomly delayed **58**, 303
pulse process **59**, 73, 289
purely random process **79**
purely random sequence **60**, 301
queueing disciplines **211**
queueing network **229**
- closed 229, **237**
- Jackson **230**
- open 229, **230**
- sequential **234**
- tandem **234**
queueing system 83, **209**
- closed **210**, 216
- insensitive **214**
- loss **212**
- multi-server **209**
- priority 211, **223**
- single-server 209, 219, **223**
- waiting 209, **215**
- waiting-loss **219**
- with feedback **233**
random experiment **1**
random integral **285**
random telegraph signal **74**, 302
random variable **4**
- discrete **4**
- continuous **6**
- independent **20**
- nonnegative **8**
- standardized **28**
random vectors **13**, **20**
random walk **141**, 142, 155, 163, 260
Rayleigh distribution **12**, 92, 93
recurrence time (of renewal processes) **111**
- backward **111**
- forward **111**
recurrence time (of a Markov chain) **153**
recurrent Markov chain **152**, 155
reflection principle **264**
regeneration cycle **131**
regeneration point **131**, 153
regenerative stochastic process **131**
renewal 89, **99**
- cycle **100, 123**
- density **102**
- emergency **94**, 133, 244
- equations **103**
- function **102**, 121
- preventive **89**, 94, 244
renewal counting process **100**
renewal process **99**
- alternating **120**, 132, 242, 244
- delayed **100**

- stationary **118**
- ordinary **100**, 242
renewal theorems **114**
- Blackwell's **115**
- elementary **114**
- key **115**
repair cost limit maintenance policy **95**
repairman problem 187, **204**
routing matrix **230**
response **59**
sample path **45**
Schottky-effect **60**
semi-Markov process **241**
sequence
- autoregressive **63**, 64, 301
- purely random **60**
- of moving averages **61**, **62**, 145
series system **246**
service disciplines **211**
shot noise **59**, **79**, 289
spectral density **299**
spectral function **299**
spectral process **309**
spectrum **298**, **299**
- continuous **299**
- discrete **298**
standard normal random variable **28**
stationary state probabilities **159**, **183**
steady state **211**
stochastic process **45**
- continuous-time **45**
- cumulative **91**
- discrete-time **45**
- ergodic **294**
- Markov **52**
- mean-square continuous **52**, 53
- second order **50**, **293**
- strictly stationary **49**
- wide-sense stationary **51**, **293**
- with homogeneous increments **51**
- with independent increments **51**
- with orthogonal increments **308**
stopping time **29**
strictly stationary **49**, 142
strike prize **274**
survival probability **8**, 199
thermal noise **46**
total probability rule **3**
traffic **209**
traffic intensity **211**
transition graph **146**, **177**
transient Markov chain **153**
transition matrix **138**, **230**
transition probabilities **138**
transition rates **174**, 176
trend function **48**, **293**
uniform distribution **6**, **9**
- continuous 7, **9**, 77
- discrete **6**, 239
waiting system 209, **216**
waiting-loss system **219**
Wald's equation **29**
Weibull-distribution **11**, 95
white noise **60**, **286**, **306**
- band-limited **308**
- discrete **60**
wide-band process **298**
wide-sense stationary **51**, **293**
Wiener process **258**
- integrated **285**
- standard **258**
Wiener process with drift **272**
- geometric **277**
z-transformation **32**
z-transforms **32**, 196, 201